TIDAL POWER

Proceedings of the symposium organized by the Institution of Civil Engineers and held in London, 30–31 October 1986

Thomas Telford, London

Co-sponsored by The Institution of Electrical Engineers and The Watt Committee on Energy

Organizing Committee: W. J. Carlyle (Chairman), A. C. Baker, D. M. Barr, R. Clare, H. J. Moorhead

British Library Cataloguing in Publication Data
Tidal power: proceedings of the symposium
  organized by the Institution of Civil Engineers
  and held in London, 30–31 October 1986.
  1. Tidal power-plants — England
  I. Institution of Civil Engineers
  627'.8   TK1457
  ISBN: 0 7277 0390 0

First published 1987

Papers or other contributions and the statements made or opinions expressed therein are published on the understanding that the author of the contribution is solely responsible for the opinions expressed in it and that its publication does not necessarily imply that such statements and or opinions are or reflect the views or opinions of the ICE Council or ICE committees

Published for the Institution of Civil Engineers by Thomas Telford Ltd, Thomas Telford House, 1 Heron Quay, London E14 9XF

Printed and bound in Great Britain by Robert Hartnoll (1985) Ltd, Bodmin, Cornwall

Contents

Opening address

D. HUNT, MP, *Parliamentary Under Secretary of State for Energy*

Renewable energy sources have an important role in the
Government's energy strategy and as we proceed into the next
century and beyond I am confident that their contribution to
the UK energy supply will increase significantly. The
Department of Energy is therefore committed to ensuring that
the renewable energy technologies realize their full economic
potential as soon as possible.

To this end the Department is supporting a major research and
development programme and in 1986 expenditure on the programme
passed the £100 million mark – an investment which is already
paying dividends. The Department is not alone in recognizing
the important role of renewable energy sources for this
country's future energy supply. An increasing contribution to
the programme is coming from the private sector and many here
are involved in collaborative projects with the Department.
This collaboration is very much welcomed.

Passive solar design and some biofuels technologies have
already been established as economically attractive and
increased effort is being placed into achieving cost reductions
and the improved performance that is necessary for those
technologies which offer promise of becoming economically
attractive in the near future. Along with wind energy and
geothermal hot dry rocks, tidal energy is one of the most
promising resources.

If two or three estuaries with the best prospects for tidal
energy were to be developed by the early decades of the next
century then they would meet about 6–8% of the current annual
electricity requirements and be equivalent to about 2% of
installed firm capacity. The best estuaries could provide
electricity at 3–3.5p/kWh and there is significant potential at
generation costs not much higher than this.

Since 1979, the Department has spent about £3 million on
tidal energy research and development, and in July 1986 the
Secretary of State for Energy, Peter Walker, announced a
further phase of work costing £5.5 million over the next 2–3
years. This further programme is aimed at reducing the
uncertainty in costs, performance, regional and environmental
issues and will bring us to the point where it will be possible
to make decisions on whether or not to plan for construction
and will include the following work.

(a) Advanced investigations and site exploration of a
    Severn barrage on a line near to Cardiff and Weston

will be undertaken. The Department of Energy, the
Central Electricity Generating Board and the Severn
Tidal Power Group will share equally the £4.2 million
cost. If a Severn barrage were to go ahead it would
provide 5% of the UK's current electricity
requirement.

(b) An in-depth feasibiity study of a tidal barrage
across the Mersey will be carried out, to examine the
technical and economic viability of the scheme, its
employment potential and its effect on the
environment. The Department is contributing up to
£400 000 towards the total cost of £800 000 for these
further studies. A Mersey barrage could provide about
0.5% of the UK's current electricity requirement if it
were built.

(c) Generic studies will also be funded by the Department.
These will include a study of the various
construction techniques for tidal barrages,
environmental issues and further work to investigate
the potential of small sites around the UK. The work
on smaller-scale sites is interesting since
indications are that some of the sites being examined
may be as cost effective as the larger ones.

This symposium, organized by the Institution of Civil
Engineers and sponsored by the Institution of Electrical
Engineers and the Watt Committee on Energy, provides an
excellent opportunity to discuss the work that has been carried
out on tidal energy in the last few years and to discuss the
way forward. Papers will be presented on all aspects of tidal
barrages including engineering, environmental impact,
organization and planning, regional issues, modelling and
economic viability. Papers will concentrate on the results of
a two-year study carried out by the Severn Tidal Power Group on
the Severn barrage, the work completed so far on the Mersey
barrage, and on small-scale barrages. The Department of Energy
has provided funds for this work, and the progress that has
been achieved on improving knowledge during the last few years
and the commitment of the Government to tidal energy will be
appreciated.

The next phase of work presents an exciting opportunity to
all workers in the field to prove that tidal energy projects in
the UK can be built to time and cost, be economically viable
and be environmentally acceptable.

It is encouraging that the tidal energy project at La Rance
in northern France is about to celebrate its twentieth
anniversary and Annapolis in Canada has achieved efficient
operation and minimal effect on the environment. These schemes
show that well-planned tidal schemes can be of significant
benefit to the UK's energy economy and the next 2-3 years will
be a vital and challenging period to both the private and the
public sectors in establishing the case for Tidal Energy in the
UK.

1. An introduction to the work of the Severn Tidal Power Group 1983–1985

R. CLARE, BSc, DIC, FICE, MIHT, *Sir Robert McAlpine & Sons Ltd*

SUMMARY The paper briefly summarises the studies into Severn Tidal Power predating those carried out by The Severn Tidal Power Group. The objectives of the STPG study are discussed and a summary given of the Group's work as an introduction to the more detailed papers to follow in this Seminar. The philosophy adopted with regard to cost estimation and contingencies is stated.

INTRODUCTION

1. The first proposal for an energy generating barrage in the Severn Estuary was made by Decoeur in 1910 at the location known as English Stones which is a little way seaward of the present bridge (Figure 1). Further schemes followed between 1917 and 1933 the last being prepared by the first Severn Barrage Committee (the Brabazon Committee). All of these were based on a barrage at English Stones. During the last war the project was reviewed again but was shelved when the war ended due to low fossil fuel prices.

2. Following the large increases in oil prices in the early seventies a second Severn Barrage Committee was set up under the Chairmanship of Sir Hermann Bondi and a prefeasibility study costing £2¼M was carried out and reported in 1981 (Ref. 1). This study considered six possible alignments for the barrage, all seaward of English Stones, and recommended in favour of a line from Lavernock Point near Cardiff to Brean Down near Weston-super-Mare. (Figure 1). The proposed installed capacity was 7200 MW, using 150 bulb turbines of 9.0 m diameter, and with an anticipated annual output of 12.9 terrawatt hours per year. The Committee recommended further studies estimated to cost £20M but confirmed that the project was technically feasible and within existing technology.

3. Following the publication of the Bondi Report the Severn Tidal Power Group (STPG) was formed by Sir Robert McAlpine & Sons Ltd., Taylor Woodrow Construction Ltd., Balfour Beatty Ltd., GEC Power Engineering Ltd. and Northern Engineering Industries plc, to be later (January 1985) joined by Wimpey Major Projects Ltd. The Group made a

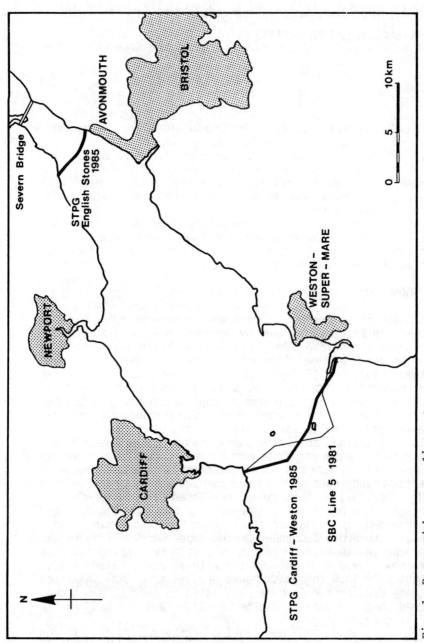

Fig. 1. Proposed barrage alignments

series of proposals to the Secretary of State to continue the work and was commissioned in 1983 to carry out a further study, based on the Cardiff to Weston line, and taking the Bondi findings as a starting point. This was estimated to cost £½M, 50% of which would be found by the Government, the balance by STPG.

4. The objectives of this study may be summarised as follows :-

(a) To assess the commercial viability of a barrage owned by a public Company established for its construction and operation.

(b) To assess environmental factors which might have an effect on the costs and benefits of the barrage.

(c) To assess the regional planning implications.

(d) To recommend on further work and stages necessary prior to a decision to proceed with construction of the barrage.

5. Within the limited funds available for the study it was not possible to commission new field work of a technical or environmental nature though much new data was assessed. Technical studies were limited to those that might result in major cost or construction time savings or increase confidence in the cost and time estimates.

6. In January 1985 the study was extended to make a comparative review of a shorter barrage on the English Stones line. The funds available were increased by £220,000 again being found 50-50 by Government and STPG. The earlier studies on this line proved of very limited use to STPG and the technical and environmental work had virtually to start from scratch. Thus the work cannot be in the same depth as that for the Cardiff to Weston line though an appreciable amount of work done on the latter is applicable to English Stones.

7. The objective of this paper is to give an overall but not detailed account of the work, results and conclusions so as to form an introduction to the more detailed reports contained in the papers to be presented during the remainder of the seminar.

Characteristics of the Schemes

8. The main characteristics of the Barrages at the two locations are given in Table 1.

CIVIL ENGINEERING (CARDIFF - WESTON-SUPER-MARE ALIGNMENT)

Introduction

9. The starting point for STPG was quite dissimilar for the two alignments. On the Cardiff to Weston line we were able to build on the work done for the Bondi Committee and select specific areas for further work.

Table 1 Data Sheet

		Cardiff-Weston	English Stones
Installed Capacity	MW	7200	972
Type of Turbine		Bulb	Bulb
No. of Turbine Generators		192	36
Availability	%	95	95
Runner Diameter	m	8.2	7.5
Generator Rating	MW	37.5	27
Generation Voltage	kV	6.9	6.9
Transmission Voltage	kV	400	400
Effective Sluice Area	m²	21600	5976
Method of T/G Erection		Single lift in situ	Single lift in situ
Weight of Maximum Lift	Tonnes	1600	1300
No. of Turbine Caissons		48	12
No. of T/Gs per caisson		4	3
No. of Sluice Caissons		62	14
No. of Sluices		186	42
No. of Plain Caissons		50	2
Total Length of Barrage	km	16.3	7.1
Annual Energy Sent Out (average year)	TWh	14.4	2.8

Alignment

10.   The alignment proposed by the Bondi Committee incorporated a severe dog leg and had an overall length of 17.9km for an estuary width of 13.4 km. (Lavernock Point to Brean Down). This has been reviewed in the light of a proposed reduction in turbine diameter from 9 to 8.2m and accepting the quantity of dredging as a variable. The revised line, shown in figure 1, has a total length of 16.3 km and impounds a marginally greater area of water than the previous proposal.

Caissons and Construction Procedures

11.   The work done in 1979/80 on caisson construction, placement and cost estimation has been reviewed by a fresh team of engineers and much of the earlier work has been confirmed. In particular it is confirmed that the preferred method for construction of the barrage is to use caissons built in shore-based facilities, floated to the barrage location and installed by ballasting down onto prepared foundations. This method is considered to have major advantages over insitu construction; firstly, there will be much less environmental impact, in particular on silt movements during construction; secondly, the majority of construction work can be carried out remote from the barrage location, thus avoiding the very large concentration of

resources which would be required for insitu construction; thirdly, it avoids much of the unacceptably long travel time for labour to mid stream locations which for much of the construction time would not be connected to the shore by road; and fourthly, a rapid programme of installation can be achieved by placing caissons on several working 'fronts' simultaneously.

12. It is now proposed to incorporate four bulb turbine sets in each caisson. The 192 turbines result in 48 caissons and in addition there will be 62 sluice caissons and 50 plain caissons which are preferred to rock embankments where the seabed is lower than -15m OD.

13. The "drive in" caisson installation method using four tugs attached to the caisson by articulated links (Ref.2) has been retained. More work on caisson placing conditions has produced slightly more favourable estimates of the times available for placement than were previously made 5 years ago.

14. A study on steel caissons in lieu of concrete (Ref.3) has been made by others and is reported later in the seminar. No significant financial advantage is thought to be offered though the use of steel in lieu of concrete in upper parts of the caisson may give advantage in reduced tow out draft.

Shipping Locks

15. The provision of shipping locks has been carefully reviewed and two alternative schemes considered. The previous proposals made provision for two locks of a sufficient size to take 150,000 ton ore carriers to Newport for the Llanwern steelworks and this again forms one of the alternatives. However these ships dock at Port Talbot and hence an alternative scheme has been considered providing locks for ships of about 10000 to 15000 tons. This would result in a saving of up to £300M and has therefore formed the basic scheme for the report, though STPG recognises that this question of shipping size needs to be finally resolved and any solution adopted for ships smaller than 150,000 tons will need existing legislation to be amended.

Construction Programme

16. An appreciable effort has been put into the sequence of construction and programme activity durations. Three factors have had a major influence on the sequence of construction. Firstly, careful note has been taken of the sedimentation studies, which will be referred to in Dr. Kirby's paper, and this has resulted in efforts to complete the barrage on the English side as early as possible, and to retain maximum water passage through the caissons until closure to avoid disturbance to the areas of mud banks in Bridgwater Bay. This aim also fits in well with the second factor which is the suggestion discussed in the paper by D. Barr, to transmit all the electricity generated to the

English shore. The third factor is the decision to construct "embankment" sections, whether rock or plain
caissons, simultaneously with the placement of power house
and sluice caissons. The result has been to shorten the
construction programme significantly from that reported in
Energy Paper 46 to 7 years to barrage closure and 9 years to
full electrical output (Figure 2). This represents a 2 year
shortening of the construction period.

CIVIL ENGINEERING (ENGLISH STONES ALIGNMENT)

Introduction
17. The proposed barrage is located just south of a large
rocky outcrop known as the English Stones, approximately 36
kilometres upstream of the Cardiff-Weston line.

Alignment
18. The location of the barrage is largely dependent on
the turbine submergence which requires a foundation level
of the powerhouse of -26m OD. The only area that
provides this depth with sufficient width for the sluices
and turbines and economical dredging is about 2km seaward of
the Shoots. The actual choice of alignment, location of
power house, sluices and locks was complex and will be
reported in the paper by Binnie and Roe.

The Scheme Concept
19. A concept similar to the Cardiff-Weston line was
adopted with turbines in caissons, sluices in caissons, a
shiplock and embankments linking to the shores. Plain
caissons were only used for the transition between embankments and sluices. The scheme is shown on Fig. 3.

Turbines and Sluice Caissons
20. Thirty six 7.5m diameter bulb turbine generator sets
in 12 caissons were selected for the power house with just
less than 6000 square metres of sluices also built into
caissons to admit the rising tide.

Embankments
21. A proportionally longer length of embankment is
required than with the Cardiff Weston line. The embankment
would consist of a rockfill control bund with sandfill used
to widen upstream, placed within quarry waste bunds.

Shipping Locks
22. As with the Cardiff-Weston line, two alternative
navigation lock sizes were chosen, one of 6,000 ton
capacity, reflecting current usage at Sharpness Dock, and
one appreciably larger capable of taking LASH (Lighter
Aboard Ship) vessels reflecting possible future expansion of
the activities at Sharpness dock by Gloucester Harbour
Trustees.

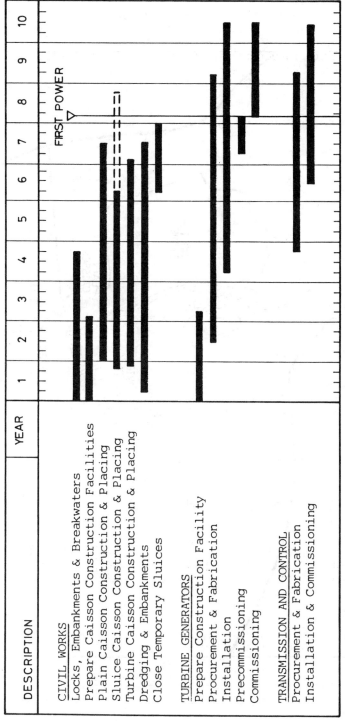

Fig. 2. Cardiff-Weston construction programme

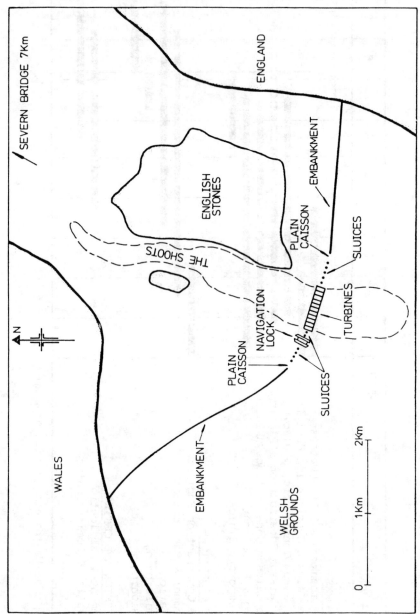

Fig. 3. English Stones – preferred alignment

Construction Procedures

23.   Caisson handling at this location must be carefully planned as upstream there is the narrow Shoots channel with high tidal currents and downstream the channel bed rises. Due to the navigational and handling problems it is proposed to use a similar fixed tug technique as intended for the Cardiff Weston line, placing the caissons during the low currents at neap tide low water.

Construction Programme

24.   In order to keep tidal velocities low, minimise scour and to maintain safe passage for shipping, it is proposed to allow as free a water passage as possible by delaying the construction of embankment founded at levels below Ordnance Datum until the installation of the navigation lock and the turbine caissons is complete.

25.   Construction of the Welsh embankment would be completed as soon as possible after this operation to allow early access for electrical transmission installation. Construction of the English embankment would be programmed so that closure coincides with completion of caisson cut off walls, permitting a small head difference for turbine testing under no-load conditions. It is anticipated that barrage closure would be completed by the end of year 5 with full power generation at the end of year 6.

MECHANICAL AND ELECTRICAL ENGINEERING

Turbine Generators

26.   It was decided to base the study on bulb types, of which the two leading European suppliers have over 125 units in service with runner diameters greater than 5m and several with diameters over 7m. A diameter of 8.2m was selected for the Cardiff Weston line that represented an acceptable extrapolation from known practice, without need for a prototype.

27.   Available water depth sets an upper limit to diameter because there must be enough submergence to avoid cavitation under all operating conditions, and dredging can be expensive. For the Cardiff-Weston line, the selected 8.2m diameter caused no special problems, but at English Stones the deep section is relatively short and dredging costs on either side are significant; so the runner diameter there was limited to 7.5m.

Electrical Transmission

28.   The sets generate at 6.9kV, and this would be stepped up to 400 kV by transformers collecting power from groups of 8 or 10 generators. The power would be exported along the barrage through single core oil filled cables in purpose-built cable ways, first to a 400 kV sub-station on the barrage at the end of the line of turbine caissons, and thence to a similar CEGB sub-station on-shore.

Installation and Maintenance Methods

29. Experience at large construction sites, especially in the nuclear industry, has demonstrated the advantage of the technique of large assemblies ex-works delivered as a single package requiring a minimum of site work; this advantage would be reinforced where men work in mid-estuary. So here each turbine generator set would be built into a 1600 tonne package with its lifting beam, and shipped direct from works to the caissons, four at a time on a North Sea flat-top barge equipped with a ballast control system to facilitate skidding onto the barrage.

30. Both turbine and generator would be designed to facilitate in-situ maintenance with parts removable through the access shafts without major dismantling of the machine. Removal of complete bulb units to a shore-side repair facility, although possible, is not envisaged.

Materials

31. Some materials are potentially critical because of the quantity and rates of delivery required. One difficulty is that some of the turbine components have very large diameters, up to 15 metres.

32. Procurement of all these materials was investigated and, in general, UK resources were found adequate for the requirements, even for the larger barrage though there is some doubt about forgings for shafts at the higher rate needed for the Cardiff-Weston barrage. These would be obtainable overseas if necessary.

ENERGY CAPTURE

33. Much effort was devoted to energy capture and this will be dealt with in a later paper by Duffett & Ward. It is sufficient to record here that advantage has been established in the addition of some pumping to the simple ebb generation concept adopted in the Bondi Report. The average annual net energy capture (gross output less pumping) established for the Cardiff-Weston and English Stones schemes is 14.4 and 2.8 TWh respectively.

COST ESTIMATES

Cost Summary

34. The cost estimates including contingencies for the two schemes are summarised below. The costs are in January 1984 money and make no allowance for financing costs (e.g. interest charges).

35. Non energy works cover essential modifications to existing work necessitated by the construction and operation of the power barrage (e.g. land drainage) and those cases where new facilities must be introduced (e.g. sewage treatment which would not have been required anyway to meet new regulations).

Table 2

	Cardiff/Weston Line £M	English Stones Line £M
Civil Engineering Works	3486	697
Mechanical and Electrical Engineering)	1629	301
Project Management & Engineering and feasibility studies)	312	77
Works not directly connected with power generation)	80	47
Parliamentary Procedures & Environmental Studies)	36	28
Total Capital Cost	5543	1150

Contingencies

36.   Contingencies were treated differently in the civil engineering cost estimate to the estimate for the mechanical and electrical works in accordance with practice in these industries..

37.   The civil engineering contingencies were estimated separately for each activity and each allowance consisted of three parts.   Firstly a percentage to allow for the preliminary state of the design and quantity take off; this contingency should be eliminated prior to start of construction as the design will be finalised and main drawings complete.   Secondly a percentage to cover the risk factor in doing the work thus this part of the contingency varies substantially from a low figure (5%) for caisson building to a high figure (40%) for caisson foundations underwater.   Thirdly there was an allowance for factors outside the control of individual contractors such as consequential effects of delays and design changes.

38.   With the mechanical and electrical work the design is well defined at the present time and the working conditions known hence a single across the board contingency percentage based on previous experience was used.   There is also an in-built contingency which it is hard to quantify as the prices were assessed on experience yet never to date has the industry had the opportunity to produce more than a handful of turbo generator units at a time.   The learning curve effect could be appreciable.

39.   The average effect of the contingencies applied was to increase the basic estimate for all works by slightly more than 20%.   This compares with the CEGB start to finish allowance of around $17\frac{1}{2}$% which they use in power station planning.

40.   It could be argued that this level of contingency is high in view of the amount of repetitive work, the extent of advanced planning, the segregation of trades and the dispersion of construction sites.   The STPG, however, believe it to be a prudent and realistic allowance.

ENVIRONMENTAL AND REGIONAL FACTORS

41.   Little funding was available to STPG to commission new   environmental   work though studies into   sediment transport were done and will be reported in this seminar by Dr. Kirby.   These studies demonstrated the importance of studying   the movement of fine sediments both   during construction and operation of a tidal barrage.   In this case a potential problem with a barrage on the English Stones line has been identified.

42.   Much useful work has been done utilising the ecological data collected for the earlier Severn Barrage Committee   and   subsequently supplementing   this   where possible, analysing it with experts and applying data from other estuaries where relevant.   Dr. Shaw's paper to this seminar comprehensively reports this work.

43.   His other paper reviews the Group's work on regional impacts and job creation both of which are of great importance.   In this work STPG received much cooperation freely given by many and varied representative and concerned organisations.

44.   Many non energy benefits, other than direct job creation,   have been identified in regional industrial development,   tourism and the leisure sector in addition to governmental   gains in taxation and reduction in social security payments.   A preliminary quantification of these benefits has been funded by STPG and the Welsh Development Agency.

PROJECT PLANNING AND IMPLEMENTATION

45.   With a project of the scale of the Severn Tidal Power Scheme,   planning and implementation need careful and detailed attention from an early stage of the feasibility studies.   STPG paid particular attention to this, considering the necessary procedures for a scheme financed, built and operated in the private sector.   This work will be reported in Mr. Buckland's paper to this seminar.

CONCLUSIONS

46.   The STPG study has confirmed that the barrage is feasible with existing technology and has resulted in a substantially reduced construction programme and reduced cost estimates for the Cardiff-Weston line compared to those previously reported by the Bondi Committee.

47.   A civil engineering solution has been devised to the

problems of constructing a barrage at English Stones and STPG believe the programme and cost estimates reported to be realistic notwithstanding the minimal resources available to carry out this part of the work. However there appears to be a potential siltation problem at English Stones and more work would be necessary on this aspect before a case for a barrage could be made at English Stones.

48. Much useful work has been achieved in mechanical and electrical engineering and in further studies into energy capture. The merits of flood pumping to supplement ebb generation deserve closer examination. The capacity of UK industry to supply the materials and to manufacture the plant, given some new facilities, has been established.

49. STPG remain convinced that in the Severn Estuary there is a major source of renewable energy that should be developed to the benefit of UK Ltd. and which will provide a most useful diversification into the UK electrical generating system and one that will be available on a much longer time scale than conventional power stations.

ACKNOWLEDGEMENT

50. The author acknowledges the permission of the Department of Energy to publish this paper and the assistance of colleagues in STPG in its preparation. The conclusions reached are not necessarily those of the Department of Energy.

REFERENCES

1. "Tidal Power from the Severn Estuary". Energy Paper No. 46, HMSO, 1981.

2. "The Towing and Positioning of Caissons in a Tidal Barrage". R. Clare and A.J. Oakley, Second International Symposium on Wave and Tidal Energy, Cambridge, 1981.

3. "Steel Caissons for the Severn Barrage". Yard/Roxburgh June 1984. Report to Department of Energy, not published.

2. Some interactions between tides and tidal power schemes in the Severn estuary

D. C. KEILLER, MA, PhD, MICE, *Senior Hydraulics Engineer, Binnie & Partners*

SYNOPSIS. This paper examines the modifications to the natural tides in the Severn estuary that would arise from the operation of a tidal power station at either the Cardiff-Weston or English Stones sites. The effects of the barrage on estuary water levels and velocities are used to calculate the energy output from the two sites including the effects of high water pumping at the Cardiff-Weston site. The same set of results are used to assess the changes to the movement of sandy sediments that might arise with the English Stones scheme.

INTRODUCTION

1. In any assessment of a tidal power scheme, predicting how tidal conditions will respond to the introduction of the barrage forms a major aspect of the study. These predictions are necessary to calculate how much energy can be extracted from the tides and to examine how the resulting changes in water level and velocity will affect the environment and other users of the estuary.

2. The predictions of the changes to tidal conditions may be made using a variety of types of physical or numerical models. Severn Tidal Power Group (STPG) commissioned Binnie & Partners to study the Severn estuary using an existing one dimensional numerical model of the estuary. This dynamic model was developed for the Severn Barrage Committee's (SBC) assessment (ref. 1) and has been described previously (refs. 2, 3). The model extends 150 km from Ilfracombe to Gloucester as shown on Fig. 1. All the assessments of Severn estuary tides made in this paper are based on this model which combines the principal gravity, inertia and friction forces with the geometric shape of the estuary to reproduce with good accuracy the observed propagation and amplification of the tides in the estuary. The hydraulic behaviour of the estuary is combined with a detailed analysis of the hydraulic performance of the turbines and sluices to predict the interaction between the barrage and the estuary tides.

WATER LEVELS AND VELOCITIES

3. Establishing the tide levels associated with the Cardiff-Weston tidal power barrage formed a major element of the

SBC studies (refs. 1, 2). These studies showed that upstream of the barrage, high water levels would be slightly reduced and considerably prolonged. Low water levels would rise to about the exisiting mean tide level between Cardiff and Bristol. In the recent studies for STPG, these results have been confirmed and extended to cover the effects of pumping. In addition tide levels associated with the English Stones proposal have been calculated. The effect of these barrage proposals on high and low water levels along the estuary is illustrated in Fig. 2.

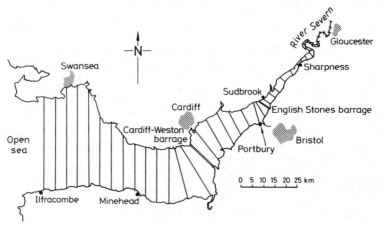

Fig. 1. The Severn estuary barrage sites and model sections

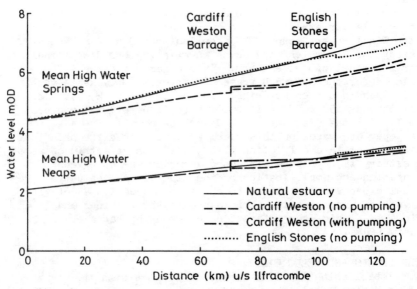

Fig. 2. Severn estuary high water profiles

4.   The effect of the English Stones barrage proposals on tide
levels downstream of the barrage is much less marked than the
effects of the Cardiff-Weston proposal. Spring tide curves
associated with the two barrage sites may be compared with the
natural tides at Portbury and Sharpness using Fig. 3. The
English Stones barrage reduces spring high water levels at
Sharpness by around 0.16 m, which is a considerably smaller
reduction than that caused by the Cardiff-Weston barrage. On a
neap tide the reduction at Sharpness caused by the English
Stones barrage is only 0.04 m. Downstream of the English Stones
barrage at Portbury there is little effect on high water levels,
though the time of the tide is advanced by about 15 minutes.

5.   The velocities in the estuary either side of the English
Stones barrage are considerably reduced by the presence of the
barrage as Fig. 3 demonstrates for Portbury and Sudbrook. At
Portbury the smaller tidal prism reduces velocities on the flood
tide. The velocities at this site on the ebb are less affected
because the turbine discharge is confined to the relatively
narrow low tide channel which prevents peak ebb velocities
dropping below their existing values.

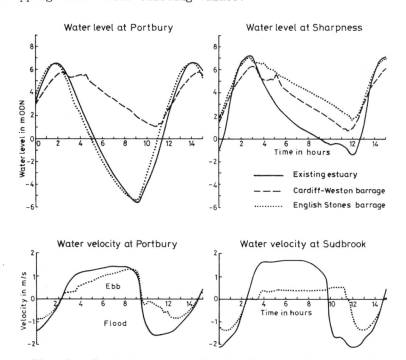

Fig. 3.   Typical spring tide level and velocity profiles

6.   Upstream of the barrage in the Shoots Channel at Sudbrook,
the barrage maintains water levels above mean tide level. This
increase in mean water depth, the reduction in tidal prism and
controlled discharge through the turbines lead to the dramatic
reduction in ebb velocities at Sudbrook during spring tides that

19

is illustrated on Fig. 3.

7. During spring tides, pumping at high water raises levels
inside the Cardiff-Weston barrage by less than 0.1 m and has
almost no effect at all on levels outside the barrage. The high
water levels inside the barrage are still lower than those that
occur at present. The effect of pumping is most clearly seen on
neap tides where the effect is to increase the level during the
high water stand at Portbury by around 0.15 m as Fig. 4 shows.

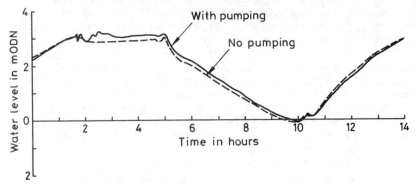

Fig. 4. Effect of pumping on a neap tide at Portbury

ENERGY OUTPUT

8. The amount of energy available from a tidal power scheme
is calculated directly by the hydrodynamic model taking account
of the modifications to the shape of the tide curve caused by
the barrage. The studies carried out for STPG examined how
different barrange locations, configurations and turbine
characteristics modified the energy output from a tidal power
barrage in the Severn estuary. This investigation was extended
to include a study of the effects of pumping extra water behind
the barrage at high tide.

Cardiff-Weston Ebb Generation Scheme

9. For a barrage on the Cardiff-Weston alignment, the five
configurations detailed on Table 1 were tested. The envelope of
the relationship between energy output and tidal range at
Ilfracombe for these schemes is shown on Fig. 5. The annual
energy output was calculated using the distribution of tidal
ranges at Ilfracombe for 1974 given on Fig. 6. This year was
used in the original SBC studies as tidal conditions were
predicted to be close to the long term average.

10. Despite the differences between the configurations on
Table 1, the annual energy output for all schemes was between
12.8 and 13.4 TWh. Although the annual energy output from all
the configurations examined showed little variation, Fig. 5
shows that the amount of energy available on tides of particular
range could be altered by changing the barrage characteristics.
This effect arises because those configurations which provide
the most energy on neap tides produce the least energy on spring
tides.

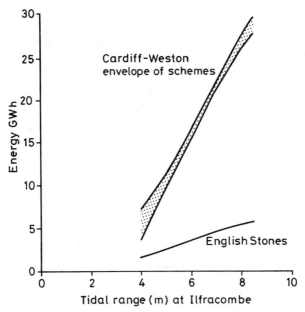

Fig. 5.   Energy output as a function of tidal range

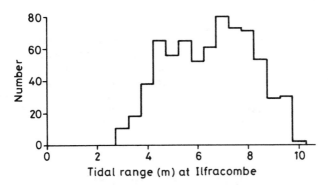

Fig. 6.   Annual distribution of tidal range at Ilfracombe
          for 1974

English Stones Ebb Generation Scheme.
   11.   The energy output for an English Stones barrage with an
operational installed capacity of 900 MW is also shown on Fig. 5
for comparison with the energy output from the Cardiff-Weston
scheme.   The principal features of this barrage are included on
Table 1.   The annual energy output from the English Stones
scheme was calculated as 2.7 TWh allowing for an electrical
efficiency of 95% between the turbine shafts and the shore
substation.

Combined Cardiff-Weston and English Stones Barrages.
   12.   One possibility investigated in the hydrodynamic model was

TABLE 1. Tidal Power Barrage Installations

| | Scheme Type | | | | | |
Barrage Location	1 C-W	2 C-W	3 C-W	4 C-W	5 C-W	6 E-S
Turbines - Number	140	144	144	182	182	36
Rated Output(MW)	60	45	45	32.6	32.6	25
Diameter (m)	9.0	9.0	9.0	8.2	8.2	7.5
Type	EWZ	EWZ	EWZ	EWZ	NEY	EWZ
Sluices - Number	160	150	150	150	150	44
Throat Area (m^2)	144	180	144	144	144	132
Annual Energy Output (TWh)	13.1	13.1	12.8	13.4	13.1	2.7

Turbine Types
EWZ - Escher Wyss
NEY - Neyrpic

Barrage Location
C-W - Cardiff-Weston
E-S - English Stones

the effect on the energy output of the Cardiff-Weston barrage if
an English Stones barrage had previously been built. For this
assessment the assumption was made that all sluice gates and
turbomachinery had been removed from the English Stones site but
that the civil structure remained. In the low velocity
conditions upstream of the Cardiff-Weston barrage, the structure
of the English Stones barrage would cause very little resistance
to flow and only a small reduction in energy output. The loss
of energy amounted to 1% of the energy output from the
Cardiff-Weston barrage.

Cardiff-Weston Ebb Generation Barrage with Pumping
 13. The benefits to be gained by pumping water through a
Cardiff-Weston barrage at high tide have been examined. A
programme of tests was carried out using turbine performance
characteristics for a machine design for which pump
characteristics were available. The tests examined the changes
in energy output associated with different amounts of pumping at
high water. The results showed that when the turbines ran in
reverse as pumps at high water, the increased energy output
achieved on the ebb tide was almost exactly balanced by the
pumping energy used at high tide. On neap tides the maximum
increase in energy output was less than 0.1 GWh and much less on
a spring tide. With this arrangement, high water pumping was
only able to increase the annual energy output by 0.3%
 14. Operating the pumps at less than the turbining speed is
possible and would improve the pumping efficiency at low heads.

Normally it would also reduce the discharge. In this case
however the pumps were operated at less than peak efficiency to
obtain the same discharge as achieved when operating at normal
speed. Driving the turbines in reverse as pumps at two thirds
the normal turbining speed, increased the net energy output from
each tide by between 0.1 and 0.3 GWh. The largest increase
occurred on a neap tide. Over a full year, the net effect of
pumping would be to increase the energy output of the scheme by
120 GWh which is just under 1%.

SEDIMENTATION
15. The sedimentation effects of constructing a barrage in the
Severn estuary near the English Stones have been examined in
outline using a potential sand transport model that uses the
velocities and levels calculated by the hydrodynamic model. The
potential sand transport is calculated at each section of the
hydrodynamic model using the Ackers-White (ref. 4) method. This
method only calculates potential transport. It assumes an
infinite supply of the sediment being modelled and does not
change the geometry of the estuary to take account of accretion
or erosion. Ackers (ref. 2) has already described the
application of this technique to other Severn barrage sites
during the SBC studies.
16. In practice, in the Severn estuary, there are inerodible
areas of exposed rock as well as extensive sandbanks.
Furthermore the modelling method is restricted to the movement
of sand sized particles by tidal action. The very different
mechanisms for mud movement are not considered nor the effect of
wave action on the location of sediment deposits.
17. The impact of the barrage has been assessed by examining
how its introduction changes the behaviour of the model. In
this way the model may be used to gain an overall impression of
how the distribution of sandy sediments in the Severn estuary
might be changed by a tidal power barrage near the English
Stones.

Sedimentation in the Existing Estuary
18. The potential sand transport model has been applied to the
Severn Estuary for an average spring-neap-spring tidal cycle.
The model has been operated with sand grain diameters of 0.5 mm
westward of the Holm Islands reducing to 0.17 mm at Denny Shoal
and 0.15 mm upstream of the Shoots Channel at Shepherdine Sands.
These sizes were chosen to reflect the natural sizes of sand on
the bed of the estuary as reported by Parker and Kirby (ref. 5).
19. The net potential sand movements in the natural estuary
which are shown on Fig. 7 are generally in agreement with the
existing understanding of long term sediment movements. The
results show a seaward drift west of Minehead and a general
landward drift to the east of Minehead. The location of this
change in the direction of net drift is consistent with the
"bedload parting" mentioned by Parker and Kirby (ref. 5).
20. Measurements by Davies (ref. 6) of sandwave migration near
Holm Sand suggested a landward movement of 500 m^3/m each year.

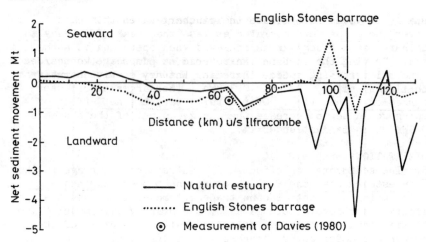

Fig. 7. Net potential sand movement over a typical
spring-neap-spring tidal sequence

If this rate were applied to the full estuary width, it would
imply a landward movement of around 0.6×10^6 tonnes over a 15
day period. This result, shown on Fig. 7, is acceptably close
to the predictions for this section. Elsewhere the high
potential transport in the Shoots at Sudbrook (chainage 109 km)
is expected, because of the high velocities at this section, but
is unlikely to result in any additional movement of sand as the
channel section is rock lined and so the potential transport
cannot be realised in practice. Overall, the model suggests the
existing net landward drift between Cardiff and Portbury is
around 0.25×10^6 tonnes over a 15 day period, or around six
million tonnes a year. This rate is a potential movement which
is likely to exceed the actual movement due to the limited
supplies available.

The Effect of a Barrage on Sedimentation
21. The introduction of a working tidal power barrage at the
English Stones reduces the peak potential sediment discharge
rates throughout the estuary as Table 2 illustrates. The
reduction in sediment discharge is most marked upstream of the
barrage on spring tides, especially during the ebb when the
potential sediment discharge becomes negligible.
22. The model results indicate that, if there is no movement
of sediment through the barrage on the ebb, the potential
upstream transport of 0.16 mm sand past the barrage could
contribute around 0.05×10^6 tonnes of sediment during each 15
day spring-neap-spring cycle. This potential upstream movement
would amount to about one million tonnes per year of sand sized
sediment.
23. The pattern of net potential sediment transport throughout
the estuary is illustrated on Fig. 7. This clearly shows that
upstream of Cardiff, the potential landward movement of sediment
is reduced and it is reversed over the 10 to 15 km immediately

TABLE 2. Maximum Potential Sand Transport Rates

Location	Tide	Sand Size mm	Max. Sediment Discharge Tonnes/Sec			
			Existing Estuary		With Barrage	
			Flood	Ebb	Flood	Ebb
Cardiff	Spring	0.30	23	14	19	7
Portbury	Spring	0.17	51	27	2	13
Sudbrook	Spring	0.16	187	52	15	0.02
Sharpness	Spring	0.15	54	8	7	0.01
Cardiff	Neap	0.30	0.6	0.4	0.5	0.1
Portbury	Neap	0.17	1.0	0.6	0.2	0.4
Sudbrook	Neap	0.16	3.3	2.4	1.2	0.5
Sharpness	Neap	0.15	0.11	0.08	0.05	0.07

downstream of the barrage. West of Cardiff the barrage appears
to cause the net movement of sediment in a landward direction to
increase. This increase arises from a greater imbalance between
the potential flood and ebb sediment discharges which both
reduce as a result of the barrage. The change in tidal balance
would also appear to move the position of "bedload parting" 25
km west from the east side of Swansea Bay to the west side.
This conclusion must be tentative as the position of parting has
been moved very close to the model boundary and will be
sensitive to any small effects the barrage has on tides at
Ilfracombe which this model does not include.

Overall Sedimentation Effects
24. The most significant feature of these results is the
dramatic drop in peak sediment discharge potential upstream of
the barrage shown on Table 2. This drop is most marked during
the ebb. This strongly suggests that once sandy sediments enter
the upper basin from either the rivers Wye and Severn or through
the sluices and turbines on flood tides, they will never leave.
These results also imply that the potential of the flood tide to
transport sandy sediments upstream will be markedly reduced in
the 20 km immediately downstream of the barrage. This will of
course reduce the rate at which the upper basin might otherwise
silt up, by reducing the quantity of sand entering the basin on
each flood tide. Sandy sediments passing through the barrage on
the flood tide will be transported upstream through the Shoots
before settling. The potential transport on the following ebb
tide will not be sufficient to move these sediments downstream
through the barrage again.
25. One feature that stands out from Fig. 7 is that the
direction of net potential sediment movement will be seaward for
10 to 15 km downstream of the barrage. At the barrage
itself, the net actual movement of sediment will be upstream,

25

since there will be little if any sandy sediment passing through the barrage when the turbines discharge on the ebb. The existence of a seaward net potential sediment transport downstream of the barrage suggests that there could be some scour of the main ebb channel past Portbury, assuming that it is erodible. Any material scoured from this channel will tend to be deposited in the channel west of Clevedon by the reduction in ebb transport potential in this part of the estuary (Table 2). Similarly the rapid reduction in flood transport potential in the same part of the estuary will tend to promote accretion of the intertidal sandbanks of the Welsh Grounds. Some of the sediments that settle in this area, either because they have been scoured from the low water channel upstream or because of a general drift up estuary from the downstream portion will be re-worked and moved upstream through the barrage on the flood tide.

CONCLUSIONS

26. The effects on tidal levels and velocities of a tidal power barrage in the Severn estuary may be predicted using numerical dynamic models which take account of the hydraulic performance of the barrage. These water levels and velocities can be used to assess the effects of the scheme on other users of the estuary.

27. The ebb generation tidal power schemes studied for STPG will generate between 12.8 and 13.4 TWh/year of electrical energy. A similar type of scheme located near English Stones will be able to generate 2.7 TWh/year of electrical energy. These energy outputs are not very sensitive to the precise numbers and types of turbines and sluices installed in the barrage. The operation of the Cardiff-Weston barrage would not be significantly affected by the presence of a barrage at English Stones which had been decommissioned by the removal of its sluice gates and turbogenerators.

28. These dynamic model studies indicate that the ability to increase the water volume stored behind the barrage by low head pumping at high water will marginally increase the net energy output from the Cardiff-Weston barrage. This increase in output is unlikely to exceed 1% of the annual energy output from the scheme.

29. The installation of a tidal power barrage at the English Stones would cause the upper basin to act as a trap for sand sized sediments. Downsteam of the barrage there may be some enlargement of the low water channel for 10 to 15 km below the barrage. There may well be some siltation of the low water channel near Clevedon and some accretion of the intertidal sandbanks between West Middle Ground and the barrage.

30. The methods of calculating the potential sediment transport described in this paper are only intended to provide a preliminary indication of the likely changes in sand movement patterns. Different techniques must be applied to determine the effects of a barrage on the muds and silts that are also present in the Severn estuary.

ACKNOWLEDGEMENT
This paper is based on work carried out by Binnie & Partners for
Severn Tidal Power Group. Their permission to publish is
gratefully acknowledged.

REFERENCES
1.   SEVERN BARRAGE COMMITTEE. Tidal Power from the Severn
Estuary, Energy Paper 46, HMSO, 1981
2.   ACKERS P. Barrage operaton, flood evacuation, surge tide
and closure dynamics. Severn Barrage, pp 65, Thomas Telford,
London, 1982
3.   KEILLER D.C. and THOMPSON G. One dimensional modelling of
tidal power schemes; paper B1, 2nd Int. Symp on Wave and Tidal
Energy BHRA, 1981
4.   ACKERS P. and WHITE W.R. Sediment transport; new approach
and analysis; J. of Hyd Div. ASCE Vol. 99 No. HY11 pp 2041, 1973
5.   PARKER W.R. and KIRBY R. Sources and transport patterns of
sediment in the inner Bristol Channel and Severn Estuary.
Severn Barrage, pp 181, Thomas Telford, London, 1982
6.   DAVIES C.M. Evidence for the formation and age of a
commercial sand deposit in the Bristol Channel. Estuarial and
Coastal Marine Science Vol. 11 pp 83, 1980

Discussion on Papers 1 and 2

MR F. IRWIN-CHILDS, Rendel, Palmer & Tritton
The report of the Severn Tidal Power Group (introduced by Mr
Clare) has covered some useful ground on the Cardiff-Weston
scheme and it is encouraging to see that the desirability of
revision of the dam alignment is being recognized. The
suggested route - with its elimination of the dog-leg - is
certainly an improvement. The line is still basically
controlled by the required depth of submergence of the turbine
generators and for caissons the amount of hand dredging needed
to achieve it. If this could be reduced, even further
consideration could be given to the alignment. What is the
intention for any special treatment at the Brean Down
landfall?

Paragraph 11 on 'Caissons and construction' contains a
statement on several advantages which caisson construction is
claimed to hold over the so-called in situ concept (such as
embodied in the Mersey scheme). It would be inappropriate for
me to proffer here a response on these matters. The answers
are largely contained in the details of subsequent papers both
on the Severn and the Mersey. For the record, suffice it to
say that when the intentions of the alternative concept are
fully understood it would be difficult to sustain the same
opinions.

Paragraph 26 on 'Turbine generators' states that the
adoption of an 8.2 m dia. runner turbine was made as an
acceptable extrapolation from existing types. What was the
axis level taken for this machine and did the saving in the
cost of dredging as against a 9.0 m runner play any part in
the decision?

Also, was any thought given to using Straflo machines?

MR R. CLARE
The exact location of the landfall at Brean Down has not been
determined and indeed the same is true of the landfall on the
Welsh shore. In both cases the shore end of the barrage is in
shallow water and will be constructed in conventional

embankment. The cost of this inshore embankment is very small
compared with the total project cost; hence the actual
location of the landfall can be controlled by local conditions
and subject to normal planning procedures.

The axis of the 8.2 m dia. turbines was taken as −20 m OD
for the study but information coming late in the study
indicated that adequate submergence would be given if the axis
were raised to −17 m OD. Paper 3 deals with the factors
affecting the alignment including dredging and raper 5
discusses the selection of turbine type.

It is intended that the alternative methods of constructing
the barrage will be fully examined in the new development
project.

MR D. KERR, MR W. T. MURRAY and MR B. SEVERN
The reduction in dredging costs between an 8.2 m dia. and a
9.0 m dia. turbine is less significant than the increase in
length of the power-house that is necessary to accommodate the
increased number of turbines (to give the same total output).
This is because the deeper parts of the channel are deep
enough for the 9.0 m turbines but it is narrow and the
increased length of power-house necessary for 8.2 m turbines
results in significant extra dredging and skewing of the
power-house line across the line of the channel.

MR I. M. WALKER, Gloucester Harbour Trustees
The ports of Bristol, Gloucester and south-east Wales have
co-operated closely with the Severn Tidal Power Group (STPG)
through the medium of the Ports Technical Panel of the
Standing Conference of the Severnside Local Authorities over
the last two years.

In particular the Ports have

(a) provided data to STPG's maritime transport
    consultants
(b) identified the detailed requirements of navigation
    and conservancy
(c) made suggestions to investigate potential adverse
    aspects of either barrage line.

The Ports are therefore extremely disappointed that the JTPG
in their published report have largely ignored the content of
the Ports' contributions and have apparently sacrificed the
rights of navigation on the altar of short-term financial
expediency.

The Ports have never discussed with the STPG the lower-bound
lock scenario costed in the report and they reject it as being
utterly and completely inadequate for present let alone future
requirements.

They invite the STPG and the Department of Energy to review
urgently this aspect of the report and to confirm their

acceptance of the upper-bound lock scenario as the only
acceptable option. They also ask that the need for
engineering solutions to investigate the adverse effects of
the port barrage regime is recognized and is processed as part
of the ongoing work in parallel with the proposed regional
studies.

MR R. CLARE
The Severn Tidal Power Group (STPG) valued the contacts with
the Ports Technical Panel throughout the study work and most
certainly have not ignored the Ports' contributions. Paper 3
has explained the study on the locks and the choice of upper
and lower bounds for the lock size. The STPG's report makes
it quite clear that no conclusion has been reached on lock
size. The STPG will enter the new study with a completely
open mind regarding lock size and will look forward to further
co-operation with the Ports Technical Panel, recognizing the
importance of shipping and ports in the regional development.
The regional studies will include the effects of the
post-barrage regime on the upstream ports.

PAPER 2

DR G. MILES, Hydraulics Research
My intention is to draw attention to some unpublished sand
transport data from the Severn Estuary which have some bearing
on the paper by Dr Keiller.

It had originally been planned to measure sand fluxes in
1980 as part of the Severn barrage studies, but a combination
of bad weather and difficulties with the equipment resulted in
a very poor recovery of data. Subsequently, in 1981,
Hydraulics Research returned to the estuary as part of their
research and development programme to prove that such
measurements were possible in the high tidal energy regime of
the estuary.

Measurements were made, using the Hydraulics Research sand
flux frame, in two sandy areas upstream and downstream of the
Holm Islands. On this occasion the experiments went smoothly
and both sites produced useful data, which have now been
processed. The non-cohesive part of the suspended load at
each site was found to consist mainly of sand in the range
0.15-0.20 mm. The data followed a well-defined trend but
there was noticeably less transport at site A (Fig. 1). This
was attributed to a shortage of mobile sediment up current
from this site and the findings emphasize the comment in Dr
Keiller's paper that the full potential transport may not
always be realized in practice.

These data should be valuable during the next stage of
investigations to help to study sand transport in more detail.

DR D. C. KEILLER
I was not aware of the sand flux data that Hydraulics
Research collected in 1981. The results of each model need
to be checked against prototype data to ensure that the model
is correctly reproducing the natural behaviour of the
estuary. These data will be of value in future studies of
the Severn Estuary. The sand size of 0.15-0.2 mm found in
the Hydraulics Research field trial is rather smaller than
the 0.25-0.3 mm that I had assumed on the basis of the
published grain size of the bed deposits. The Ackers-White
sediment transport theory used in my model gives a similar
sand flux over the velocity range 0.7-1.5 m/s. This sand
transport theory seems significantly more sensitive to
velocity than the observed data are. One possibility is that
the continual changes in velocity during a tide cycle reduce

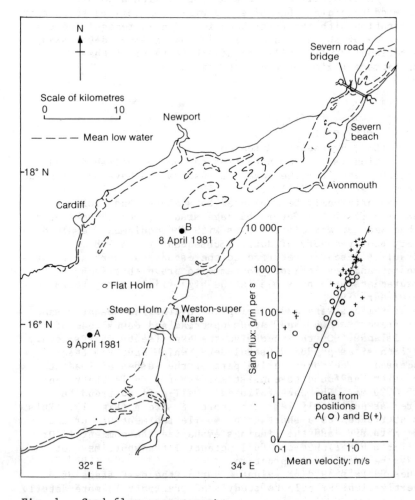

Fig. 1.   Sand flux measurements

the sensitivity of the sand flux to the local velocity which is assumed to be steady in the theory. This is an area where further research into the mechanics of sand transport would be beneficial.

DR N. R. C. BIRKETT and DR N. K. NICHOLS, University of Reading, and DR D. A. C. NICOL, Cambourne School of Mines

DYNAMIC MODELS FOR OPTIMAL CONTROL OF TIDAL POWER SYSTEMS

The calculation of operating strategies which maximize energy output from tidal power schemes has previously been carried out with static flow models. Computational procedures are described which couple optimal control techniques with dynamic models of the estuarine flow and the hydraulic performance of the power plant to maximize revenue from the derived energy. Tests using data for a proposed Severn tidal barrage establish that the incorporation of dynamic effects leads to new operating strategies and increased revenue predictions.

INTRODUCTION
The economics of a tidal power scheme depend ultimately on the efficiency of plant operation. To predict the maximum revenue which may be derived from any scheme, the best operating strategies must therefore be assessed. In previous studies of tidal power schemes, operating regimes for turbines and sluices have been optimized using static, 'flat' basin models and estimates of power output have then been determined by numerical simulations using dynamic models of estuarine flow under the predicted regimes (refs 1-3).

This approach requires an <u>a priori</u> estimate of the hydraulic effect of the barrage on water levels and velocities at the barrage site. It fails, moreover, to allow the dynamic interaction between estuarine flow and flow controlled through the barrage to be exploited.

Prompted by these considerations, the Science and Engineering Research Council and the Central Electricity Generating Board have co-funded an investigation at the University of Reading into the use of dynamic models for the optimization of control strategies. As a result of this study, new models have been developed which combine the techniques of mathematical control theory for determining optimal operating strategies which maximize revenue from the derived output with stable numerical procedures for simulating non-linear fluid flow in a full estuary of variable geometry containing a barrage with specified hydraulic characteristics. This approach has been found to be both computationally feasible and flexible in treating a variety of schemes including both ebb and two-way generation, pumping and schemes with two barriers (refs 4-8).

MODEL

The optimal control problem considered is that of maximizing the average tariff-weighted power derived over a tidal period (or periods), subject to the hydrodynamic equations of flow in the estuary. In the initial stages of the study, simple static and linear dynamic equations for the tidal flow were used (refs 4, 6-8). In the current model (ref. 5), flow in an estuary of variable geometry is simulated by the non-linear, one-dimensional shallow water equations for the conservation of fluid and momentum. The tidal elevation is imposed downstream of the barrier and the hydraulic performance of the barrage is described by the discharge characteristics of the turbines, sluices and pumps, defined as non-linear functions of head difference. The total discharge through each type of device is controlled between zero and a maximum availability value. Controls are determined to satisfy necessary conditions for optimality.

The computational procedure couples an iterative method for maximizing the revenue functional with finite difference techniques for solving the equations of flow and the associated adjoint equations for the control problem. A conditional gradient method, which can be shown to converge, is used to determine the optimal controls. An efficient and accurate explicit method for approximating the flow is applied, together with special implicit boundary conditions at the barrier, which are selected to guarantee stability of the numerical procedure.

RESULTS FOR SEVERN BARRAGE

Results for various schemes have been obtained with this model using data for a proposed Severn tidal barrage. The model extends 120 km from Ilfracombe to Sharpness. With the specified geometric data, the predicted tidal ranges in the estuary, without the barrier in place, show good agreement with measured data.

For comparison, data from an early intensive study of a barrage on the Cardiff-Weston line carried out by Binnie & Partners (refs 1 and 2) have been used to describe the plant hydraulics. It is assumed that there are 140 working turbines (corresponding to 155 turbines at 90% availability) and 160 sluices (12 m^2). The turbines are rated at 57 MW. Discharge and derived power characteristics are the same as those for the early study and, essentially, follow the line of maximum efficiency until the maximum power limitation is reached.

Energy output

In Table 1 results are shown for an ebb generation scheme at two tidal states - spring and neap - where average power has been optimized over one tidal cycle of 12.42 hours. With model NLD1, sluicing is constrained to take place only on the flood tide - a strategy which arises naturally from optimization using a static, flat basin model. These results

are comparable with those of the earlier study, shown as
model BP in Table 1, which were obtained with the less
sophisticated optimization process.

In model NLD2 sluicing on both flood and ebb tide is
permitted and a significant increase in energy is obtained.
The explanation for this somewhat unexpected result lies in
the dynamic effects of the additional sluicing on flow in the
tidal basin. Figs 2 and 3 show the tidal elevations on the
upstream and downstream sides of the barrier and the
operational state of the plant over one tidal cycle for
models NLD1 and NLD2 respectively. With model NLD2, the
sluices are fully opened at the end of the generation period
when the head difference is too low for the turbines to
operate, but the flow is still on the ebb. The sluices are
closed when the head difference becomes zero and are then
opened again at the optimal time on the flood to obtain
maximal elevation in the tidal basin.

The advantage of this strategy is that a larger head

Table 1. Average energy output for an ebb scheme – one tidal
cycle

Model	Spring (8.5m): GWh	Neap (4 m): GWh
BP	29.6	4.1
NLD1	30.2	5.5
NLD2	32.0	7.6

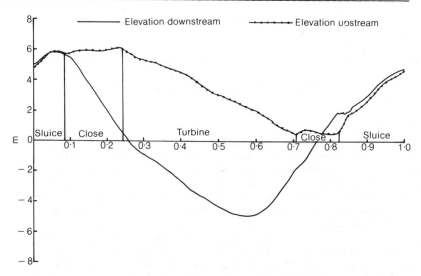

Fig. 2. Model NLD1 (flood sluicing only): tidal elevations at
the barrier over one tidal cycle – spring (8.5 m)

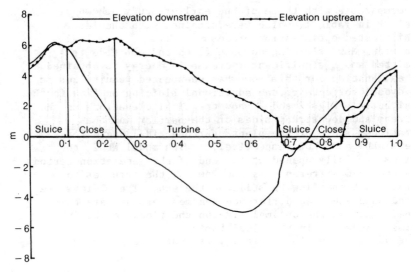

Fig. 3. Model NLD2 (two-way sluicing): tidal elevations at the barrier over one tidal cycle - spring (8.5 m)

difference is created for exploitation during the generating cycle than with the strategy of the earlier models. This difference in behaviour is able to the propagation of a wave upstream of the barrier and to subsequent llations in the basin, and the result confirms previous predictions that es in power output could be achieved by exploiting the dynamic effects of flow.

Two-way generation

The model also permits two-way generation strategies to be optimized. For the Severn data, if the restriction to ebb generation only is removed, it is found that the model converges to a scheme in which sluicing is carried out on both ebb and flood tides, but power is generated only on the ebb, i.e. the ebb schemes already studied are optimal over all strategies.

Maximum revenue

For full 14 day lunar cycles, both power output and revenue have been optimized for the Severn using model NLD1 (flood sluicing only) during a typical winter period. For maximum power output a unit tariff per unit energy is taken, and for maximum revenue a tariff based on winter electricity rates is used. Table 2 gives the predicted average power output and total revenue computed for both schemes. The result shows that by a careful scheduling of the flow over a 14 day cycle an extra profit of about 4% can be achieved at the expense of a slight reduction in the average power generated. Figs 4 and 5 show the instantaneous power generated in each case and the corresponding tariff weighting function.

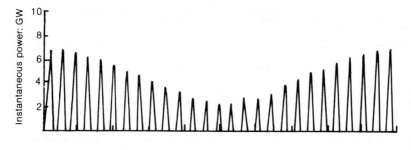

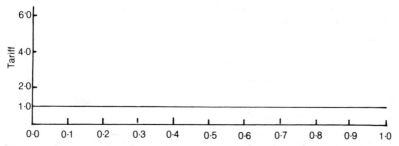

Fig. 4.   Power output with unit tariff for one lunar (14 day) cycle

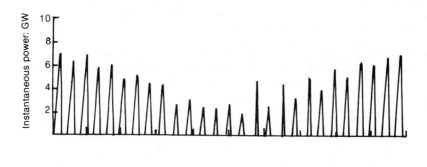

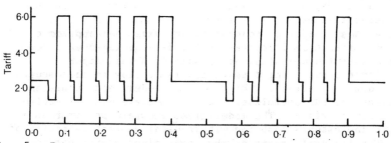

Fig. 5.   Power output with winter tariff for one lunar (14 day) cycle

Table 2. Comparison of power output and revenue for an ebb scheme - one lunar (14 day) cycle

Tariff	Average power: GW	Total revenue: units
Unit	1.42	5.05
Winter	1.38	5.25

Table 3. Average power output for an ebb scheme with pumping - one tidal cycle, spring (8.5 m)

Model	No pump: GW	With Pump: GW
NLD1	2.44	2.68
NLD2	2.58	2.75

Pumping

To test the pumping option in the absence of machine data, a simple model for the discharge characteristics of a pump has been taken. It is assumed that the pump discharge is constant at 800 m^3/s and that the power expended is directly proportional to the head difference. The net revenue is maximized over one tidal cycle of 12.42 hours, assuming a unit tariff weighting, i.e. assuming that the pumping cost per unit of power consumed is equal to the revenue per unit of power generated. (The model permits the application of different tariff structures to pumps and generators). The results obtained using model NLD1 (flood sluicing only) and model NLD2 (two-way sluicing) are shown in Table 3. For this simplified data, a significant increase in net derived power is obtained with pumping, and again two-way sluicing considerably improves the predicted output.

CONCLUSIONS

Techniques of optimal control theory have been applied to develop computational procedures for determining operating strategies which maximize net revenue from the output of a tidal power plant using dynamic models of flow in the estuary and across the barrage. The numerical models are stable and efficient and the optimization techniques are cost effective and highly flexible. The model gives accurate predictions of flow in the estuary combined with a detailed representation of the hydraulic performance of the power plant at the barrage. Schemes with two-way or ebb-only generation and schemes with pumping can be treated. Different tariff structures can be applied

Tests using geometric and hydraulic plant data for a proposed Severn tidal barrage have established that the incorporation of dynamic effects in the optimization procedures leads to significantly different control strategies and increased power output. Maximizing tariff-weighted power output leads to an increase in net revenue with only a small loss in generated power.

A major advantage of the optimal control formulation for the tidal power model is that additional constraints on operating strategies can be directly incorporated. In particular, integration into the national grid may impose restrictions on the rates of shedding or loading power which could be treated by this approach. It is concluded that the optimization of operating strategies for tidal power schemes using dynamic models of flow coupled with control techniques is both desirable and feasible.

DR D. C. KEILLER
The main strength of Dr Nichols' approach is that it allows an optimum to be found that might not be apparent from the normal engineering approach to the problem. The main weakness in the Reading approach to date is that there has been greater interest in obtaining an optimum solution than in the details of the hydraulic performance of a real barrage. Dr Nichols might quite correctly reply that these details can easily be included into her solution technique once that is fully developed. Nevertheless the omission of detail in the specification of the hydraulic behaviour of the estuary and the barrage can cause important errors in the energy assessment.

Dr Nichols' estimates of energy output are optimistic mainly because she has underestimated the hydraulic expansion losses downstream from the barrage. Her assessment overestimates the turbine head by about 0.25 m, which will increase power output by 300 MW on neap tides. The effect on sluice discharge is more dramatic as the omission of expansion losses will overestimate sluice discharge by 67%. The situation may not be as bad as this since Dr Nichols' model omits sluicing flow through the turbines, which will halve the overestimation of the sluicing capacity. If reverse sluicing is allowed, Dr Nichols' model, which assumes identical performance in both directions, will have overestimated the sluice capacity by a factor of 1.9 if her model does not allow reverse sluicing through the turbines. The large apparent sluice capacity in the Reading model is likely to alter the dynamic response of the estuary and will increase energy output.

The effect of delayed sluice opening was checked during the Severn Barrage Committee's studies and it was found that it had little effect until the sluicing head reached 0.5 m and then caused a reduction in energy output with higher starting heads. It would be surprising if the extra energy predicted by the Reading model can be achieved when the appropriate reductions in sluicing capacity are included.

Despite these deficiencies, which can be overcome by further model development, Dr Nichols' model is a useful addition to the armoury of models that are available for tidal power assessment. It is important to realize that the traditional hydrodynamic model like the model described in the paper, with its emphasis on the correct reproduction of the prototype behaviour of both estuary and barrage, still has a role to play in assessing the correct energy output from the Severn or other estuaries.

REFERENCES
1. SEVERN BARRAGE COMMITTEE. Tidal power from the Severn Estuary. Her Majesty's Stationery Office, London, 1981, Energy paper 46.
2. KEILLER D. C. and THOMPSON G. One dimensional modelling of tidal power schemes. Proceedings of the 2nd International Conference on Wave and Tidal Power. British Hydromechanics Research Association, 1981.
3. WILSON E. M. et al. Tidal energy computations and turbine specifications. Severn Barrage, Thomas Telford, London, 1982, 79-84.
4. BIRKETT N. R. C. Optimal control in tidal power calculations. PhD thesis, University of Reading, 1985.
5. BIRKETT N. R. C. Non-linear optimal control of tidal power schemes in long estuaries. University of Reading, 1986, NA Report 9/86.
6. BIRKETT N. R. C. et al. Optimal control problems in tidal power. Journal of Dam Construction and Water Power, 1984, 37-42.
7. BIRKETT N. R. C. et al. Optimal control problems in tidal power generation. Applied optimization techniques in energy problems (ed. H.J. Wacker). Teubner, Stuttgart, 1985, 138-159.
8. BIRKETT N. R. C. and NICHOLS N. K. Optimal control problems in tidal power - a case study. Industrial numerical analysis (eds S. McKee and C. M. Elliot). Clarendon, Oxford, 1986, 53-89.

3. Civil engineering aspects of the Cardiff–Weston Barrage

D. KERR, MA, MICE, *Sir Robert McAlpine & Sons Ltd,*
W. T. MURRAY, BSc, MICE, *Taylor Woodrow Construction Ltd,* and
B. SEVERN, MA, MICE, *Balfour Beatty Engineering Limited*

SYNOPSIS
STPG consider that the barrage should be constructed using concrete caissons as recommended by SBC. A barrage line of 16.3km length is proposed making maximum use of caissons to increase flexibility in construction sequence and minimize redistribution of settled sediments. STPG studies on caisson design, towing and emplacement, dredging, foundations, locks and embankments are described. Further studies of caisson design are recommended to improve equipment layout and reduce draught. A construction programme with a duration of 7 years to barrage closure is considered feasible. The potential advantages to the project programme and cost from provision of smaller locks than proposed by SBC are discussed.

INTRODUCTION

1. The studies carried out for the Severn Barrage Committee (SBC) covered all the main civil engineering aspects of the barrage. These studies resulted in preliminary designs and cost estimates for the major components of the barrage and concluded that the project is feasible using proven technology. The objective of the civil engineering part of the Severn Tidal Power Group (STPG) Interim Study was to examine the key engineering and technical factors which affect the cost of construction of the barrage. The STPG study had a limited budget and thus the work concentrated on those aspects of the civil engineering design and construction methods which have a significant influence on capital cost and programme - the latter because it determines the cash flow.

2. The main principles for construction of the barrage that were proposed by the SBC - based on the use of large concrete caissons built in shore-based construction facilities and floated to the barrage location - have been accepted by STPG. The majority of the outline designs have also been accepted as adequate for the present level of study. However, improvements to barrage alignment, construction sequence and programme are proposed and in addition aspects of the designs which merit further study have been identified. Previous cost estimates have been updated and revised to suit the amended alignment and programme.

3. It must be emphasised that none of the component designs can be regarded as optimised designs and all would benefit from a thorough review of alternatives during the next stage of study.

FORM OF CONSTRUCTION

4. Caissons have been adopted as the most suitable method of constructing the turbine house and sluices in the barrage. The method offers flexibility in programming the work, since it allows barrage installation to proceeed on several fronts simultaneously. The majority of the caisson construction work can be carried out remotely from the barrage location, thus avoiding the very large concentration of resources with associated logistics and management problems which would be required if in situ construction were adopted.

5. It will be possible to determine the sequence of installation of the barrage to minimise silt movements during construction, and the environmental impact of construction work will be minimised by the relatively low material requirements associated with caisson construction. The risks of storm damage to caissons during barrage construction are relatively low, since they can be placed and ballasted during fair weather periods, thus being secure against damage in subsequent storms.

BARRAGE ALIGNMENT

6. A study of barrage alignments has been carried out by STPG, using capital cost as the principal criterion for evaluation of the alternatives considered.

7. The SBC proposed a barrage line from Lavernock Point to Brean Down which was 'dog legged', resulting in an overall length significantly greater than the straight line crossing distance. This alignment was chosen for various engineering and environmental reasons, including an assumption that dredging for turbine caissons would be difficult and costly and should be minimized. However, STPG believed that a more direct line of lesser overall length might be expected to reduce the cost. In order to establish whether savings can be made, the alignment study was carried out retaining the same general crossing position and installed capacity but with other parameters considered variable.

8. The barrage alignment is determined primarily by the powerhouse, which must be founded on rock and has a foundation level determined by turbine submergence requirements. The length of powerhouse for 192 turbines of 8.2 diameter is 3.5 km and the foundation level was taken as – 29 m O.D. minimum. The only areas sufficiently deep for the powerhouse are the main channel (rockhead typically – 30 m to – 35 m O.D.) and the buried channel between the main channel and Steep Holm. The required powerhouse length of 3.5 km can

only be obtained on a skewed line across the channel width. After studying the rockhead topography three basically different powerhouse alignments were chosen for comparison as shown in Fig. 1.

9. For each powerhouse alignment two landfalls at each end were considered in order to encompass the likely variations: Lavernock Point and Ball Rock on the Welsh coast and Brean Down and a point south of Brean Down on the English coast.

10. The main conclusions of the alignment study were:-

(i) 75% of the barrage cost is for items which are largely independent of alignment: turbines/generators, electrical transmission, turbine caissons, caisson construction facilities and locks. Thus change in alignment affects only 25% of total cost and cannot be expected to produce a major change in cost.

(ii) The lowest cost alignments (A and C) show a reduction of about £200M from the SBC scheme (updated January 1984). This is 4% of total barrage cost and is 7% of total civil engineering cost.

(iii) The barrage should pass north of Steep Holm. Alignment B, which passes south, is significantly more expensive than the other alignments due to its extra length and the deeper water through which it passes between Steep Holm and the English coast. Also, the inclusion of Steep Holm within the basin is less environmentally attractive than leaving it outside.

(iv) There is little cost difference between different landfalls for each alignment, so the choice of landfall will not influence the choice of alignment and can be based on environmental and local planning considerations.

(v) Alignment A should have the least effect on existing fine sediments as it is aligned roughly normal to existing flows.

(vi) Alignment A is preferred on present evidence. However, confirmation of the proposed alignment and optimisation of the line will only be possible when, more geological information is available and, further engineering studies are carried out to evaluate cost variations more accurately.

11. Following the alignment study a barrage arrangement was

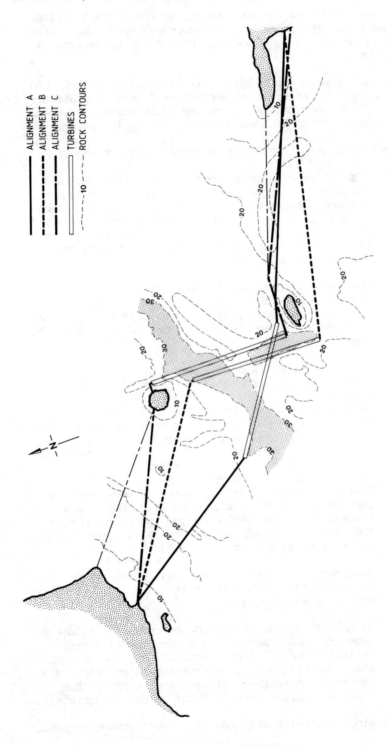

Fig. 1   Alternative Barrage Alignments

prepared based on Alignment A. (See Fig. 2). The barrage length is 16.3 km and is made up as follows:

- 48 turbine caissons (total length 3.5km).
- 62 sluice caissons (total length 3.8km).
- 50 plain caissons (total length 2.7km).
- 3.8km length of main embankment.
- 2.0km length of embankment alongside Brean Down.
- Main shipping lock complex, occupying 0.5km length of barrage, and associated breakwaters
- Small craft lock situated at English end of the barrage.

CAISSON DESIGN

12. Reinforced concrete caissons are proposed for the construction of the barrage, to give long maintenance free life. The conditions of exposure will be very severe but adopting modern standards of design and construction it is possible to achieve a life in excess of 100 years.

13. A separate study of steel caissons was undertaken in 1983-84 by YARD/ Roxburgh and Partners. Although cost comparisons for directly comparable designs have not so far been made, STPG consider it is unlikely that steel caissons would offer construction cost savings. Furthermore, they would carry a continuing higher maintenance cost than concrete.

Turbine Caissons

14. The extent of deep water in which turbine caissons can be placed without excessive rock excavation is limited. It is therefore desirable that the turbine generators should be placed at the closest possible centres to minimise the overall length of the power house. For this reason the box caisson developed by Binnie and Partners for the SBC study was adopted as a basis for the study. Budget restrictions have precluded the development of alternative designs and work was directed mainly towards a check on structural adequacy and the weighting and ballasting of the caisson.

15. It was necessary to modify the caisson design to accommodate the 8.2m diameter turbine generator in place of the previously proposed 9.0m machine. Changes have also been made to simplify the construction and ballasting by reducing the number of diaphragms. (See Fig. 3).

16. Consideration was given to caissons accommodating 2,3,4,5 and 6 turbine generators. Large caissons are advantageous from the construction and programming viewpoints; however, sea towing and placing considerations limit the practical maximum size. It is accordingly proposed to adopt caissons holding four turbine generators.

17. Towards the end of the study, during consideration of

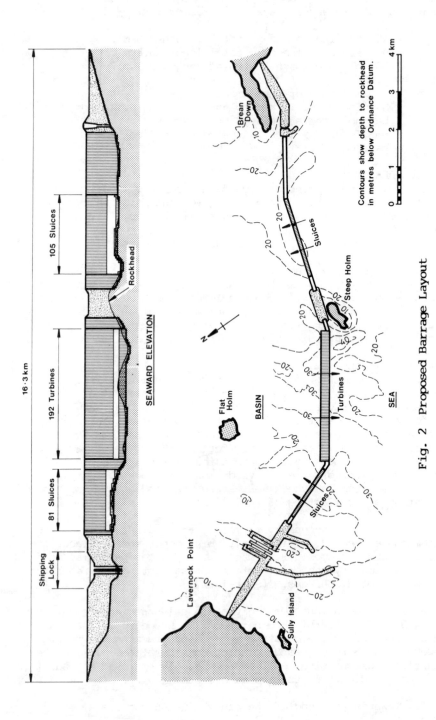

Fig. 2 Proposed Barrage Layout

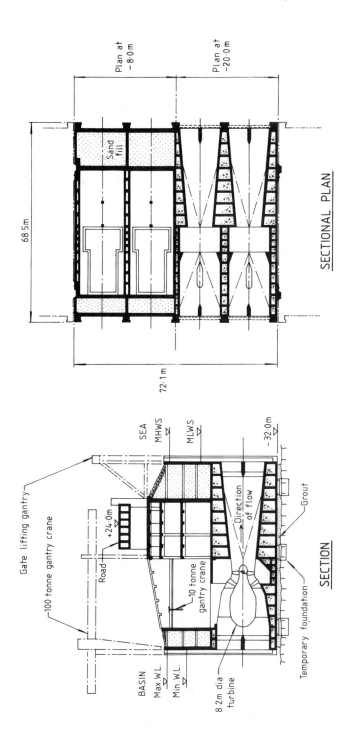

Fig. 3 Turbine Caisson Layout

the English Stones scheme, the turbine caisson design was further developed, mainly to reduce the draught. Two principal modifications were made. First, cylindrical walled cellular construction was adopted above the level of the top of the draught tubes. Second, the compartmentation was modified by the addition of a longitudinal wall separating the turbine access areas from the electrical equipment area. It was then possible to remove the lateral walls subdividing the electrical equipment area and thus to facilitate much more flexible equipment layouts and also provide greater security against flooding for the electrical equipment. These developments, which are illustrated in Paper No. 4 will be equally applicable to the Cardiff – Weston scheme and will produce a cost saving in both caisson and construction facilities.

18. There is a need, during the next stage of work, for a detailed evaluation of alternative structural concepts for the turbine caissons. The internal layout should be reviewed with a view to reducing the space allowances for electrical equipment and possibly simplifying the cranage provisions. If a simpler structural form can be devised, there would be both programme and cost benefits. Consideration should be given to the use of structural steelwork for the suspended floor construction, which is likely to result in reduction of overall weight and therefore draught.

Sluice and Plain Caissons

19. The sluice caisson design incorporating vertical lift gates prepared by Binnie and Partners for the SBC study was accepted by STPG as a basis for study. (See Fig.4). A limited amount of work was undertaken to verify the behaviour of the caissons during towing and installation. It emerged from this work that, whilst the deeper caissons can be placed at low or high water, the shallower units can only be placed at high water, since there is otherwise insufficient depth.

20. STPG propose that considerable lengths of the barrage should be constructed using plain caissons instead of embankment as previously assumed. This has both programme and cost advantages and also permits the introduction of temporary sluices in the plain caissons if required. A design for plain caissons has been prepared during the study (see Fig.5) This design, which has vertical sides, will present a uniform feature in the barrage regardless of bed level and will subsequently present a suitable berthing face for other caissons before they are placed.

CAISSON CONSTRUCTION

21. The method of construction of caissons proposed in the SBC report was reviewed and is regarded as satisfactory and economic. Caissons will be built in a number of separate stages. Initially, they will be constructed in dry docks to a

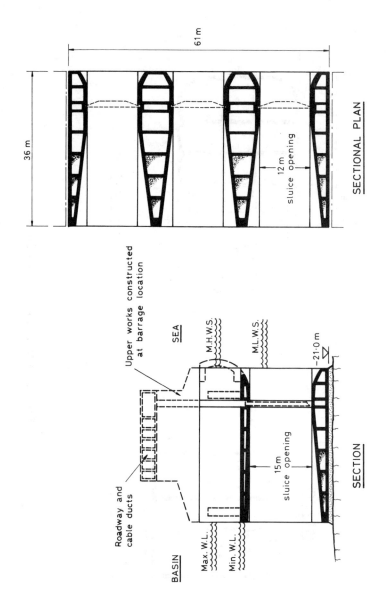

Fig. 4 Sluice Caisson Layout

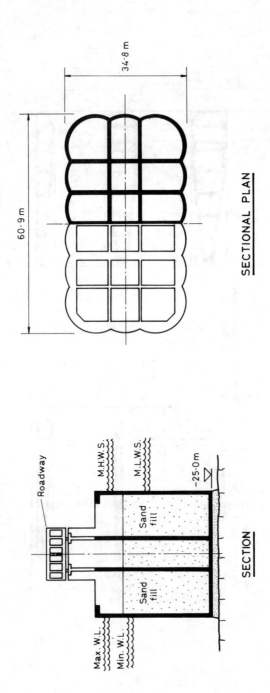

Fig. 5 Plain Caisson Layout

sufficient height to enable them to be floated out.
Subsequent construction will take place at wet berths, where
the partly completed caissons will be ballasted down onto
prepared beds. For turbine caissons, first stage and second
stage wet berths, with different bed levels, are proposed
because of the large height of the caisson. The construction
periods envisaged for individual turbine, sluice and plain
caissons are in the region of 18 months, 8 months and 4
months respectively.

22. The construction planning undertaken indicated that
production rates of 14 turbine caissons, 16 sluice caissons
and 12 plain caissons per year could be achieved and would
match the proposed rate of installation.

23. Purpose built construction yards will be required: for
planning and estimating purposes allowance has been made for
two yards for turbine caissons and two for sluice and plain
caissons. The arrangement of a typical yard is outlined in
Fig.6. However, these matters must be regarded as
provisional pending confirmation of caisson sizes and detailed
surveys of possible sites.

24. During the study a preliminary assessment of possible
sites for construction yards was made, in order to provide a
basis for cost estimating and to establish whether sufficient
potential sites exist. Subject to further consideration of
the wider environmental and social issues it was
concluded that sites for construction of caissons could be
found within the Severn Estuary and Bristol Channel. However,
sites elsewhere should also be considered in future studies.

CAISSON TOWING AND PLACING
25. Strong currents occur naturally in the Estuary,
especially during the spring tide cycles. As construction of
the barrage proceeds, still stronger currents will be
experienced close to the works. The manoeuvring and placing
of caissons needs to be done with some precision and a minimum
of risk : it follows that these operations must take place
during the relatively slack-water periods at about the times
of high water and low water. The proposed limiting conditions
are with currents not exceeding 1.0 m/s, wave heights not
exceeding 2.0m, and wind speed not exceeding 10.0 m/s, for a
duration, or "window", of at least one hour.

26. In the later stages of closure the available
"windows" become significantly shorter and scarcer and in
considering the last closure phase, programmed to take two
months, only tides of 6m range or less will allow such
opportunities (see Fig. 7). In this period four caissons
are to be placed. Occurrence of the required small tides is
irregular, although predictable. Typically, from 2 to 24 such
tides may occur, in different periods of two months. Having

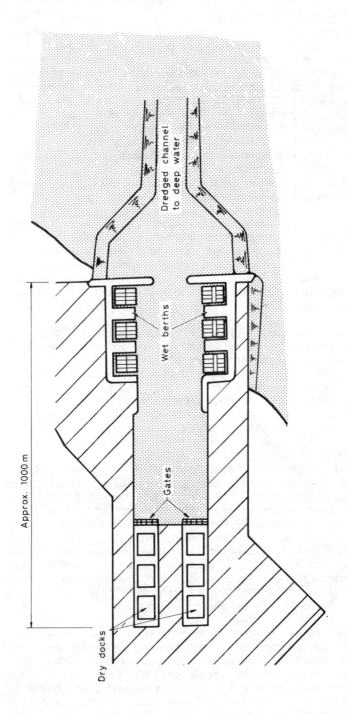

Fig. 6 Typical Caisson Construction Facility

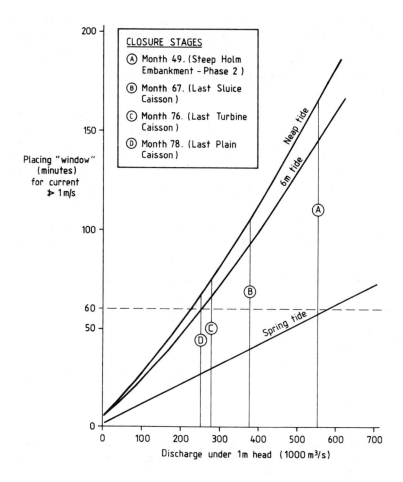

Fig.7 Caisson Placing "Windows"

regard also to the weather criteria, it will clearly be necessary to give attention to the tidal predictions for the specific calendar months when these last caisson placings are proposed.

27. The methods and procedures envisaged for tug attachment and caisson placing are the same as proposed in the SBC report. For the four-unit caisson now proposed for the rather smaller turbine-generator units, the holding force needed in a 1 m/s current is assessed at about 250 tonnes, still safely within the capacity of the four 10,000 h.p. tugs proposed.

28. The sequence of manoeuvres is illustrated in Fig. 8. Just prior to manoeuvring a caisson into position for placing, it will have been water ballasted to about 1m above final level. Ballasting will continue as soon as the caisson is in position, and when the caisson rests on its foundation, rapid general flooding of main waterways and remaining under-water spaces would secure it there.

29. Once the positioning is accepted as satisfactory, underbase grouting and addition of further permanent ballast will follow.

30. Depending on the locations of caisson construction yards, towing to the barrage may occupy between 5 and ·30 hours. Longer tows would include periods towing or holding against tidal currents. A pair of 20,000 h.p. tugs are proposed for these duties. In order that the timings of caisson production and caisson placing can be independent of each other and of the intermediate towing, holding areas are proposed near each end of the journey. There would be a deep water mooring, at least near each turbine caisson yard, and temporary berthing areas beside the already - constructed part of the barrage.

31. In the course of detailed design and planning for these operations it will be appropriate to examine more closely the towing forces and floating behaviour of caissons, by use of both computer and physical modelling. It will also be prudent at some stage to perform some full-scale trials, perhaps on early production caissons, to confirm or refine the arrangements for attachment of tugs, for ballasting, for possible emergency actions, etc. The earlier part of the caisson production programme would allow opportunities for such checks, and appropriate finalization.

DREDGING
32. Dredging is required to provide a level foundation on sound rock for caissons and to provide navigation channels to the shipping locks. The turbine caissons must be founded at - 29m O.D. or lower in order to meet submergence requirements. The plain and sluice caisson foundation levels will be varied

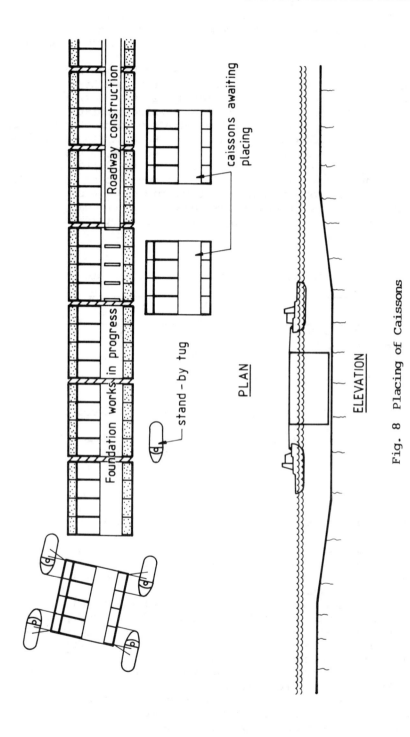

Fig. 8 Placing of Caissons

to suit existing rock levels and so to minimise dredging. Sufficient rock will be dredged to ensure the removal of weathered and unsuitable material and create a level surface.

33. There are three geological strata in the barrage location: Carboniferous Limestone, Trias and Lias. It is anticipated that the Carboniferous Limestone will require blasting whereas the Trias and Lias, which consist of beds of mudstones, shales and limestones with some gypsum veins, can probably be dredged without blasting. Rock has therefore been divided into 'hard' and 'soft' categories, the former including the Carboniferous Limestone and the latter Trias and Lias. Approximate quantities to be dredged are shown in Table 1. (It must be emphasised that the division into 'hard' and 'soft' rock is a preliminary judgement which can only be confirmed when rock strengths and strata thicknesses are known.)

Table 1
Dredging for Caisson Foundations - Approximate Quantities

	Sediment M.m^3	Soft Rock M.m^3	Hard Rock M.m^3
Turbine Caissons	3.2	1.9	0.2
Sluice & Plain Caissons	0.1	0.8	0
Total	3.3	2.7	0.2

34. In addition to rock type the choice of dredging methods is affected by wave conditions, current and water depths. The deepest level of rockhead along the barrage line is approximately - 35 m O.D. If dredging is to continue at all states of the tide, for the majority of tidal ranges, this implies a maximum water depth of 40 m. However, it may be more economical to restrict dredging for the deepest areas to half the tidal cycle rather than provide a dredger which can operate under the worst combination of tide and depth.

35. During the SBC studies the view of most dredging contractors was that a purpose-built, jack-up cutter suction dredger should be used for 'soft' rock because of the severe tidal conditions and the large water depth. Now, the consensus of opinion favours the use of large ship-type cutter suction dredgers modified for increased depth and supplemented with large grab dredgers for the deepest water and for removal of blasted material. This change of view is due to the development of much larger ship-type cutter suction dredgers coupled with the lack of success with the only jack-up dredgers that have been built due to a combination of economic circumstances and development problems. In addition, there would be an unacceptable risk to the project in relying on a single purpose-built dredger of new design.

36. It is proposed that drilling and blasting is carried out from a large jack - up platform and the blasted rock removed by a large grab mounted on a floating pontoon or jack- up barge. It is anticipated that a secondary blasting operation on a close grid pattern will be required under the caisson locations to minimise overbreak.

37. A difficult problem is how to deal with loose rock below formation level. A cutter suction dredger leaves a layer of disturbed material 1 to 2 m deep depending on cutter head diameter. Also, there will be rock fragments left by blasting operations. In both cases a large grab could be used to clear the foundation over the width of the caisson.

38. Alternatively, the loose or disturbed rock could be left in position and the voids filled using recently developed underwater grout or concrete materials.

CAISSON FOUNDATIONS

39. A foundation system is required which will permit safe and reliable installation of caissons in the difficult operating conditions in the Severn Estuary and provide a sound and durable foundation during the life of the barrage. The system chosen must permit rapid placing of caissons and allow them to be levelled with reasonable accuracy. In addition it must accommodate tolerances in caisson placing, support the caisson temporarily and incorporate an adequate system for grouting the gap between caisson and seabed.

40. Three foundation schemes have been reviewed by STPG (See Fig.9):
   - the use of 6 m diameter concrete stub piles formed by drilling within a steel casing, with flat jacks to provide even load distribution and permit levelling of the caisson. Flexible grout bolsters would be used to form a seal around the caisson edges.
   - the use of steel piles jacked down from the caisson and driven into the rock surface under the weight of the ballasted caisson, with grout bolsters as for the stub pile scheme.
   - the provision of downstand beams along the edges of the caisson which seat in predrilled trenches in the rock. This system was used by Taylor Woodrow for Dubai Dry Dock.

41. The downstand beam proposal is not favoured at present mainly because of the additional draught due to the beams beneath the caissons. Also it might prove to be an expensive method in view of the requirement for continuous and very accurate pre-drilled trenches at all downstand beam locations. Nevertheless, it retains the advantage of providing a reliable edge seal for the grouting of the void

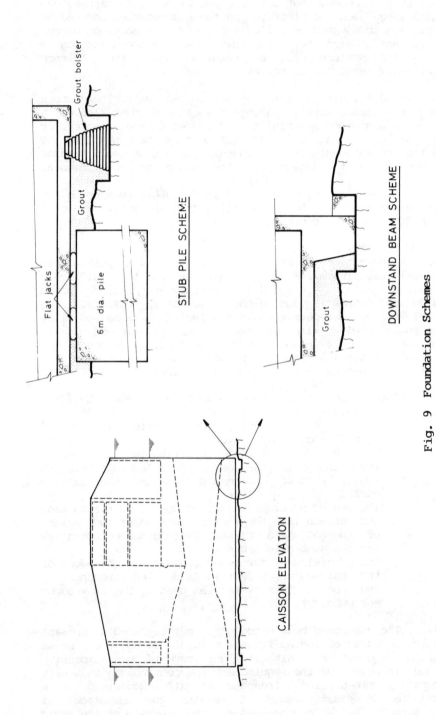

Fig. 9 Foundation Schemes

beneath the caissons. Of the other two proposals, that
involving concrete stub piles forming temporary pad
foundations is preferred on account of its relative simplicity
and greater reliability. However, the grout bolster method of
sealing the caissons to grout the foundations needs
considerable development work. It will have to be reliable in
the comparatively rough seabed conditions which will exist
after dredging. It is anticipated that further preparation of
the seabed on the lines of the bolsters will be needed in some
areas after dredging, in order to ensure an adequate seal.

EMBANKMENT DESIGN AND CONSTRUCTION
42.   STPG have adopted the traditional embankment design
proposed by the SBC.   A Typical embankment cross-section is
shown in Fig. 10 .   The embankment consists of a rockfill
structure to gain initial control over tidal currents, sand
fill, filters and armouring of rock and concrete units.

43.   The design of the embankment is governed by water
levels, tidal velocities, wave climate, foundation stability,
and the space required for cables and service road.

44.   The total quantities and rates of supply of rock and
sand required for the embankments, breakwaters and the sand
island for lock construction are shown in Table 2.   The
maximum rock sizes are 6 t for breakwaters and 4 t for the
embankment.

Table 2
Materials for Embankments, Breakwaters and Lock Sand Island

Material	Total volume M.m^3		Maximum Supply Rate M.m^3	
Rockfill	9.6)	14.1	2.7)	4.0
Rock Armour	2.5)	(total	0.7)	(total
Filters	2.0)	rock)	0.6)	rock)
Sand Containment	2.1		0.5	
Sand	24.1		16.0	

Note: 1)   Concrete armouring units not included.
      2)   Quantities based on locks for 150 000 dwt vessels.

45.   Rock represents about two thirds of the cost of
embankment construction and thus STPG studies were directed
primarily at investigating sources of supply, transport
methods and costs of rock.

46.   It was concluded that the requirement for rock can
probably be met by the limestone quarries of South Wales and
the Mendips.   The larger sizes might be supplied without
causing much disruption to the normal flow of graded materials
by selecting blocks from several quarries before primary
crushing.   The most efficient and environmentally preferable

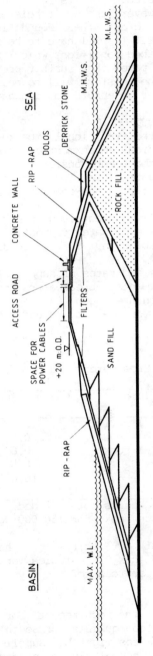

Fig. 10 Typical Embankment Cross-Section

methods of transporting rock are by rail and by sea. There are rail-connected quarries in South Wales and the Mendips at present and in addition there are possible sources of supply outside the U.K. The supply of rock will clearly be a major operation and will require detailed study in conjunction with both major suppliers and local authorities to establish the best overall scheme.

47. There is a plentiful supply of sand in the Severn Estuary and the 24 M.m³ required represents about 1% of existing deposits. Sand could be dredged and either pumped into place or bottom dumped, depending on location, very economically.

48. STPG's studies suggest that the lengths of embankment should be minimised because the alternative, which is to use plain caissons, uses much smaller volumes of materials than embankment. The environmental and logistical problems of transport of materials for embankments, rock in particular, will be eased if maximum use is made of caissons. Furthermore, the use of caissons gives greater flexibility in planning the construction sequence and allows the introduction of additional temporary sluice openings.

49. The cost estimates prepared by STPG show that embankment becomes cheaper than plain caissons for a rockhead level above about -20 m O.D. However, STPG propose the use of plain caissons below a level of -15 m O.D. because the benefits outlined above are judged to outweigh the relatively small extra costs of caisson for levels above -20 m O.D.

NAVIGATION LOCKS
50. STPG have accepted SBC's proposal to provide two shipping locks sited at the Welsh end of the barrage. The sizing of the locks depends on projections of shipping usage over the life of the barrage and will be subject to interests other than those of STPG.

51. For the purposes of the study, STPG have considered upper and lower bound lock schemes which would accommodate vessels up to approximately 150,000 dwt and 10,000 dwt respectively. The sizes of the locks considered are as follows:

	Upper Bound Scheme	Lower Bound Scheme
Length	370m	140m
Width	60m	25m
Sill Level	-19m O.D.	-10m O.D.

Also, a small craft lock is proposed, located at the English end of the barrage.

52.   Most of the development work carried out during the
study was for the upper bound scheme.   For this scheme it is
necessary to construct the locks in the deep water of the
channel to Cardiff, which means that they must be operational
before the only other deep channel, which passes between Flat
Holm and Steep Holm, is obstructed by the advancing line of
turbine caissons.   (Navigation routes are shown on Fig.11 and
the upper bound scheme on Fig.12).   The study showed that this
programme requirement would not be met by constructing the
locks using caissons.

53.   The proposed construction method is diaphragm walling
within an artificial sand island.   This technique permits
construction to commence early in the programme.   The island
would be constructed by pumping sand from the estuary into a
progressively raised protective rock rubble bund.   Diaphragm
walling would be constructed through the sand island and into
the underlying rock to form self stable substructures similar
in concept to those constructed at the Royal Portbury Dock in
Bristol.

54.   The lower bound lock scheme (see Fig.13) can be
constructed closer to the Welsh shore.   Since the deep channel
to Cardiff can then be used by shipping after the main channel
has been blocked, lock construction is no longer on the
programme critical path and thus there is the possibility of
reducing the overall construction period.

55.   The lower bound scheme has the added advantages of
needing much less dredging for navigation and permitting an
additional length of the barrage to be constructed using
caissons, with corresponding benefits in terms of increased
construction flexibility and reduced requirement for
materials.   The net effect of these changes is a cost saving
to the project from adoption of the lower bound lock scheme of
about £300m.   However it must be emphasised that the size of
locks to be provided needs further study and discussion with
all interested parties.

56. Diaphragm walling construction has been assumed for the
lower bound scheme.   Whilst this appears to be a satisfactory
solution, it is proposed that further consideration is given
to the use of caissons for the locks, as there may be spare
capacity in the construction yards manufacturing barrage
caissons.

57.   Whatever size of shipping locks is adopted,
breakwaters will be needed to shelter the seaward approaches
to the locks.   The rubble mound breakwater layouts adopted in
the study were based solely upon judgement.   Clearly model
studies will be necessary in order to determine the
requirements.

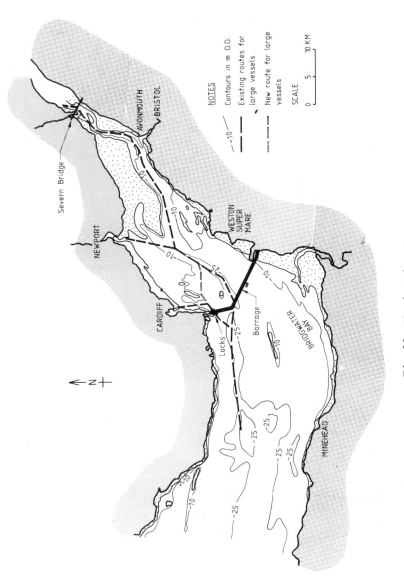

NOTES

−10   Contours in m O.D.

       Existing routes for large vessels

       New route for large vessels

SCALE

0   5   10 KM

Fig. 11 Navigation Routes

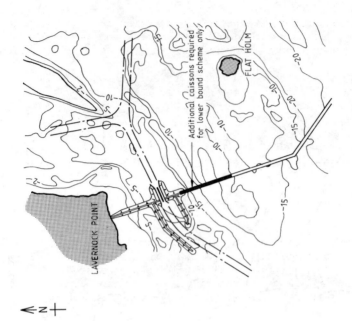

Fig.13 Lower Bound
        Locks Scheme

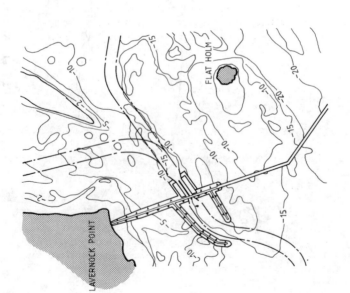

Fig.12 Upper Bound
        Locks Scheme

BARRAGE CONSTRUCTION SEQUENCE AND PROGRAMME

58.   The   proposed   programme gives an overall duration   of
construction   to   barrage   closure of   7   years   (84   months)
compared with 9 years to closure proposed by the SBC.   The
principal   reasons   for the reduction in duration are,   firstly
the use of plain caissons in lieu of embankment and, secondly,
closure by caissons rather than embankment.

59.   The   proposed programme and construction sequence   are
shown   on Figs. 14 and 15   (upper bound lock scheme).   A
construction   sequence   which gives priority to completion   on
the   Welsh side of the barrage was also considered by STPG.
However,   the sequence shown in Fig. 14   is preferred because
it   provides   earlier   access   from   the   English   shore   for
transmission   installation   and gives   a   shorter   overall
duration.

60.   The main features of the programme are as follows:

   (i) Caisson   construction   and installation rates   are
       matched   to give an installation rate of 28 caissons
       per year initially,   increasing to a peak of 42   per
       year - when turbine,   sluice and plain caissons   are
       all   being   placed - reducing to less than   one   per
       month   during   installation of the final   plain
       caissons.

  (ii) Caissons   are placed on four working fronts.   The
       proposed   sequence of installation would be   adjusted
       if   studies   show   this to be necessary to control
       sediment movements Also, temporary sluices could be
       provided in plain caissons if necessary.   The barrage
       is   completed by installation of a special plain
       closure caisson on the corner at the Welsh end of the
       turbine caissons.

 (iii) The   timing of embankment construction is determined
       on the Welsh side by the need for access to construct
       the   locks   and on the English side by   the   adjacent
       caisson   installation and the access requirements for
       electrical transmission installation.

  (iv) The   average   rate of   progress   on   dredging
       will probably exceed the rate of caisson installation
       and   thus   the critical aspect of the dredging and
       foundation   preparation programme will be to meet the
       requirements   for   installation of the first   few
       caissons.

   (v) The   majority of sluice   and   turbine waterways
       must be kept open until the last caisson is placed to
       minimise   water velocities in the remaining gaps.
       Following the installation of the last caisson,   the

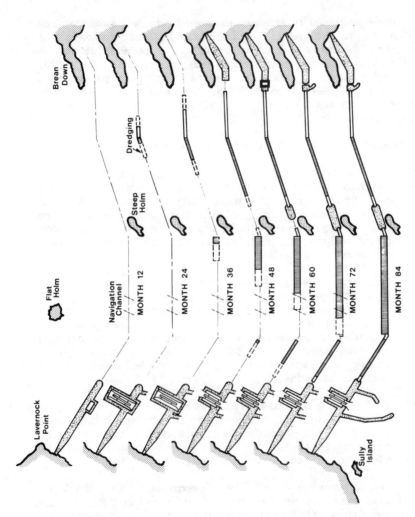

Fig. 14 Proposed Barrage Construction Sequence

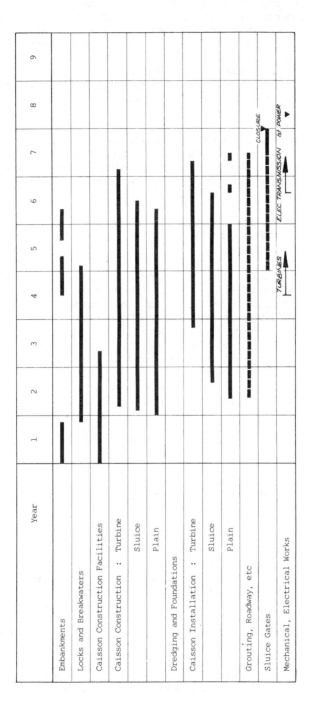

Fig. 15   Proposed Barrage Construction Programme

barrage must be closed as soon as possible to permit
generation and revenue earning to commence. The
large number of waterways to be closed will make the
logistics of this exercise very complex and a
detailed programme has not been attempted at this
stage. However, a period of 6 months has been
allowed after the installation of the last caisson
(Month 78) to achieve closure.

(vi) Access for installation of the first 8 turbines (i.e.
access to the first two caissons) will be available
at Month 42. Initially access for men and equipment
will be by water until Month 69 when access will
become available along the barrage from the English
shore.

(vii) Access for electrical transmission installation
will be available along the barrage from the English
shore at Month 69.

61. The proposal to provide only small locks has been
referred to earlier. Apart from the significant cost
reduction this proposal would have advantages to the barrage
layout, construction sequence and programme. It has been
assumed at present that the overall programme duration to
closure of 84 months would be unaltered by the substitution of
small locks but this could be reduced if the rate of caisson
construction and installation can be increased. In theory
caissons could be constructed at any desired rate by providing
an appropriate number of construction yards. However an
increase in the rate of caisson installation is limited by
availability of suitable tidal conditions. Thus the minimum
overall barrage construction period will be determined
primarily by the maximum rate at which caissons can be
installed. It is considered that alternative rates of
installation should be studied further in future studies to
determine the optimum overall programme for project economics.

COST ESTIMATE
62. The cost estimate prepared by STPG for the civil
engineering works is shown in Table 3. The contingency
includes allowances for uncertainties in quantities and
pricing, for design development and for problems in
construction and installation. The contingency for each item
was assessed separately and this resulted in an overall
allowance of 20%. This figure would be re-assessed when
further site data is available and design has progressed
further.

63. Project management and engineering has been allowed for
separately in the overall project cost estimate.

Table 3
Cost Estimates for Civil Engineering Work   (£M)

Shipping Locks (Inc. Breakwaters)	393	(144)
Caisson Construction Facilities	343	
Caisson Construction - Turbine	864	
- Sluice	281	
- Plain	158	
Dredging for Caisson Foundations	159	
Caisson Installation - Turbine	305	
- Sluice (inc. gates)	359	
- Plain	90	
Embankments	285	
	3237	(2988)
Contingency	647	(598)
Total	£3884M	(3586)

Note: Figures in brackets are for lower bound lock scheme

64. The main changes from the cost estimate prepared by the SBC (£4300M up dated to 1984) are due to the substitution of plain caissons for embankment, the alteration in barrage alignment and the re-assessment of contingencies. An important aspect of the latter is the STPG intention that outline design should be complete and detailed design well advanced before construction commences so that changes during construction are minimised.

ACKNOWLEDGEMENT
65. This paper is based on work carried out for the Severn Tidal Power Group Interim Study which was part-funded by the Energy Technology Support Unit on behalf of the Department of Energy. Permission to publish the paper is acknowledged.

4. Civil engineering aspects of an English Stones barrage

C. J. A. BINNIE, MA, DIC, FICE, FIWES, *Director, W S Atkins &
Partners,* and D. E. ROE, BSc, MICE, *Group Engineer, Wimpey
Major Projects Ltd*

SYNOPSIS. This paper describes the civil engineering aspects
of a barrage generating electricity at the English Stones.
Following an introduction to the scheme, the estuary at English
Stones and arrangement and alignment of the barrage are
discussed. Each of the elements of the barrage is described
with regard to design and construction and cost estimates are
presented. Finally conclusions are given.

INTRODUCTION
History of English Stones Scheme.
1. The first scheme for generating electricity at the
English Stones site was proposed over seventy years ago, in
1910. The Severn Barrage Committee appointed in 1933
recommended construction of an 800 MW scheme. The scheme was
reviewed again in 1943 but no action was taken. In 1977 A V
Hooker of W S Atkins & Partners proposed another scheme at the
English Stones. This was considered by the Severn Barrage
Committee and reported in Energy Paper Number 46 (Ref 1). The
scheme was discounted mainly due to its small energy output in
comparison with the larger schemes further down the Estuary.
2. In 1983, Wimpey Major Projects Limited and W S Atkins and
Partners formed a Joint Venture to promote the
scheme. Proposals for a pre-feasibility study were submitted
to the Department of Energy in November 1983 (Ref 2).
Subsequently the Wimpey Atkins Joint Venture represented by
Wimpey Major Projects Limited, joined the Severn Tidal Power
Group who had been appointed to undertake Interim Studies of
the Cardiff Weston scheme. These studies were then expanded to
cover the English Stones Scheme.
Interim Study.
3. The main objective of the interim study was to assess the
commercial viability of a privately owned and operated barrage
selling power to the public grid system. This paper describes
the civil engineering aspects of the English Stones
Barrage. The work was undertaken between February 1985 and
April 1986 as part of the Severn Tidal Power Group's Interim
study.

Availability of data.

4. The studies of the barrage have been completed using data collected during previous work and from other relevant sources. This data was more limited than that available for the Cardiff Weston Scheme as this part of the estuary has not been covered in as much detail during previous studies.

THE ESTUARY AT ENGLISH STONES

Location.

5. The English Stones is an area of rock outcrop located just upstream of the proposed barrage site. The location of the English Stones barrage in relation to the Cardiff Weston scheme and the Severn Estuary is shown in Figure 1.

Topography and Bathymetry.

6. The main feature of the area is The Shoots channel through the English stones. This channel is 3 km long, 350m wide at its narrowest point, with bed levels varying from -20 to -25m O.D. Each side of the channel, rock outcrops are exposed at about 0.0m O.D. Some 2.5 km downstream of The Shoots the main channel widens and the bed level rises to -15.5m O.D. About 1.5 km north of The Shoots the bed levels rise steeply to above -10m O.D. Large areas of sand and mud are exposed at low tide. Approximately 75% of the proposed barrage line is above mean low water level of spring tides.

Geology and Soils.

7. Underneath the main channel a thick layer of sandstones, referred to as the Pennant Sandstones,is likely to be found. Either side of this channel, a thick layer of Carboniferous Sandstone is overlain unconformably by the Triassic strata comprising red or red-brown thinly bedded sandstones, argillaceous siltstones and mudstones which are calcareous at depth (Keuper Marl).

8. Only sparse data were available to interpret sediment thickness. On the western side of the channel sediment thickness varies from 1m to over 10m adjacent to the Welsh shore. Some small tidal channels and creeks cross the intertidal flats and these typically have an upward sequence of basal gravel overlain by clay, silt and crossbedded sands, followed by muds and sands at the surface. In the main channel the thickness of sediments increases downstream from a thin veneer up to 3m thickness.

Tides and Currents.

9. The existing tide regime at the site will be modified by the introduction of a barrage. The effect of the barrage depends on the design and numbers of turbines and sluices installed and on their mode of operation. A one-dimensional model of the estuary was used to predict the changes in tidal regime. The main changes can be summarised as follows:

- immediately seaward of the barrage the tide range would be slightly reduced, in particular the low water level would be raised by 0.5m.
- landward of the barrage the minimum water level would be about +2m O.D.

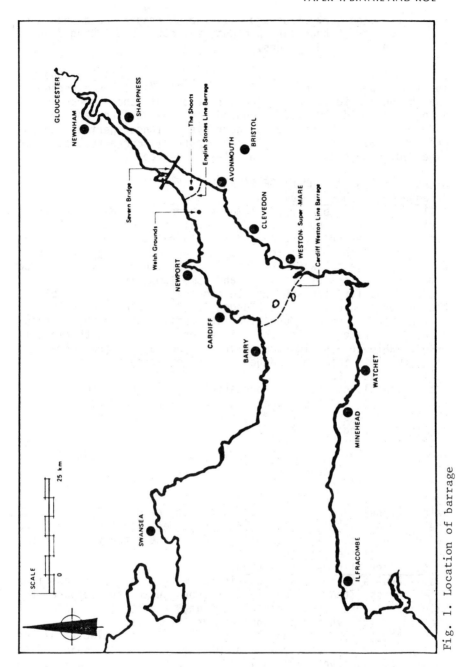

Fig. 1. Location of barrage

-      the tide curve landward of the barrage would be more asymetric than at present, the ebb phase taking 7 hours and the flood 3 hours.

10. Due to the narrowness of the channel at The Shoots, localised peak velocities in excess of 5m/s can be expected on the highest tides. After construction of the barrage, ebb velocities would be much reduced as the controlled release through the turbines would occur over a longer period than the present ebb. During flood, the upstream currents would be expected to be reduced, but to a lesser degree.

ARRANGEMENT AND ALIGNMENT OF BARRAGE
Arrangement of barrage elements.
11. The main barrage elements are turbines, sluices, navigation lock and embankment. From the energy evaluation work the preferred turbine generator arrangement was determined as 36 units with 7.5m diameter runners and each with an output of 27 MW together with 42 sluices of the submerged venturi type some with openings of 12m x 11m and some 12m x 14m.

12. To satisfy the submergence requirements of the turbines, the power house must be founded at -25.5m O.D. The only area that provides a good foundation at sufficient depth, without excessive dredging, is the main river channel. Furthermore it is desirable to position the powerhouse normal to the direction of flow to minimise changes to the existing flow pattern and sediment movements.

13. The sluices require foundation levels of -15m O.D. and -18m O.D. In addition the sea bed upstream and downstream of the caissons must be dredged to slopes of   1 on 10 rising to meet existing sea bed levels.

14. A shiplock will be required to allow passage of commercial shipping to and from the ports upstream of the barrage and to cater for pleasure craft passing up and down the Estuary. Regarding the location of the shiplock there are conflicting requirements between keeping the lock within the main channel and locating it away from the water flows through the turbines and sluices. The preferred location for the navigation lock is on the Welsh side of the estuary, within the section of sluice caissons.

Criteria for alignment selection.
15. Four alignments were considered within a 1.5km corridor at English Stones. The main factors used to compare the alignments were cost, energy output, amount and ease of dredging, hydrodynamic and environmental impacts.

16. About 80% of the barrage cost is for electrical and mechanical works, caissons and navigation locks, which were considered constant for all four lines. Thus a change in alignment affected only 20% of the total. The estimated amount of energy available from each alignment varied by about 3.5%, due to the difference in basin areas.

Preferred Alignment.
17. The preferred alignment was chosen because it has the
lowest cost, least dredging for sluices and turbines, lower
current velocities during construction and short navigation
channel although it has the longest embankment. A plan and
part longitudinal section of the preferred alignment are shown
in Figs 2 and 3. The main features are listed below:

- 36 No 7.5m diameter turbines housed in 12 concrete
  caissons located in the main river channel, founded at
  -25.5m O.D.
- 30 No 12m x 11m sluices and 12 No. 12m x 14m sluices
  housed in 14 concrete caissons, founded at -15m O.D. and
  -18m O.D. These are located either side of the turbine
  caissons.
- A single navigation lock 140m x 20m for ships of 6000
  DWT located in the Welsh side sluice section and founded
  at -15.5m O.D.
- Plain caissons will be provided only between the
  embankment and the sluices.
- Transmission cables will be routed on the Welsh side,
  with a substation on the plain caisson adjacent to the
  sluices.
- The embankment is likely to be a composite rock and sand
  fill structure.

CIVIL WORKS DESIGN
Concept
18. The structural elements for a barrage at English Stones
would be similar in many respects to those for the Cardiff
Weston scheme. The main difference lies in the smaller scale
of the English Stones scheme, as may be seen in the following
comparisons:-

	English Stones	Cardiff Weston
Overall length of barrage	7.1 km	16.3 km
Average water depth at MHWS on line of barrage	10 m	22 m
Number of turbines	36	192
Number of sluices	42	186

19. Similarities are evident between the physical
requirements for the civil works elements of the two schemes.
The design concepts developed for Cardiff Weston are therefore
in general applicable to English Stones. In particular it is
considered that, for the main structural elements of the
barrage, the caisson method of construction is most suitable.

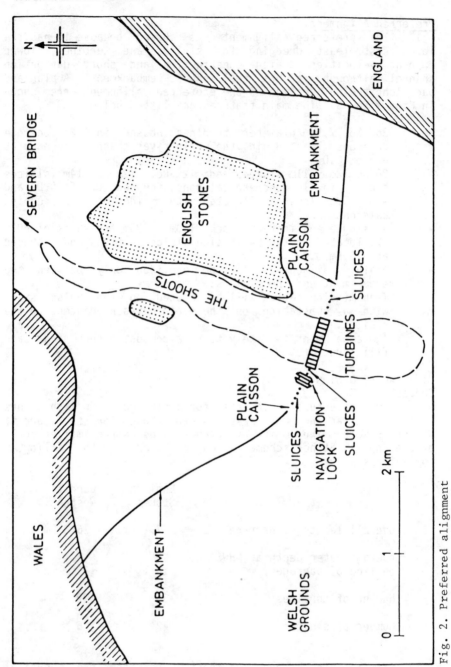

Fig. 2. Preferred alignment

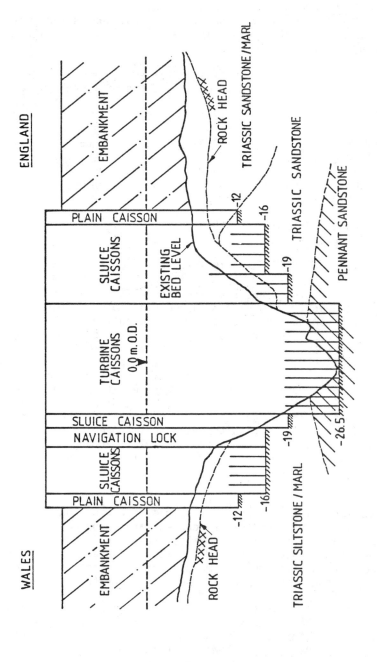

Fig. 3. Part of longitudinal section

Caisson Structures

20. Principal data for the various caisson structures are given in the table below and described in more detail in the following paragraphs.

	Turbine Caissons	Sluice Caissons	Plain Caissons	Ship Lock
Number required	12	14	2	2
Width along barrage (m)	50.4	58	65	30
Length across barrage (m)	63	34	40	160
Foundation level (m O.D.)	-25.5	-15 & -18	-11	-15.5
Towing draft (m)	18.7	11	8	7

21. The turbine caissons are designed to house three generators each. It is considered desirable to limit the size of each caisson to minimise towing and handling forces and to increase manoeuvrebility bearing in mind the strong currents and limited searoom at the site. Smaller caissons, for less than three sets, would suffer from stability problems during installation. Caisson designs were developed from the Cardiff Weston model, modified for the smaller turbine diameter and shallower foundation level. The modifications included the addition of a bulkhead wall to separate the turbine chambers from the transformer and switchgear areas to reduce the risk of flooding. Also an alternative cellular arrangement of the upper ballast chambers and a reduction in the number of diaphragm walls were introduced to reduce the weight and towing draft of the caisson and to simplify the construction. It is desirable to keep the towing draft as low as possible to allow maximum flexibility in the operations of towing and placing the caissons. Two 100 tonne gantry cranes are provided to handle the temporary steel limpet gates required for dewatering and maintenance. Each turbine cell would have a 10 tonne travelling crane for routine maintenance. Transmission cables would be installed in cable ducts incorporated in the precast roadway units.

22. For the sluices, vertical lift gates operating in conjunction with a submerged venturi-type water passage are envisaged, as for Cardiff Weston. The length of caisson parallel to the water flow is determined by the required exit configuration for optimum discharge rather than by structural stability requirements. There are three sluices per caisson.

23. Plain caissons are used only at the two junctions of embankment and sluice sections. They are of cellular construction, sand filled for stability. The plain caisson on the Welsh side supports the main electrical substation

building. This allows electrical installation work in the substation and interconnection with the generation section to be commenced before completion of the Welsh embankment.

Embankment Design
24. The embankments fulfil three functions:

- to retain water for power generation
- to carry an access road to the turbines and sluices
- to carry ducts for the transmission cables.

25. The lowest natural foundation level of about -6m O.D. is near to mean low water spring tide level of -5.2m O.D., so that most of the embankment is above minimum tidal level. Over most of the intertidal zone the foundations consist of a few metres of sand and gravel overlying bedrock. Close to either shore there are mud deposits.

26. A design wave height of 2.5m has been adopted, taking account of the partially sheltered conditions at the English Stones site. The barrage crest elevation has been selected at +11.5m O.D. with a wave wall on the seaward side projecting 2m above this level. These levels are determined from predicted water levels, wave heights and wave run-up. A crest width of 17m provides adequate space for the access road and for power transmission and other cables.

27. The main design features and proposed method of construction can be summarised as follows:-

- An anti-scour blanket of suitable rock placed ahead of main embankment construction.
- A rockfill control bund placed on the seaward side of the embankment.
- Quarry waste bunds placed in stages along the upstream face.
- Filter layers placed on the inside slopes of the rockfill and quarry waste bunds.
- Dredged sand filling placed in stages between bunds.
- Rip rap protection to upstream and downstream faces of embankments, placed in stages to suit embankment construction.

Ship Lock
28. Existing port facilities above the proposed barrage site are mainly at Sharpness and Gloucester and to a lesser extent at Chepstow and Lydney. The docks at Sharpness can accept vessels up to 16.8m beam and 6.55m draft, on suitable tides. These parameters would imply a maximum overall length of vessel of about 125m. The proposed size of the ship lock through the barrage is therefore 140m in length between gates, 20m width, with a sill level of -5.50m O.D. Proposals by Gloucester Harbour Trustees to cater for LASH (Lighter Aboard SHip) vessels travelling to Northwich Roads to offload, would require a much larger lock.

29. The favoured location for the lock, for optimum

navigation conditions, is on the Welsh side of the main channel, within the sluice section. Commercial vessels travelling upstream would generally lock through an hour or more before high water; those travelling downstream would lock through after high water. Pleasure craft would use the same lock at different times from the commercial traffic and special penning facilities are provided to ensure their safety during locking.

30. A suitable foundation level at the proposed lock site is some 10m lower than the required sill level. This feature lends itself to the use of a cellular base to the lock caisson, providing buoyancy and a convenient space for housing rising sector gates. Unlike vertical sector gates these do not require housings in the lock walls and therefore allow the use of a relatively narrow caisson structure. This would allow the option of constructing the caisson in an existing dry dock, instead of in purpose-built facilities, with consequent advantages to the construction programme.

31. Use is made of the cellular base to accommodate cable tunnels allowing the power transmission lines to pass under the lock.

32. Upstream and downstream of the lock entrances, lead-in jetties are planned to assist vessels entering and leaving the lock and provide moorings for those awaiting passage through the lock. These jetties comprise lines of piled dolphins, suitably fendered, with mooring facilities and interconnecting walkways.

CIVIL WORKS CONSTRUCTION
Dredging

33. Dredging is required;

- to provide a level foundation for the various caisson structures
- to allow unobstructed flow of water to the turbines and sluices
- to provide manoeuvring space for caisson installation
- to provide a safe parking area for caissons being moved to the barrage site.

Dredging is also required for the caisson construction yard and to form a channel through the "Denny Bar" area off Avonmouth to enable the caissons to be towed to the barrage site.

34. Excavation in the Pennant Sandstone and possibly some of the harder strata in the Keuper rocks would require drilling and blasting. Excavation in the softer rock, most of which it is expected can be excavated without blasting, would be carried out using a heavy duty cutter-suction dredger.

Caisson Prefabrication

35. Not including the shiplock, there are 28 caissons to be prefabricated. This could conveniently be carried out in a single caisson construction yard with two "production lines",

one for turbine caissons and one for sluice and plain caissons. Ideally the yard should be located near to the main navigation channel of the estuary and within easy reach of the barrage, allowing the tow to the barrage site to be accomplished on a single flood tide. Each "production line" consists of a dry dock with removable gate, in which the first stage construction would be carried out, and a wet basin with fitting out berths for completion of the caissons to full height and fitting out with limpet gates and equipment for controlling the flooding and installation of the caissons.

Caisson Foundations

36. Foundation details for the caissons of a barrage at English Stones and their methods of construction are similar to those proposed for the Cardiff Weston Scheme. Stub piles about 5 metres in diameter would be constructed by drilling a short distance into bedrock using a full face drilling machine supported on a jack-up barge. An oversize casing would be used to protect the drilling shaft from waves and tidal currents. The stub pile would be concreted and finished to a plane surface before removal of the casing. The caisson would be landed onto a system of linked flat jacks installed on the tops of the stub piles. Collapsible bolsters would be lowered around the perimeter of the caissons to retain grout pumped in around the stub piles to fill the gap between the rock surface and the underside of the caissons.

Caisson Towing and Placing

37. The width of the main channel at Chart Datum level (-6.5m O.D.) is about 1200m at the barrage, narrowing to about 300m in The Shoots channel some 2000m upstream. The existing tidal currents near the proposed site of the barrage and particularly in The Shoots channel are already fast and turbulent. The tidal range at the site is greater than elsewhere in the estuary. The operation of towing-in and positioning the caissons, and sinking them onto prepared foundations, must be carefully planned and carried out using methods and equipment to cater for all eventualities. The construction sequence for the barrage has been arranged to keep tidal currents as low as possible while the biggest caissons, for the turbines and shiplock, are being placed. Also part of the embankment construction in the deeper water would be delayed until all caissons have been placed. Current velocities and the lengths of time when currents in the remaining gap do not exceed values of 0.5 and 1.0 metres/second have been estimated using a one-dimensional mathematical model, for various stages of construction and tidal states. These are summarised in Table 1.

38. To make use of the longest periods of low current velocities, and to allow the caisson placing operation to be readily aborted if necessary, it is planned to place the majority of caissons at low water slack of a neap tide. Sufficient over-dredging must be done to enable the floating caissons to be safely manoeuvred into position in these low water conditions. Also the rate of ballasting after touchdown must be fast enough to ensure positive contact

TABLE 1
WATER LEVEL DIFFERENCES AND VELOCITIES DURING CLOSURE

Closure Stage	0	1	2	3	4
Area Blocked (%)					
Flood	0	15	45	60	73
Ebb	0	9	39	55	70
Peak Av. Velocity					
Caisson Gap (m/s)					
Spring Flood	2.0	1.90	2.1	2.97	-
Spring Ebb	1.3	1.53	1.8	2.75	-
Neap Flood	0.8	0.98	1.1	1.43	-
Neap Ebb	0.75	0.90	1.0	1.58	-
Peak Av. Velocity					
Embankment Gap (m/s)					
Spring Flood	1.7	1.01	2.0	3.3	4.86
Spring Ebb	1.3	0.81		2.83	5.15
Neap Flood	0.8	0.47	1.0	1.38	-
Neap Ebb	0.75	0.43		1.46	-
Time Gap Velocity					
0.5 m/s (mins)					
HW Spring	34	36	35	16	18
LW Spring	32	24	15	12	126
HW Neap	90	93	70	51	-
LW Neap	105	97	70	51	-
Time Gap Velocity					
1.0 m/s (mins)					
HW Spring	102	81	74	42	23
LW Spring	288	120	81	45	156
HW Neap	not exceeded	not exceeded	225	114	-
LW Neap	not exceeded	not exceeded	330	148	-

Note: Closure Stages are defined as follows:-

Stage	Number of caissons placed		Embankment completion	
	Turbine	Sluice	Welsh	English
0	NIL	NIL	NIL	NIL
1	6	NIL	30%	NIL
2	11	7	55%	NIL
3	12	13	95%	30%
4	12	14	100%	95%

between caisson and foundations during the subsequent tidal rise.

39. Current velocities of 0.5 m/s and 1.0 m/s are considered to be the desirable maximum and limiting values respectively for caisson placing. Durations at mean low water neaps when currents do not exceed these values vary from 105 to 51 minutes for 0.5 m/s and from unlimited duration to 148 minutes for 1.0 m/s, depending on the extent of construction already completed. Placing caissons within these times is considered to be achievable, bearing in mind that the shorter periods occur when considerable experience in placing caissons would have been built up.

40. For the Cardiff Weston scheme caisson placing would be carried out using powerful purpose built tug units attached, one at each corner of a caisson, with rigid arms capable of vertical articulation. For the English Stones scheme serious consideration was given to alternative methods in order to avoid the considerable expense of such units, which would have to be spread over a smaller number of caissons. No other scheme could be devised, however, which offered the same speed of operation and flexibility as the fixed tug method; qualities essential for ensuring safe towing and placing operations. Allowance has therefore been made for the use of four tugs (plus one reserve) each of 6000 to 7000 HP and capable of developing a force of 60 to 70 tonnes, permitting control in currents up to 1.5 m/s.

41. It is considered necessary to provide a safe parking place for caissons arriving at the barrage site and awaiting the right conditions for placing. Such an area would have to be specially created, as at present the only areas with water deep enough to prevent grounding at any state of the tide are in the main channel, where current velocities in excess of 3 m/s would be experienced, requiring mooring forces approaching 2000 tonnes. A deep water refuge would be created off the main channel, immediately to the south of the English Stones rocks, where individual caissons can be safely held awaiting placing and to which they can be returned if the placing operation is aborted.

Embankment Construction

42. The rock for the anti-scour blanket would be placed partly at low tide by land based equipment and partly, where sea bed is below Ordnance Datum level, by bottom dump or side discharge barges. Similarly, placing of the rockfill control bund would be partly by barge but mostly by end tipping from dump trucks. Placing of the quarry waste bunds and filter layers would be mainly by barge, but the higher levels would be transported by dump truck.

43. It is proposed to stockpile dredged sand at the root of either embankment. Placing the sand fill in the embankment would be by a combination of direct pumping from dredger and transport from stockpile to embankment by dump

truck. Rip rap would be placed by a combination of end tipping, use of rock trays, and by grab or grapple.

Closure

44. Closure of the barrage would be effected in the embankment on the English side. The closure section is in an area off the main channel where the necessary rock foundation is expected to be found at a level of about -7m O.D. i.e., about 1.5m below mean low water springs. Closure would be carried out by placing the control bund in horizontal layers across the gap, initially by barge only and later by a combination of barge at high water and dump truck when the bund is uncovered.

Programme

45. The proposed construction sequence is shown in Fig. 4. First power generation is programmed 5 years after starting construction with full power generation after 6 years. The programme provides for the ship lock and new navigation route to be operational at month 30, before placing the majority of caissons in the present navigation channel. Turbine caissons must be ready to receive the first three sets, delivered in month 40. The substation would be installed on the Welsh side plain caisson, giving sea access for installation of electrical equipment by month 41. Completion of the welsh embankment by month 53 allows the last 1500m of transmission cables to be installed.

Cost Estimates

46. Cost estimates at 1984 prices have been prepared for the civil engineering works for the barrage. Estimates, for a barrage incorporating the ship lock described in paragraph 30 above, are given in Table 2. These estimates are based on quantities of the principal materials and work items derived from preliminary design studies of the works. Prices have been estimated for all significant elements and operations involved in the construction of the barrage. Advice from specialist suppliers and subcontractors has been sought where appropriate.

47. An overall contingency of 20% of the construction cost has been included and this is intended to allow for the following:-

- Take-off and estimating; an allowance for miscellaneous items not measured and for uncertainties in basic prices.
- Design; an allowance for the preliminary state of designs.
- Construction; an allowance for the preliminary state of construction planning, for unforeseen foundation conditions, for excessive weather downtime, labour disputes and other factors.

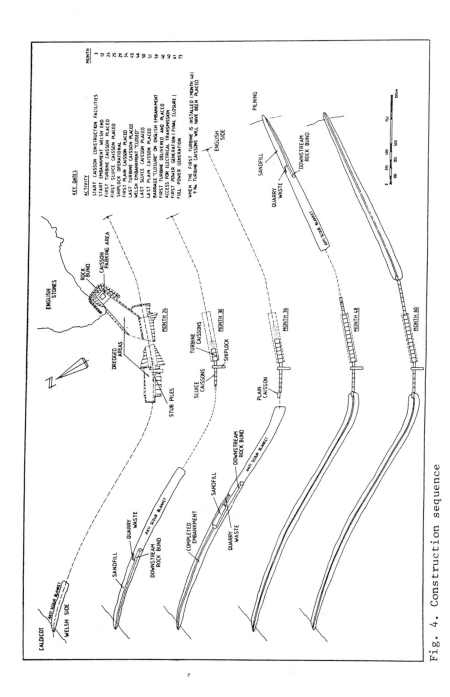

Fig. 4. Construction sequence

TABLE 2

CIVIL ENGINEERING COST ESTIMATE

Item	£M
Navigation Lock	18
Caisson Construction Facilities	76
Caisson Construction	
- Turbines	127
- Sluice	81
- Plain	7
Dredging for Caisson Foundations	30
Caisson Installation	
- Turbine	68
- Sluice (incl. gates)	75
- Plain	6
Embankments	93
	581
Contingency (20%)	116
	697
Management, Design, (7%) Inspection	49
TOTAL	£ 746 M

CONCLUSIONS

Feasibility.

48. On the basis of work carried out during the interim study construction of a barrage at the English Stones to generate electricity is technically feasible. The scheme would consist of a power house with turbine generators in the deep water channel, flanked by sluices on each side and connected to the shore by embankments. The preferred alignment is about 2km downstream of the English Stones. There may be a problem of basin siltation with this scheme; this is discussed in the paper by Dr. R. Kirby.

Cost and Programme.

49. The estimated capital cost of the Civil works is £746 million at January 1984 prices. The construction duration would be five years to barrage closure and first power generation with full power generation after six years.

ACKNOWLEDGEMENTS
The paper is based on a study carried out by STPG and funded by

the Department of Energy and STPG.

The authors acknowledge the permission of the Dept. of Energy to publish this paper and the assistance of colleagues in the STPG in its preparation.

REFERENCES

1. Department of Energy. Energy Paper No. 46 Tidal Power from The Severn Estuary. 1981.

2. Wimpey Atkins. Wimpey Atkins Severn Tidal Barrage - Prefeasibility Study Proposals. November 1983.

Discussion on Papers 3 and 4

MR R. N. BRAY, Livesey Henderson
Dredging and pretreatment of rock is called for in exposed sea
conditions where the tidal currents in spring tide conditions
may exceed those normally associated with acceptable operating
conditions.

Have the study team considered the current regimes in the
areas where dredging and pretreatment are to be carried out
with a view to establishing the down time incurred due to
increases in tidal currents caused by the partial closure of
the estuary for the construction works?

The study team has suggested that floating dredgers should
be employed and has implied that more than one would be
desirable - to avoid the risk of a critical failure which
slows down the construction programme. I fully agree with
this. However, I do not necessarily agree that the dredgers
should be employed in the bottom half of the tidal cycle as
suggested in paragraph 34 of Paper 3 since this eliminates one
of the slack water periods when currents are low.

Large cutter suction dredgers of the floating type could
cost of the order of £300 000 per week to operate. With two
of them (or possibly even three) and with an extended
programme caused by down time, a large proportion of the
dredging budget would be used.

With respect to pretreatment of rock, large jack-up pontoons
have to be raised to a height well in excess of the highest
anticipated wave at high water conditions - this is for
insurance reasons. This means that the platform could be
50-60 m above rock head. In the exposed sea conditions and
the current regimes predicted it will be extremely difficult
to position blast holes accurately due to the bending and
movement of the casing tubes.

Secondary blasting as suggested, on a close grid pattern,
would be all the more difficult to achieve. In addition it
should be noted that drilling blast holes for secondary
blasting in areas which have already been blasted in the
primary blasting process will be made considerably more
difficult because the rock will be full of fractures.

In view of these remarks, do the Authors of Papers 3 and 4 feel that they have an adequate allowance in their cost estimates for the dredging and pretreatment works?

MR D. KERR, MR W. T. MURRAY and MR B. SEVERN
The dredging works required have been considered by two dredging companies and significant allowances have been made for down time due to various factors including tidal currents. The dredging works should be complete before the later stages of caisson installation when the increase in tidal currents will be most significant. The point of the comment in paragraph 34 of the Paper was to note that it may not be economic to provide a dredger that is capable of dredging the deepest areas at all states of all tides.

Mr Bray's comments regarding the difficulties of accurate drilling for blasting are accepted. Methods of achieving the required end results have not been considered in detail yet but will be an important topic for investigation in the further studies. It has been assumed that a large jack-up of the type developed for North Sea oil operations will be required and that some form of guide frame and/or heavy protective casings may be required on the sea bed to overcome • the effects of currents.

Our cost estimates include preliminary allowances for dealing with these difficulties. However, the areas that are anticipated to require blasting are only a small proportion of the total dredging works (see paragraph 33 and table 1 of the Paper).

MR M. HORDYK, Steel Construction Institute
Installation operations offshore strive for absolute simplicity, and for independence from tides and weather. Most importantly, the installation operations must be integrated into the design of the structural components to be used offshore.

The installation of the foundations for the Gaviota platform in the Bay of Biscay presented a similar problem, the foundation conditions being similar to those likely to be found across the Severn Estuary. The design of the chosen foundation system used stud piles within holes pre-drilled in the rock and their use to support the caisson units could result in simpler offshore operations.

It would be possible to use standard sheet pile sections to form a seal between the caissons and the rock head, and the Rosenstock shock blast process would allow sheet piles to be driven into rock. An example of the use of the technique can be found in Liverpool Docks.

MR D. KERR, MR W. T. MURRAY and MR B. SEVERN
The two suggestions raised will both be considered in future

studies. The use of sheet piles driven to form a seal around
the caissons has been considered briefly for soft rock but the
experience in their use quoted by Mr Hordyk will enable
considerations for a range of rock conditions.

MR D. E. ROE
A report in New Civil Engineer on the use of the technique at
Brunswick Lock, Liverpool, indicated that some problems had
been experienced on one of the two coffer-dams. Difficulties
with pile driving had necessitated a second blasting
operation.

DR H. R. SHARMA, Central Electricity Authority, Government of
India
Was the decision to have a four-turbine caisson for the Severn
barrage based on an optimization study? Could the Authors of
Paper 3 elaborate on the decision to have four turbines in one
caisson?
  What is the dry weight of the four-turbine caisson?
  Is it self-buoyant?
  How much time has been estimated for the construction of
each four-turbine caisson?
  What will be the total time period of construction for all
the power-house caissons?
  Could the stages of caisson construction be elaborated: will
the turbines be installed before sinking the caissons into
place?
  What is the latest thinking about the foundation treatment
for the Severn barrage project?

MR D. KERR, MR W. T. MURRAY and MR B. SEVERN
The Severn Barrage Committee report recommended the use of
caisson units for three turbines. The feasibility of
increasing the size of the caissons to accommodate four
caissons was studied and it was found that the tug towing
capacity required was not significantly increased and that the
tug placing capacity proposed was adequate for the larger
caisson. The advantage of reducing the number of caisson
placing operations is that the number of periods of good
weather required is reduced. An optimization study was not
carried out as it was not considered prudent to consider the
use of even larger caissons at this stage of development. The
weight of a four-turbine caisson is approximately 100 000 t
and the caisson is self-buoyant floating with a draught of
about 22 m.
  Details of the construction and placing periods for turbine
caissons are given in fig. 15 in the Paper, and the answers to
the questions on caisson foundations will be found in
paragraphs 39-42. The turbogenerators will be installed after
the caissons have been placed in the barrage.

MR F. IRWIN-CHILDS, Rendel, Palmer & Tritton
The reason given for considering the smaller size of lock is
that 150 000 t ore carriers could unload at Port Talbot
outside the barrage. The navigation study for the Severn
Barrage Committee reported that ships of 75 000-100 000 t
could use Portbury Dock at Bristol and that deep water
riparian berths could be established as at Cardiff - thus
pointing to the desirability of opting for the larger size
locks.

MR D. KERR, MR W. T. MURRAY and MR B. SEVERN
The Severn Tidal Power Group (STPG) considered various lock
sizes to understand how each could best be provided, how it
could be integrated into the overall project and what it would
cost to provide and operate. No decision on the lock size
suited to commercial shipping operations past the barrage from
such time as it could be commissioned have been considered by
the STPG. The data available to do the rational grounds
remain to be collected. The STPG's forthcoming study with the
Department of Energy and the Central Electricity Generating
Board will start to assemble these data. A possible change in
commercial shipping demands in the Severn Estuary with the
barrage in place has already been identified by the STPG. The
possible scale of this and its impact on existing operations,
on the demand and case for investment, and on overall regional
economic activity have to be judged. The appeal in principle
of having permanently deeper water and much reduced currents
is self-evident to ship operators. This does not necessarily
mean, however, that their routes would on economic grounds
sensibly include the Severn Estuary.

MR E. GOLDWAG, GEC Turbine Generators Ltd
Both Mr Clare and Mr Kerr have suggested that each
turbogenerator set would be built into a 1600 t package with
its own lifting beam and shipped directly from the works to
the caissons, where it would be installed in a single lift.
This approach has been chosen because experience gained
elsewhere, mainly in the nuclear industry, has suggested that
the module built demonstrated the advantage of the technique
of large assemblies delivered in single packages requiring a
minimum of site work.
  While the principle of the module built was widely adopted
in the construction of both nuclear and thermal
power-stations, as far as rotating machinery is concerned the
size of modules has historically been limited to a single
turbine cylinder, a mere 150 t or so, and in any case no part
of the module depended on embedment in concrete to maintain
its shape and integrity.
  In the hydroelectric field, only very small units have ever
been dealt with on the single-module principle.
  Thus units of the size and bulk under discussion require a

more conventional approach during erection, with the embedded parts installed separately from the rest of the equipment, which should as far as possible be dealt with as a minimum number of modules.

What is the state of the art?

MR D. KERR, MR W. T. MURRAY and MR B. SEVERN

It was proposed in the Severn Barrage Committee (SBC) report that the turbogenerator sets should be placed in single units, on the basis that current lifting capacity used in the construction of North Sea oil platforms would be more than sufficient to handle and place complete units in the Severn barrage. This principle offered considerable advantages in that complete units could be prefabricated and tested at the works elsewhere and towed and placed at the barrage. It was also possible to separate the civil construction in the precasting yards from the mechanical and electrical trades. There is a need to minimize the caisson draught and it is an advantage to avoid placing turbines in the caissons before leaving the construction yards. The placing of turbine caissons can also proceed ahead of the turbine generator manufacture which offers programme advantages. For all these reasons the Severn Tidal Power Group (STPG) decided to adopt the SBC proposal.

To the authors' knowledge there is no precedent to placing such large units as single packages but the mechanical and electrical engineers in the STPG have not suggested that it is not feasible to place complete units. They have been advised by the turbogenerator companies that it should be feasible to design the structures to be sufficiently stable for transport and installation, although no detailed designs have been carried out.

MR W. J. CARLYLE, Binnie & Partners

Is the experience of previous tidal power projects in relatively sediment-free estuaries relevant to the Severn which is an estuary choked by sediment? The loss of tidal volume after closure of the barrage would surely alter the regime of the downstream/seaward channels. Problems of sediment choking would be expected in the outlet channels which will have to be rather wider than the natural deep ebb channels.

Finally as a dam engineer I feel that careful attention will have to be paid to the sealing of the caisson foundations not to prevent loss of water but to prevent the risk of scour in the foundation.

MR D. KERR, MR W. T. MURRAY and MR B. SEVERN

Tidal power projects are of necessity attracted to estuaries with high tides, which means that their estuaries tend to be long and, because of tidal asymmetry and strong currents, muddy towards their head. (The Rance estuary might be regarded as an exception, but this is only because it is not at the head of the tidal system which is a considerable distance up estuary in Mont St Michel Bay.)

The Severn Estuary is certainly not choked in sediment. The existing regime is surprisingly stable compared with, say, the Mersey Estuary, as historic Admiralty charts clearly show. Fresh input rates of muds and sands are very slow relative to the water volume of the estuary, so that no overall sediment problems are expected.

Much of the bed of the estuary seawards of the barrage is exposed rock, and most of the remainder is veneered with sand. The barrage would variously alter the strength and distribution of currents over this area, less change occurring seawards of the turbines and further down the Bristol Channel where conditions would be progressively nearer to the present conditions. Away from the turbines, the movement of material is more likely to be determined by flood flows through the sluices, by wave action and by any circulations set up by flows from the turbines. It is not immediately obvious why these would lead to choking of the outlet channels, because velocities close to the barrage will generally be stronger than further from it. Material moved to the structure is therefore more likely to be carried through and well past it, a possible process to which the Severn Tidal Power Group has already given attention because of the various environmental implications.

In response to the need for prevention of scour beneath the caisson, this matter has not been considered in detail yet but an allowance has been made in the cost estimate for some injection grouting beneath the caissons, in addition to the bulk grouting to the interface between caisson and prepared rock bed. It is possible that there could be large fissures or voids in some of the rock strata, for example the Keuper marl, and hence these would require grouting. However, bearing in mind the relatively low maximum and average head of water across the barrage compared with large river dams, it is considered unlikely that extensive rock grouting will be required.

MR M. J. GRUBB, Imperial College of Science and Technology
A 500 kW oscillating water column (OWC) wave power generator, grounded in a cliff face, has recently been demonstrated with claimed costs of under 4p/kWh.

It is possible that OWC generators could be incorporated in the seaward face of many of the barrage caissons. The additional civil and electrical costs arising from such machines would presumably be much less than those for machines sited on a cliff face, for which all civil and electrical work would have to be installed from scratch. The Cardiff-Weston barrage also provides a natural and lengthy cliff-type face directly on to deep water, so that frictional dissipation of waves on the sea bed would presumably be smaller than for many shore-based sites.

I have been unable to find any relevant wave data, but it is possible that the incorporation of wave generators could add between 3% and 8% to the barrage output, in a form which would not worsen any problems of system integration.

Has this possibility ever been seriously investigated?

MR D. KERR, MR W. T. MURRAY and MR B. SEVERN

It is probably correct that some form of wave energy device could more easily be incorporated into the seaward face of a tidal power barrage than into a cliff face, but this does not necessarily mean that it would give a more economic wave energy station.

The Severn barrage (and tidal power-stations in general) are not constructed at the seaward end of estuaries because that is not where the tide range is highest nor the economics of a tidal power scheme most attractive. The Severn Barrage Committee (1981) illustrated this point very clearly by their decision to favour a relatively up-estuary location for the barrage from those considered.

The strength of the case for moving a tidal power scheme seawards to allow it to perform two duties can be judged by comparing the amounts of energy which it would harness in each. Depending on the location in the outer Bristol Channel, the available mean annual wave energy density will not exceed about 20 kW/m. Even assuming 100% energy capture over a 30 km barrage, this would give 5 TWh per year, or about one-third of the output of the preferred tidal barrage near Cardiff.

The move seawards to this location might add a little to its potential to harness tidal power, but its cost would increase several fold. The net result would be a much less economic project.

The alternative to installing wave energy devices on a Cardiff barrage would be to add only a fraction of 1% to the annual energy output of the scheme and again would not be justified. Mr Grubb's suggestion that the gain would be 3-8% is questionable.

The addition of these amounts of wave energy would not affect system integration because this is not a problem for tidal power anyway and any additions would at most be only small.

MR F. P. LOCKETT, Coventry Polytechnic Energy Group

It is unlikely that there is sufficient wave energy at proposed UK barrage sites to merit exploitation. However, the Coventry Polytechnic Energy Systems Group is actively researching ways to use the air turbine power take-off systems developed for harnessing wave energy in low head hydropower applications, including tidal schemes. The simplest device conceptually consists of an air box with water inlet and outlet valves and a Wells (bidirectional) turbine in the top. Suitable operation of the water valves over a cycle of say 20 s then produces an oscillating water column driving the air turbine. Such a device could offer economic power take-off in small tidal schemes, where the high power/barrage area ratios of water turbines are not appropriate, and fits well into caisson barrage design.

MR D. KERR, MR W. T. MURRAY and MR B. SEVERN
Mr Lockett does not offer data to substantiate the suggestion that air turbines offer a more economic form of power conversion than water turbines for small tidal power schemes.

It is not immediately obvious why there should be a change in the balance between the economic viability of these two types of machine with the size of the scheme. The difference between large and small schemes constructed on estuaries with similar tide ranges is essentially in the number of power units installed. The unit price of generation can, as usual, be expected to be less for the larger schemes, for reasons of economy of scale.

Sensible comparisons of the two systems must await the availability of equally reliable and comprehensive cost and performance forecasts for each. However, it is clear that air turbines are not an alternative to water turbines for large tidal power schemes, such as the Severn scheme, because they are not proven in long-term operation at present.

PAPER 4

MR C. J. A. BINNIE
The English Stones site is about 7 km downstream of the existing Severn Bridge at the point where the Severn Estuary widens (Fig. 1). This is also the site selected for the second road crossing.

The spring tide range at this point is about 12 m with a high water area of greater than 100 km^2. There are two important features: one is the English Stones which is a large area of rock exposed at low water; the other is the deep water channel called the Shoots, which narrows to 350 m and through which there are very high tidal velocities.

A cross-section just downstream of the English Stones shows the deep water channel in the centre. At the base is the hard Pennant sandstone. Overlying this are rocks of the Triassic series. Above those are gravel deposits and close to each shore is a layer of mud.

The English Stones is a rock outcrop in the estuary about 2 km by 3 km and is a largely horizontal mass of Triassic Keuper marl protected by a capping of thin sandstone beds.

The English Stones is the site which has been considered traditionally for tidal power generation. A scheme was

proposed in the 1940s with vertical axis turbines, sluices and
an embankment dam in the deep water section. This scheme
appears to have been intended to be constructed in a series of
coffer-dams with final closure in the deep water section. It
is interesting to consider the advances in turbine generator
design and construction techniques since then.

The next major tidal power study of the Severn Estuary was by
the Bondi Committee in about 1980 but this was aimed at
maximizing power output from the estuary and did not seriously
consider the English Stones site.

The Severn Tidal Power Group (STPG) considered that with
modern methods of tidal power generation this site might have
the advantage for private sector investors of a lower capital
requirement and a higher rate of return than the Bondi line.
It was therefore studied by the STPG.

Several alternative barrage lines were considered using
similar elements to those adopted for the Cardiff-Weston line
described by the STPG. Although a line straight across the
Shoots has the advantages of being the shortest length, the
Shoots deep water channel is too narrow for the number of
turbine generator caissons required. It also suffers from very
high tidal velocities during construction and short caisson
placement time windows. In addition, there was the risk that
any caisson that was not ballasted down at low water could be
carried out of control through the Shoots by the tide and
damaged. A site about 2 km downstream was therefore selected,
where the deep water channel was wider but shallower. This
gives sufficient submergence for 7.5 m turbines without an
unacceptable amount of dredging.

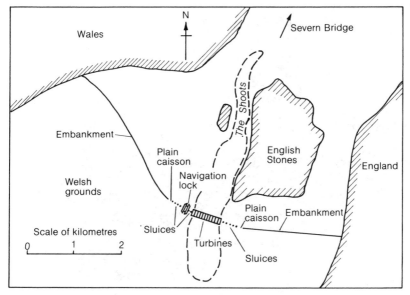

Fig. 1.   Layout of the proposed English Stones scheme

The turbine generators are in the deep water channel, with the sluice caissons on either side. The navigation lock is located on the general navigation route but away from the turbine discharge flows, to minimize ship-handling problems.

Almost all the embankments are founded above spring tide low water level. The embankment design varies for different foundation conditions but in general consists of a rockfill closure section followed by dredged sand infill contained within quarry waste bunds.

One of the problems for the designer at this site is to select the correct amount of sluice area. Smaller sluices reduce capital expenditure. For construction, though, it is necessary to have the maximum sluice area to minimize construction tidal velocities. Once operational, a large sluice area would maximize power output and would allow high tidal levels inside the barrage to be close to existing levels, which is important for Sharpness Docks. It would also provide operational flexibility to reduce sedimentation.

For the scheme, there would be 36 7.5 m dia. turbine generators housed in 12 turbine caissons. Each turbine would be rated at 27 MW to provide an installed capacity of 972 MW and an annual power output of about 2.8 TWh.

MR H. J. MOORHEAD, EPD Consultants Ltd
Figure 2 shows the interpreted rock head contours in the area of Flat Holm and Steep Holm. The foundation of the turbine caissons needs to be around 30 m below ordnance datum level for

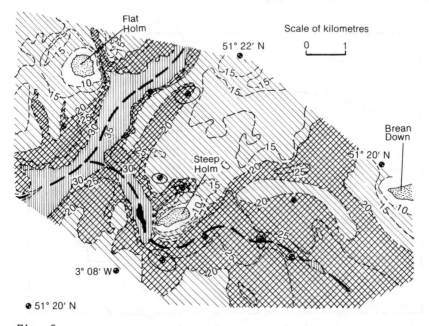

Fig. 2

the turbines to have adequate submergence. The rock head
contours as currently interpreted from sonar investigations
present certain geomophological anomalies.

As recently as about 10 000 years ago the sea level in this
area was about 50 m lower than at present. The rock surface in
the Severn Estuary is generally thought to date back to about
late Tertiary age. During the last ice age an ice sheet
extended into parts of the estuary and some of the old
preglacial river channels have been partially infilled. The
preglacial Severn and Parrett rivers had their confluence
somewhere between Flat Holm and Steep Holm as shown by the
broken lines on the diagram. Both the alignments proposed by
EP46 (Bondi) and more recently by the Severn Tidal Power Group
take advantage of this old river channel junction feature.

The further joint programme of work currently being
considered will be attempting to obtain a more positive
appreciation of the rock head contours in this area by using
sonar with borehole correlation. Experience elsewhere has
shown that it can be very difficult with geophysical methods
alone to distinguish the interface between certain types of
alluvium and the lower strength rocks. As can be seen from
Fig. 2, on the present interpretation, the old river channels
run uphill in certain areas.

MR D. E. ROE
One of the first priorities in the next phase of study will be
to carry out a geophysical survey and to sink a series of
boreholes. The boreholes will provide data, lacking at
present, which should clarify these issues.

5. Power generation studies of a barrage on the Severn

G. L. DUFFETT, MA, FIMechE, *Product Planning Manager, Northern Engineering Industries plc,* and G. B. WARD, BSc, ACGI, *Business Development Executive, GEC Energy Systems Ltd*

SYNOPSIS: Bulb turbines are proposed for both Cardiff-Weston and English Stones lines. The paper outlines the factors affecting the selection of the plant design parameters and indicates the method of installation. Studies are reported on energy outputs as they are affected by the number of turbines and sluices, plant availability, generator rating and the mode of operation. There is consideration of operation with ebb generation, double-effect generation and ebb generation plus flood pumping. The relationship of energy output to energy value is considered and reasons are given for the preferred schemes selected by the STPG.

INTRODUCTION

1. This paper is concerned with the two barrage lines investigated by the STPG, namely the Cardiff-Weston and the English Stones lines.

2. The electrical output depends primarily upon the tidal characteristics, the design of plant and the operating regime selected. The financial benefit depends upon the capital cost and also upon the tariff structure for the supply, and when pumping, the purchase of electricity.

3. The hydraulic characteristics of tidal power are dominated by a variable and relatively low head. Bulb turbines were selected because they meet the requirements for efficient power generation under these conditions and because there is a large amount of service experience with these machines.

4. Bulb turbine generator manufacturers' technologies are available to the STPG from Neyrpic/Alsthom and Sulzer-Escher Wyss/Elin Union through their association with GEC and NEI respectively. No model tests of designs specifically for the conditions of the Severn have yet been undertaken. Existing data for other modern machines have been used.

5. The operating regime chosen for the first detailed studies was that of ebb generation only, followed by a less detailed study of the alternatives.

TURBINE GENERATOR DESIGN

6. Within the constraint imposed by the economic depth of channel, a smaller number of larger turbines and associated caissons is less expensive than a larger number of small turbines with the same total capacity, although the optimisation is relatively flat. However, for a barrage on the Cardiff-Weston line, the diameter is determined by the maximum extrapolation in runner diameter from existing experience that can be proposed with confidence. The need for a prototype is then avoided. Over 120 bulb turbines with runner diameters of 5m or greater have been built by, or to the design of, Neyrpic and SEW. For example, there is operational experience of 7.4m diameter runners in turbines with a continuous rating of 53MW and of 7.7m diameter runners in 25MW machines. Consequently, a runner diameter of 8.2m was selected for an output which was known to be in the range 30MW - 40MW at peak, and considerably less under most tidal conditions, and for which cavitation would not prove to be a severe limitaton.

7. The varying head involved in any tidal scheme suggests that a significant increase in efficiency, and hence in output, could be achieved if the turbine speed were adjusted to match the operating conditions. If there is a DC link with a rectifier at the input and an inverter at the output, then the turbine generator can operate at its most appropriate speed. The system also provides built-in isolation of the generators from the transmission system which has advantages in the event of electrical faults. However the electrical losses are increased. With synchronous machines connected to the AC grid, losses vary from about 2% at full output to 20% at 20% output, while the comparable DC losses are 3.5% and 25% respectively. Nevertheless, it is calculated that 2.3% more energy would be sent out with variable speed operation. However, in the case of the longer barrage, the capital costs would be increased by about £300M and these would not be recovered by the increased revenue. Similar considerations apply to plant at English Stones.

8. The relationship between runner diameter and the economically available channel depth is determined by the caisson design and by the minimum submergence which the turbine can accept without cavitation. Although a reduction in submergence mainly limits maximum power and so leads to little loss except during maximum spring tides, cavitation is a potential cause of serious damage. The minimum safe submergence is determined by model tests and,

in the studies, it has not been reduced below the minimum figure recommended by the designers. During part of a high spring tide generating cycle the selection of a particuar submergence can cause the power output to be limited to below the installed generator capacity. Whilst the two design companies have different cavitation criteria, the difference in optimised nett output over a cycle of tides is negligible in the range of practical submergence.

9. For a barrage at English Stones, the runner diameter is limited by the channel depth. Early work was carried out assuming 6m diameter runners but 7.5m diameter was found to be practicable and this size was later adopted.

10. The best rotational speed is determined by using hill charts and cavitation data derived from model tests from existing designs. The annual energy production from ebb generation only was determined for rotational speeds varying from 50 - 70 rpm for a Neyrpic turbine with fixed distributors and variable blades and for a Sulzer - Escher Wyss turbine with fixed blades and variable distributors at different settings. The optimum for both machines with 8.2m diameter runners was at, or very close to, a synchronous speed of 55.56rpm (108 poles) and this was selected for all subsequent work on the Cardiff-Weston barrage. The selected rotational speed for the 7.5m diameter turbines at English Stones was 57.69rpm (104 poles).

NUMBER OF TURBINE GENERATORS

11. At both sites, the number of turbine generators is determined by the width of the deep channel that can be made available economically. Theoretical calculations show an increase in energy output if the number of turbines could be increased beyond this practical limit.

12. The output is also dependent upon the quantity of water transported into the basin during the flood tide which is determined by the effective total flow area of the sluices and the turbines when acting as sluices. Even if flood pumping is used, total sluicing area remains the dominant parameter. Up to, and beyond, the ranges of practical interest at both Cardiff-Weston and English Stones, the energy sent out per annum increases with the number of turbines and, to a lesser extent, with the number of sluices. This is illustrated in Fig. 1 which presents the results of calculations for a barrage at English Stones. Over the ranges considered, doubling the number of turbines increases the energy sent out by about 68% but doubling the number of sluices produces an increase of only about 11%. It may be necessary to select the number of sluices to obtain adequate tidal control during

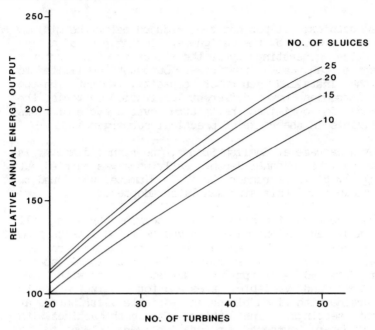

FIG.1. Output Variation with Increasing Number of
TG's and Sluices at English Stones.

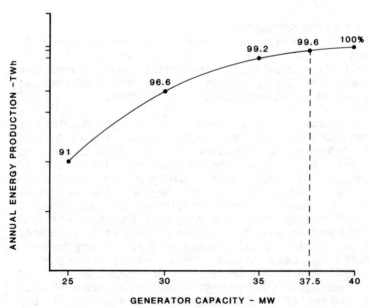

FIG.2. Effect of Generator Rating on Total Annual
Energy Production.

construction rather than for reasons of performance, and this is the case at English Stones.

13. These engineering considerations and the extra costs of accommodating much increased numbers of turbines within the restraints of the actual estuary led to 192 machines with 8.2m diameter runners on the Cardiff-Weston barrage, and 36 units, 7.5m diameter at English Stones. The longer barrage contains the equivalent of 150 sluices of the submerged Venturi type of 12m x 12m, while there are the equivalent of 42 sluices in the shorter barrage with a flow area of 5976m², there being variations in vertical dimension depending upon the available local depth of water. The initial studies assumed a discharge coefficient of 1.5. Test data show this figure to be low and a more realistic figure was used in the final assessment.

14. It is not the purpose of this paper to describe the method of manufacture and erection of the turbine generators in any detail, but some comment is relevant because there is a direct impact upon the timing of start of electrical generation and upon energy output during the early years of operation. The rate of construction of the barrage must not be limited by the rate of manufacture of the turbine generators and the time for installation and commissioning on site must be kept to a minimum.

15. With the large number of machines, particularly in the case of the longer barrage, these objectives are achieved by manufacturing and assembling complete turbine generator units in the works. It is necessary to consider manufacture at least six years in the future and it would not be prudent to rely on capacity which is currently spare. Business evolution, the refurbishment or dismantling of works, and advances in production technology occur continuously over shorter timescales. The longer barrage requires the delivery of three turbine generator sets per month, so the STPG cost estimate includes for two parallel manufacturing facilities. One would suffice for the smaller barrage. This approach also allows comprehensive pre-commissoning testing before the complete assembly is transported to site by barge. Each turbine generator is assembled with a special lifting beam and loaded onto a barge and off-loaded into the caisson cell in a single lift. In the case of the larger machines this involves a single lift of 1600 tonne. The turbine caisson is already in its final position and the lifting beam remains part of the permanent structure. Access is provided for the removal and replacement of damaged or worn components and most routine maintenance operations will be carried out in situ.

16. The electrical output depends not upon the number of turbine generators installed, but upon the number operating. A figure for availability can be based only upon experience. At first the STPG studies assumed the same availability, 90%, as in EP 46 (Ref. 1) but further consideration of operating experience gave confidence that 95% can be assumed.

GENERATOR RATING

17. The selection of the optimum generator rating is not entirely straightforward. A large capacity can be utilised during only a few spring tides in the year although the calculations carried out by the STPG are for an average year, 1974, while the potential energy capture varies by +7% to -6% over the 18.6 year tidal cycle. The capital cost implications of increased generator rating go beyond the costs of the turbine generators and local electrical plant as CEGB grid transmission requirements must also be taken into account. The transmission system reinforcement is discussed elsewhere (Ref. 2). The effect of generator rating on the total annual energy production by ebb generation is shown in Fig. 2 for a barrage at Cardiff-Weston.

18. The generator rating finally selected for machines on this site was 37.5MW, giving an installed capacity of 7200MW. It may be noted that the 7200MW recommended by EP46 referred to the turbine output. There is therefore an increase in the maximum output of about 3% due to the exclusion of generator losses.

19. The rating selected for the turbine generators on the English Stones barrage was 27MW output, giving an installed capacity of 972MW. The value selected would have been slighly higher if the increased output had not involved a significant step change in the cost of ancillary electrical equipment.

DETERMINATION OF ENERGY OUTPUT

20. The determination of energy output is a complex matter, not only because of tidal variations, but also because basin levels depend upon tidal, turbine and sluice characteristics. Hence it is essential to have access to adequate computer programs.

21. STPG made great use of the computer models available from Salford Civil Engineering Ltd. (SCEL) with work, primarily confirmatory, also carried out by Binnie & Partners and by the Engineering Mathematics Department of the University of Bristol. The SCEL flat basin, non-dynamic, model was used for the main energy output evaluations. This is considered adequate for the present

purpose and more complex models are very expensive to run. However in the next phase of assessment a complex dynamic model will have to be used, especially for the detailed considerations of the real benefits from pumping.

22. Uncertainty in the present results could be up to +/-10%, of which up to +/-7% is associated with potential inaccuracy in the calculation of the changes in tide level caused by construction of a barrage across the estuary which are data input to the SCEL performance programs. IOS values for tidal range were used as input to the Salford model.

23. The SCEL computer model simulates the operation of a tidal power scheme during an infinite series of identical tides. Hence the energy output can be predicted for any tidal range. The annual energy production can then be estimated by multiplying the energy output for each tidal range by the number of annual occurances obtained from a histogram of the tidal ranges that are expected at the barrage.

24. The turbine performance is derived from hill charts which have been determined from model test results with corrections for majoration in accordance with the Hutton formula which takes account of differences in dimensions and operating temperatures between model and full-scale plant. Practical constraints, such as limits on generator output and on cavitation, can be fed into the computer model.

GENERATION REGIMES
25. The reasons for discarding variable speed operation have been explained. Hence fixed speed machines were assumed when investigating other possible generation regimes. Differences between maximum energy output and maximum value benefit depend upon the tariff structure for electricity generation and the 1984 Bulk Supply Tariff was used in all the studies. Most of the work was carried out for turbine generators on the Cardiff-Weston line but the conclusions can be assumed to be valid for a barrage at English Stones.

26. Although double effect generation reduces the energy output, it offers the potential for a reduction in the installed capacity and hence in the cost of the transmission system. It should also be possible to demonstrate a better firm power contribution. However, double effect generation requires double regulated and, hence, more expensive turbines and produces lower basin levels which is undesirable from an ecological point of view. It is concluded, as did EP 46, that on balance,

double effect generation is less attractive than ebb generation only.

27. The effect of retiming generation on the value of energy sent out using ebb generation only was examined by varying the starting times after high tide and the duration of generation. Its effect was found to be small, amounting to only 1% increase in value for the 1984 BST structure. However this retiming would be done in practice as a matter of normal operating procedure. Even for maximum energy production the generation start time after high tide varies with the tidal range.

28. Most of the performance work for the STPG study was based upon ebb generation only because preliminary calculations suggested that there would be only a small gain from flood pumping. However, to assess the level of the gain a minor study was undertaken.

29. To avoid major modifications to the SCEL computer models, a simplified approach was used when investigating flood pumping, which included assumptions that the generator and motor efficiencies were the same, that pumping commenced at the peak of the tide, and that transmission losses need not be included if a datum ebb generation case was recalculated without such losses. The increase in energy sent out at constant energy prices was calculated to be 2.25% using the first available machine characteristics, although the energy value could increase much more if the input energy cost was much less than the output energy value. In practice this condition only occurs for a limited number of tides.

30. This gain was considerably less than that reported by EdF for La Rance which was 11% in output when the plant operation is optimised for energy production.

31. To test the effect of the differences in basin and machine parameters, the calculations were repeated using the La Rance data. The energy gain was calculated to be 10% which compares well with that reported for La Rance and gives confidence in the SCEL model despite some known limitations.

32. Later in the study an improved turbine/pump characteristic for a fixed distributor/variable blade machine became available, and further calculations were carried out which resulted in an increase in annual energy gain over ebb generation of about 5%. Some conservative approximations were made in this limited study and no attempt was made to re-optimise the design parameters. For example, there is little doubt that the selected speed is

no longer the optimum.

33. Obtaining maximum value for the electricity exported while minimising the cost of pumping, leads to a reduction in the net energy capture. It is not economic to pump on all tides but, in some circumstances, there may be a value benefit in starting pumping before high tide. The present SCEL computer model cannot simulate this latter feature. An assessment of the overall situation suggests that a 7% increase in value by using ebb generation with flood pumping on appropriate tides over that obtained by using ebb generation alone can be regarded as a reasonable estimate. As already noted, a more complex, dynamic model will be required to confirm any real benefit of flood pumping.

34. The reduced basin area behind the English Stones barrage, compared with the total flow area through the turbines, suggested initially that there would be a significant benefit from the addition of pumping. The benefit in terms of the incremental gain in energy sent out was first assessed using the same approach as that adopted in the early investigation for the Cardiff-Weston barrage. With the tidal regime assumed, the energy gain was found to be about 2% but the gain would be increased somewhat with the reduction in tidal range resulting from construction of the barrage. Taking into account the results obtained from the later studies of the Cardiff-Weston barrage, it is suggested that a gain in energy value of 7% over ebb generation might be expected here also.

35. Table 1 shows the development of calculation of energy output from the Cardiff-Weston line for turbine generators operating at fixed speed. Initially the assumed barrage alignment differed from that finally proposed which enclosed a slightly larger basin area. Information from the potential generator manufacturers gave confidence that 97.5% generator efficiency at full load would be achieved although it should be noted that the calculations of losses at part load were more accurate than in earlier studies.

36. The effect of generator rating upon the annual energy production was discussed above (see Fig. 2). 30MW machines produce about 96.6% of the potential energy and 37.5MW sets 99.6% on the basis of ebb generation only. This increase is worth, capitalised, over £200M.

37. Table 2 summarises the corresponding variations in value based upon the 1984 BST. For the average year, retiming of the ebb generation increases the annual value by 1.1% but the improvement with flood pumping can be considerably greater if the periods of pumping and

TABLE 1 - Summary of Findings Relating to Energy Sent Out - Ebb Generation Fixed Speed - 192 TGs Installed

CASE	GENERATOR EFFICIENCY %	AVAILA- BILITY %	GENERATOR RATING MW	CYCLIC VARIATION ON APPROX 18.6 year CYCLE* %	ENERGY SENT OUT TWh	VARIATION FROM REF. SCHEME %	INSTALLED CAPACITY GW
Reference Scheme for Computer Surveys	95	90	30	+7.0 0 -6	13.1 12.2 11.5	0	5.76
Increased Generator Efficiency	97.5	90	30	+7.0 0 -6	13.4 12.5 11.8	+2.6	5.76
Increased Turbine Generator Availability	97.5	95	30	+7.0 0 -6	+13.8 12.9 12.1	+5.8	5.76
Increased Generator Rating to meet capacity limit set by 4 lines	97.5	95	31	+7.0 0 -6	13.9 13.0 12.2	+6.3	5.952
Addition of flood pumping to the above scheme	97.5	95	31	+7.0 0 -6	14.1 13.2 12.4	+7.9	5.952
Increased Generator Rating to give same installed capacity as EP46	97.5	95	37.5	+7.0 0 -6	14.2 13.3 12.5	+9.1	7.2
Addition of Flood Pumping to the above with better pump characteristic	97.5	95	37.5	+7.0 0 -6	15.0 14.0 13.2	14.8	7.2

*1974 assumed as base year

TABLE 2 – Summary of Findings Relating to Factors Affecting Energy Value

CASE	GENERATOR EFFICIENCY %	AVAILABILITY %	GENERATOR RATING MW	CYCLIC VARIATION ON APPROX 18.6yr CYCLE %	VALUE VARIATION FROM REF. SCHEME %	INSTALLED CAPACITY GW
Reference Scheme for Computer Surveys	95	90	30	+7 0 -6	0	5.76
Reference Scheme plus allowances for increased generator efficiency, and increased availability	97.5	95	30	+7 0 -6	+13.2 + 5.8 - 0.5	5.76
As above but with generator rating increased	97.5	95	31	+7 0 -6	+13.5 + 6.3 - 0.3	5.952
As above plus retiming of generation	97.5	95	31	+7 0 -6	+14.9 + 7.4 + 1.0	5.952
As above plus flood pumping	97.5	95	31	+7 0 -6	+20.1 +12.2 + 5.5	5.952
Reference scheme plus allowances for increased generator efficiency and increased availability; plus generator rating increased to 37.5MW	97.5	95	37.5	+7 0 -6	+16.7 + 9.1 + 2.5	7.2
As above plus retiming of generation	97.5	95	37.5	+7 0 -6	+17.9 +10.2 + 3.6	7.2
As above plus flood pumping and better pump characteristic	97.5	95	37.5	+7 0 -6	+24.9 +16.7 + 9.7	7.2

TABLE 3 – Data Sheet for Bulb Turbine Generator Sets

LINE OF BARRAGE		CARDIFF–WESTON	ENGLISH STONES
Runner Diameter	m	8.2	7.5
Maximum Discharge per set	m^3/s	650	550
Maximum Head on Sets	m	9.2	10
Level of Centre–Line	m OD	-16.9	-14
Operating Regime		Ebb generation with flood pumping	
Capacity per Set	MW	37.5	27
Generation Voltage(AC)	kV	6.9	6.9
Generator Efficiency	%	97.5	97.5
Rotational Speed	rpm	55.556	57.69
Number of Poles		108	104
Method of Installation:		Single lift in situ on barrage	
Weight of heaviest lift	tonnes	1600	1300
Number of Sets		192	36
Number per caisson		4	3
Total Installed Capacity	MW	7200	972
Time Availability	%	95	95
Average Annual Net Energy Output	TWh	14.4 *	2.8

*Based on preferred base case with conservative enhancement for improved sluice discharge coefficient.

generation are carefully selected on the basis of the cost/value of electricity.

38. While the operator, on any tide, is free to pump or not, there is only a limited period for pumping which varies with the tidal range.

CONCLUSIONS

39. A summary data sheet of the recommended schemes is presented in Table 3 for turbine generators at barrages on the Cardiff-Weston line and on the English Stones line.

40. The runner diameter is selected, for the longer barrage, on the basis of safe extrapolation from existing plant to avoid full-scale prototype build and testing and, for the shorter barrage, on the maximum diameter that can be used in the available deep channel.

41. The levels of the centre lines are determined by the cavitation characteristics of the turbines as derived from model tests.

42. The operating regime finally selected is ebb generation with flood pumping, although much of the work was carried out assuming ebb generation only, and thus some of the parameters will require re-optimisation during the course of any future more detailed assessment of pumping which will also require a more sophisticated program than any used to date.

43. The requirements and cost of reinforcing the CEGB transmission lines need to be reviewed together with the philosophy on the required number of spare lines to meet the occasional periods of high output from the Cardiff-Weston barrage. It may be that some reduction in generator capacity would have economic benefits but the whole spectrum of tides over the 18.6 year cycle needs to be taken into account.

44. The computer models of pumping with ebb generation need to be improved to obtain realistic assessments of energy output and net value. It is suggested that this should be a subject for collaboration within the UK to make maximum use of limited resources.

REFERENCES

1. Severn Barrage Committee "Tidal Power from the Severn Estuary",
   Energy Paper No. 46, 1981
2. Gardner, G.E., "Transmission system reinforcement"
   Paper No. 9: ICE/IEE Symposium on Tidal Power, 1986

3. Banal, M and Brichon, A, "Tidal Energy in France: the Rance Tidal Power Station; some results after 15 years of operation"
BHRA, 2nd International Symposium on Wave & Tidal Energy, 1981

ACKNOWLEDGEMENTS

Many individuals contributed to the studies summarised in this paper. However the authors wish to acknowledge, in particular, the contribution by R. Potts of International Research & Development Co. Ltd. and to thank the STPG and the Department of Energy for permission to publish this paper.

6. Electrical stability of the Severn Barrage generation

W. P. WILLIAMS, *GEC Transmission and Distribution Projects Ltd*

SYNOPSIS

The paper briefly reviews the phenomena of transient instability in an a.c. power system and the factors that affect it. It describes the assumptions made for and the results of transient stability calculations for both barrage schemes. It is concluded that for the Cardiff-Weston proposal the transient stability is marginal and a fuller investigation is required. With its lower loading the English Stones scheme appears stable though again further studies are necessary.

INTRODUCTION

As part of the 1981 prefeasibility study an examination of the choice of generating plant type and the manner of coupling them to the grid was made (ref. 1). In this it was recognised that direct a.c. coupling of the Barrage generation to the CEGB system would present problems of both dynamic and transient stability. It was concluded that the electrical characteristics of the generators would have to be carefully chosen to ensure adequate damping to prevent the low frequency oscillations symptomatic of dynamic instability and on the basis of simple transient stability studies it was suggested that because of the comparatively low rotational inertia of the turbine/generator set special measures might be necessary to overcome instabilities following system faults. In the study it was envisaged that the power would be distributed to a number of points on the CEGB 400 kV network on both the English and Welsh sides of the Severn Estuary.

As the more recent study by the Transmission and Control team of the STPG proceeded, the preferred arrangement evolved that all the power should be exported through a single 400 kV substation on the English shore. In addition the decision to base the study on a bulb generating set confirmed that the rotational inertia of the turbine generator would be on the low side. In view of these factors, but particularly because of the decision to concentrate all the Barrage output through one point it was considered desirable to carry out further checks on the stability of the Barrage generators. Only transient stability was studied since dynamic stability studies would require final design parameters of the Barrage generators to be meaningful or produce new information and in any case control techniques are available to deal with such instabilities. It must be stressed that the studies to be described in this paper were based on a installed Barrage capacity of 5580 MW.

TRANSIENT INSTABILITY

To assist non-specialists in the appreciation of the significance of the various factors to be discussed a brief understanding of transient instability is desirable. In any a.c. power system all synchronous generators and motors must operate at absolutely identical frequencies, apart from transient variations around the common value. In a simple two machine system of a generator supplying a synchronous motor load, power is transmitted between the two machines due to a difference in electrical phase angle between their internal e.m.fs (Figure 1). The maximum power that can be transmitted occurs when their phase angle difference reaches 90° electrical, its value being limited by the magnitude of the e.m.fs. and the inductance between them. This is expressed by the equation

$$P = \frac{E_1 E_2}{X} \sin \theta$$

where P = transmitted power,

E_1, E_2 and θ = generator and motor internal e.m.fs. and the phase angle between them.

X = system reactance.

In this simple system if a three phase short-circuit occurs at an intermediate point then no power can be transferred. The impedance of an electrical network is pre-dominately reactive hence the generator although its currents may be many times normal, does not supply any electrical load. Its mechanical power input initially remains constant hence it accelerates at the rate of

$$\frac{P_m \times 9000}{H \cdot S_r} \text{ degrees/sec}^2 \text{ (for 50 Hz system)}$$

where P_m = mechanical power MW

116

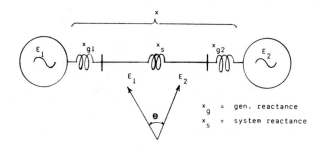

Fig. 1. Power transfer in simple system

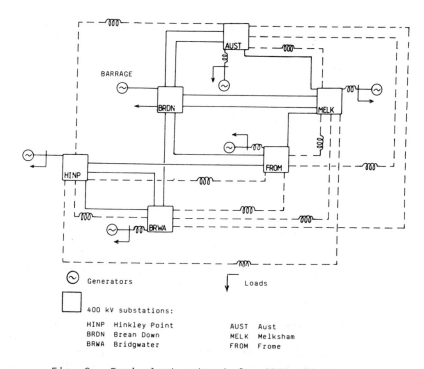

Fig. 2. Equivalent network for CEGB 400 kV system

117

H = turbine-generator inertia constant MW secs/MVA.

S_r = generator rating MVA.

Conversely the motor deprived of its source of input power decelerates.

If the three phase short-circuit is cleared rapidly, before the phase angle between the two machines has increased too much, then power will again flow from the generator to the motor, its value dependent on the angles between them. The rotational inertia of the machines will cause them to oscillate and their phase angle difference may transiently exceed 90°. If at the point of maximum angular swing the generator can still transmit more than its pre-fault power then the oscillations will die down and stable conditions will be resumed. If however this critical angle is exceeded then the oscillations will increase, synchronism will be lost and a stable power condition will not be attained. This condition is defined as transient instability.

The important factors to the maintenance of system stability are:

- the rotational inertia of the turbine-generator sets; the larger the value the slower will the machines change speed

- the duration of the fault; a shorter fault time keeps the machines from swinging further apart and to this end reliable high-speed fault clearance is an absolute essential in a transmission system

- the speed of response of the generator excitation system; by boosting the excitation in the post fault period it is possible to increase the power transfer thus strengthening the ties between the machines

- the post-fault network impedance; fault clearance by switching out a section of line usually produces a higher network impedance between swinging machines and aggravates the instability

- the pre-fault phase angle between the generators; it is necessary to allow an adequate margin for the machines to swing and still remain below the critical angle. This is achieved by ensuring the pre-fault angle is significantly less than 90°.

One special technique for maintaining stability in critical cases is the use of braking resistors. These are large short-time rated resistors that can be rapidly switched on to the generator terminals during the period of the fault and switched out immediately the fault is cleared. By so loading the generator its rate of acceleration during the fault period can be reduced and the subsequent swinging also reduced.

REPRESENTATION OF THE SYSTEM

Within the limited time and financial resources available for studying the transient stability of the Barrage scheme, it was only possible to attempt to obtain a feel of possible problems. It was decided therefore to restrict the investigation to cover the "worst case condition" that the Barrage generation would have to ride through and continue to maintain full supply to the CEGB system. The worst loading condition was seen as that in which the Barrage generation formed the highest proportion of the total system generation and the worst fault condition as that in which no Barrage power would be exported during the fault, the clearance of which would increase the impedance of its link with the main system to above its pre-fault value.

In a multi-machine power system there is a tendency for machines which are electrically close and not separated by a fault point to swing together hence it is possible to simplify a large system by combining generators.

During and after fault clearance the Barrage generators would obviously swing together hence could be represented by one large equivalent generator with suitably scaled parameters. The representation of the CEGB system was more difficult, but after discussion the Technology Planning and Research Division CEGB derived an equivalent network with generation (Figure 2). This is the network that would be seen looking into the system from the Barrage generators at Brean Down substation. It is a reduction of the complete UK system to six main nodes around the South West peninsular. The nodes are interconnected by actual or projected lines with additional interconnections to represent the remainder of the system. At four of the nodes equivalent loads and equivalent generators representative of the remainder of the system are connected; the generation and loading at Hinkley Point represent the actual generators at the power station and the load local to the substation; at the sixth node the Barrage generation is connected.

The parameters used for the Barrage and the Hinkley Point power station generators were those appropriate to the particular existing or projected machines; as seen later the Hinkley Point generators are at a sensitive point·of the system and accurate modelling is essential. The other generators are equivalent machines based on the parameters of a typical 660 MW set.

The loading conditions proposed were those corresponding to an "average summer week night trough demand" for 1995, i.e. a light load. However it was considered excessively pessimistic to assume this load coincided with all Barrage machines in operation at maximum output. It had previously been agreed with the CEGB that their system reinforcement investigation should be based on an available peak Barrage output of 5340 MW and this value was also used for the study. The Barrage output was then 29% of the total system generation.

CALCULATIONS

The "worst case condition" for the fault was taken as a three-phase fault on the double-circuit Brean Down - Melksham line, cleared in 0.1 seconds by switching out both circuits, this fault coinciding with one of the Brean Down - Aust circuits being out for maintenance. These lines were chosen since they are the most heavily loaded lines from Brean Down, each double circuit carrying about 40% of the Barrage output under normal operating conditions, whilst a three-phase fault at Brean Down reduces the transmission of power from the Barrage to zero for the duration of the fault. With the single circuit maintenance outage on the Brean Down - Aust line the remaining circuit takes 28% of the Barrage power and the double circuit line to Melksham 44%.

The initial transient stability studies were carried out using a simple mathematical representation for the generators to keep computational time to a minimum. These showed that with the premises chosen the Barrage generators continued to accelerate after fault clearance, losing synchronism in only a further 0.2 seconds, followed by the Hinkley Point generators approximately 0.2 seconds later. This is illustrated in Figure 3(a). (In this and the other graphs in Figure 3 the generator rotor angles are all plotted with respect to the Melksham generator).

Before starting the transient stability calculations load flow and fault level studies were carried out. The load flow studies gave acceptable system voltages, and power and reactive power flows. The fault levels, neglecting the contribution from the Barrage generators, show reasonable correlation with the expected levels due to the CEBG system. These results gave confidence in the equivalent grid network.

Possible ways to overcome the instability are the use of high speed of response excitation systems, an increase of inertia of the sets and the use of braking resistors. Since the first two only involved minor adjustments in the data they were tried out. With high-speed-of-response excitation on the Barrage generators the margin of instability was reduced but both the Barrage and Hinkley Point generators

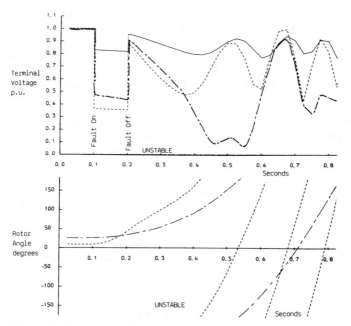

(a) Simple generator models, no fast excitation,
     preferred Barrage inertia

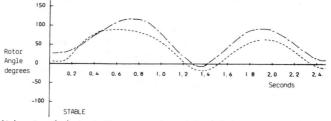

(b) As (a) but Barrage inertia high
     - preferred value +60%

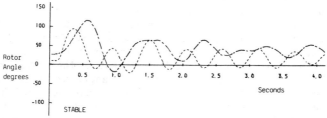

(c) All generator models include sub-transient effects.
     Barrage with fast excitation and preferred inertia.

----- Barrage   ———--——Hinkley Point   ———————Melksham

Fig. 3.   Cardiff-Weston Scheme
          Transient stability plots

121

still become unstable after approximately 1.3 seconds. By increasing the inertia of the Barrage sets by approximately 60% stability was achieved (Figure 3(b)), but with only a 36% increase they still lost synchronism. In a further experiment one of the six 900 MW Barrage generating groups was also tripped at fault clearance but the remainder of the sets still became unstable at the same point in time.

It is known that by using a more accurate and detailed mathematical model of synchronous generators the response of the generators and the damping of the system is better represented. A further series of studies was therefore carried out with the sub-transient reactances and time constants included on all generators; this introduces the effect of the machine damper windings. The high-speed-of-response excitation system on the Barrage generators was still included. In this series although the Barrage and Hinkley Point generators swing significantly they do not lose synchronism hence stability is maintained (Figure 3(c)).

ENGLISH STONES SCHEME

A brief study was carried out for the English Stones alternative. The same loading conditions for the CEGB system were used. In order to allow for additional CEGB generation the number of equivalent sets connected to Melksham were increased from 12 to 19 and their interconnecting reactance reduced in the inverse ratio. This gave satisfactory system voltages, load flows and fault levels that correlated with those expected. The Barrage generation is now only 5% of the total. To allow the best comparison with the previous study the same fault contingencies were used with the same fault clearance time.

Due to the higher loading taken by the equivalent generator connected to Melksham the rotor angles of the other machines with respect to it are different in the pre-fault condition, although their spread is of a similar magnitude. There is an increased angle between the Barrage and Melksham generators but of an opposite sign. The first swings of the critical machines i.e. Barrage and Hinkley Point are however significantly reduced and subsequent swings show reasonable damping. The English Stones arrangement therefore shows itself to be more stable than the Cardiff-Weston barrage.

USE OF HVDC LINK

During the investigative stages of the Cardiff-Weston transmission scheme the possibility of using a d.c. link between the Barrage generators and the 400 kV CEGB network was examined. There are certain technical advantages in the use of such a link that gives an asynchronous and controllable connection. The main advantage is that the generators can operate at a variable speed with a range of possibly 1.0 : 0.6 which will allow additional energy recovery. More importantly from this paper's aspect the possibility of asynchronous operation also overcomes any instability problems. Firstly the Barrage machines cannot get out of step with the system on the initial swing and secondly its power controllability can be used to actively damp out the grid system swinging after fault clearance and during steady-state conditions. No studies were carried out with a d.c. link solution.

CONCLUSIONS

It was appreciated at the start of the brief investigation that it would provide little more than a feel for the stability of the projected scheme. It has shown that the inertia of the generators is of prime importance in the stability of the a.c. interconnected scheme and that with the preferred values the stability of the Cardiff-Weston scheme is marginal under a reasonable worst case condition. Stability was only achieved by the damping introduced by machine/system interaction. Without relying on this damping the increase in inertia required is significant, being greater than 36% but less than 60%.

It is known that the use of braking resistors to slow down the Barrage generators during the fault period would certainly assist stability but their application was not studied. Equally well it is known that the introduction of a d.c. link would ensure stability but in view of the high cost of such a scheme it was not considered worthwhile to demonstrate this.

A brief examination of the English Stones scheme shows it to be the more stable.

Although the studies provide a reasonable assurance that stability will be achievable, it is essential that a full investigation is carried out examining other possible worst case conditions to arrive at an optimum solution. This is particularly important if the maximum capacity of the Cardiff-Weston scheme is increased from 5580 to 7200 MW.

REFERENCE

1. GARDNER G.E. and JERVIS W.B. Generator design and coupling to the grid. Severn Barrage, Proceedings of Symposium, I.C.E., October 1981.

7. Transmission and control

D. M. BARR, BSc, *Chief Electrical Engineer, Balfour Beatty Engineering Limited*

SYNOPSIS. The arrangement of the electrical connections from generator to shore for a tidal power station evolves from factors similar to those for other generating stations. These factors include the rating and number of generators, the length of the connections and the magnitude of power loss which can be accepted due to a fault. Also to be considered are the ratings available for switchgear, cables, transformers and busbars. The security level is important and physical restrictions such as weight and access. The most important consideration however is cost. The paper describes the way in which these factors have contributed to the choice of connections for the Severn Barrage between Lavernock Point and Brean Down.

PREVIOUS PROPOSALS (ref. 1)

1. In a previous report an AC system was proposed (FIG 1) which had a 72MVA 6.9/132kV transformer per generator and ten 132kV circuits in parallel connected to each 720MVA 132/400k transformer. Two of these 720MVA transformers were then connected in parallel to the shore by 400kV cable.

2. An alternative DC system was also proposed (FIG 2) to allow more efficient operation by variable speed turbines. In this case fourteen generators were connected in each group and ten sets of rectifiers connected to five separate 400kV grid substations.

GENERATOR GROUPS

3. There are 192 x 37.5MW generators in the present proposal. It has been requested that to match the spinning reserve criterion of the CEGB a fault on the barrage should result in the loss of no more than 1000MW. The generators can conveniently be arranged in groups of 24 giving a maximum capacity of 900MW per group and a total of 8 groups.

4. The most economical generator voltage was given as 6.9kV resulting in a current of approximately 3,500 amps per set which can be handled by standard switchgear. By splitting the low voltage windings of the generator transformers into two it is feasible to connect eight generators on to one 6.9k /400kV 333MVA step-up transformer. Three of these transformers can then be connected to a single 1000MVA (900MW) 400kV circuit.

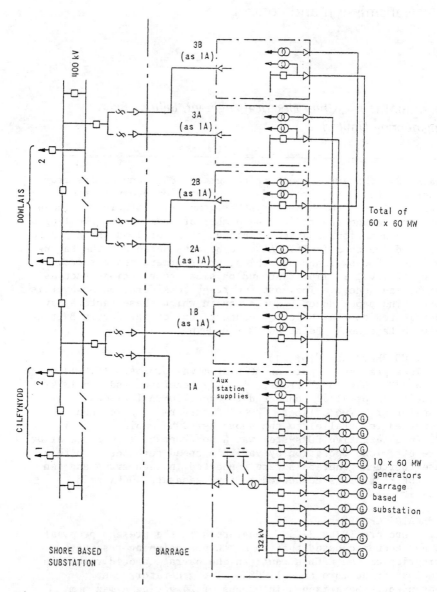

Fig. 1. Three-voltage system

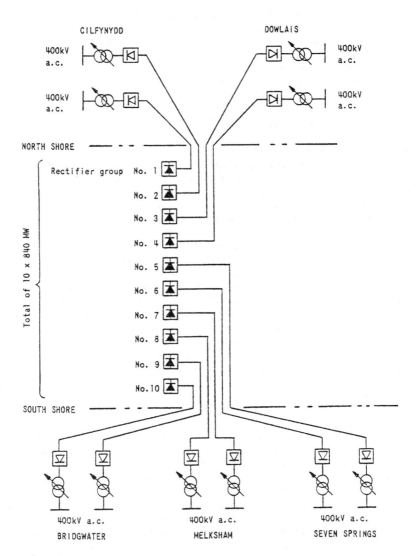

Fig. 2. DC system

CONNECTIONS TO SHORE

5. The 400kV system has eight (1000MVA) generator transformer circuits. However the availability of generation has been taken as 0.95 giving a total maximum output of 7600MVA. This can be carried by five 400kV cable circuits each of 2500 sq mm cross section rated at 1660MVA. One spare circuit has been provided in case of maintenance or other prolonged outage. In the event of a fault coinciding with a circuit outage, four circuits will sustain the full output for all but a few hours of the year. This is because the full output is only available at maximum spring tides and because its duration will infrequently exceed the thermal time constant of the cables.

BARRAGE SUBSTATION

6. To enable the eight generator groups to be connected to the six cable circuits to shore a substation is proposed on the barrage. Alternative layouts considered were a double busbar, a circuit-breaker-and-a-half and a mesh. Since the mesh is most economical and gives an acceptable level of security it was adopted. Also the major objection to a mesh namely the difficulty of future expansion does not arise.

7. It will be noted that unlike previous proposals the connections to shore are in one direction only. This avoids the possibility of overloading circuits by transmitting power from one shore to the other. It also reduces the expense of having standby capacity in both directions. Since Wales has a surplus of generation the connections are to the English side.

8. The main electrical connections are shown in FIG 3. Note that since the 400kV barrage to shore connections are 10 km long compensating reactors are switched in with the cables.

GENERATOR CONNECTIONS

9. Each generator has a 6.9kV circuit breaker for isolation but the electrical protection relays will operate the 400kV circuit breakers. The connections from generator to switchgear will be by air insulated phase segregated busducting suitable for the current of 3500 amps. From the switchgear to the transformer the current will be 14,000 amps. The connections will also be busducting. The 6.9kV breakers will be able to withstand the through fault current and after a fault is cleared by the 400kV circuit breakers the 6.9kV breakers will be able to quickly and permanently isolate the faulty component so that the remainder of the plant can be re-energized. Thus only 37.5MW may be out of commission while the fault is repaired.

GENERATOR TRANSFORMERS

10. The 6.9kV/400kV transformers are arranged as single phase units to limit their size and weight but also to enable a spare unit to be provided at minimum cost. The 6.9kV bus-ducting will be flanged and bolted where necessary to enable a faulty single phase unit to be readily replaced. Neutral earthing will be arranged with high impedance transformers to limit the earth fault and circuit breakers will be metal clad between phases.

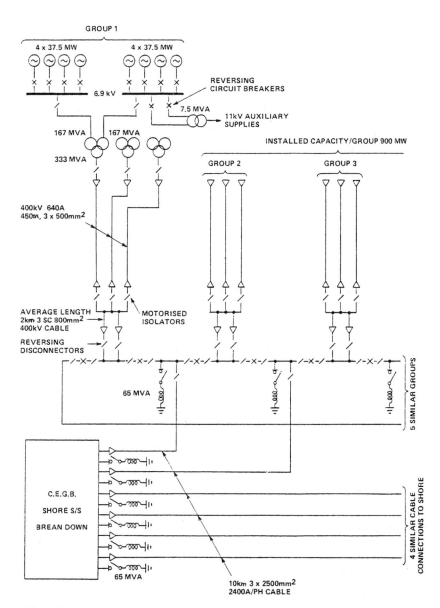

Fig. 3. Two-voltage system

GENERATOR SYNCHRONIZING

11. The synchronizing of the generators can be effected by the local 6.9kV breakers on a predetermined automatic run-up sequence in a conventional manner. An alternative rough synchronizing method can be adopted without the use of governing equipment and entails operating the machines as induction machines up to near synchronous speed and then energizing the field which pulls them into synchronism. However the requirements of system operation under forced outages may require governing action on the output of the tidal barrage as a whole. It is also possible that the stresses produced by rough synchronizing twice per day 365 days per year would create unacceptable high levels of forced generator outages. These are points which should be more fully investigated during the final definitive design.

PUMPING CYCLE

12. During the course of the studies on the energy optimization it was found that a significant increase in energy production might be possible by including a pumping cycle. To enable the turbines to act as pumps the machine rotation is reversed and the generator used as a synchronous motor. To reverse the rotation requires the transposition of two phases, ie Red Yellow Blue to Blue Yellow Red. In addition the machine has now to be started not as a turbine is started by opening the waterways but as a pump started by applying electricity to the pump motor. For small machines attached to large electrical systems it is possible to start them direct-on-line as induction motors changing to synchronous operation by energizing the excitation when the machines are at near synchronous speed. However there are two sets of shocks to the windings and the mechanical structure which on a repeated basis cannot be anything but deleterious. A less onerous method is to provide variable frequency starting equipment which accelerates the pump motor to just above synchronous speed. As the machine decelerates through 50 Hz it is synchronized by the appropriate by-pass breaker on to the normal grid busbar, which also has had its phase rotation reversed. The machine voltage is matched to the 6.9kV grid voltage by its AVR.

GENERATOR CONTROL

13. The machine controls are otherwise conventional with excitation derived from static inverters for rapid response and adequate range. Normal governing action would be modified by the necessary damping to avoid hunting with the wave action. Starting and stopping will be manually initiated but will have rapid automatic sequencing control to enable the 192 machines to start within the minimum time allowed by the CEGB for shutting down the equivalent thermal plant. This time will vary according to the tidal range but will be of the order of half an hour. The rate at which the machines may be unloaded ie 70MW per minute would mean a shut down time of about an hour and a half. The CEGBs thermal plant will easily be able to accept this rate of loading.

PROTECTION

14. The protection envisaged for the barrage plant and circuits will take into account the peculiarities of the arrangement. For example it is not considered essential to have a fault isolating circuit breaker for each generator and the cable compensating reactors will form a whole with their corresponding cables for fault isolation. However detail protection studies will be required before implementation engineering is carried out.

STATION AUXILIARY SUPPLIES

15. The auxiliary station supplies when the generators are not running will be taken from the local area board at 33kV. In the running mode unit supplies will be derived from the generator busbars via 24 6.9/11kV 7.5MVA transformers. The common station services will continue to derive their supplies from the 33kV system. In addition if the 33kV supplies are out of commission three 1.5MW diesel generators provide a standby facility. It should be noted that the unit auxiliaries supplies are derived from a source which may be subject to reversal of phase rotation. To cater for this condition a phase reversal switch must be provided on the auxiliaries board feeder. Where the load is sensitive to interruption it will be fed from the 33kV station supplies. An alternative to the unit transformer system would be to derive the auxiliaries supply from a pair of 400/33kV transformers of say 80MVA rating. This could save some transformer costs but increase cable costs and involve two additional expensive 400kV circuits. Detail studies on the auxiliaries systems to optimize the arrangement will be required when the location and rating of the loads is established.

DIRECT CURRENT TRANSMISSION

16. An alternative method of connection to the shore which would allow variable speed operation of the generator/turbines is by direct current link and associated rectification and inversion equipment. There are a number of significant advantages over the alternating current system namely
a) improved stability
b) output less susceptible to wave action
c) the barrage electrical system can be isolated electronically rather than by operating the circuit breakers
d) the fault level of the grid is not increased by a contribution from the barrage
e) increased turbine efficiency due to variable speed operation
17. The rectifiers would operate at say ±300kV and be arranged in 4 banks which would each be capable of handling 1800MW in the bipolar mode. Temporary 900MW monopolar operation could be maintained in the event of the failure of equipment or cable in one pole. This would require the use of adequately rated earth return paths. Theoretically an electrode system using the sea as an earth path could be used but a metallic path is preferred.

The main problem with a DC link is that the cost of the terminal equipment must be recovered by increased output, better efficiency, reduced circuit costs, improved reliability or reduced indirect expenses. To achieve a net gain from the DC link compared with an AC link the length of the conductors is critical. If for example the output from the tidal barrage were carried by direct current overhead lines to the load centre of England somewhere between London and the Midlands the saving on the main overhead line costs could be substantial. However there would be additional distribution costs since the DC link would not have facilities for interconnecting with the 400kV grid except at a converter station. In addition the DC link would occupy valuable way-leave routes which are required for the 400kV grid. Comparing the additional cost of the short direct current system to a terminal on the shore with an alternating current system shows that the DC system costs approximately 220 million pounds more than the AC system. It is not shown elsewhere that this additional expenditure can be justified by additional benefits and there is a reduction in security because of the special conversion equipment.

STABILITY

18. The system studies reported elsewhere indicate that it may be necessary to improve stability margins. This could be done by increasing machine inertia or by using artificial loading resistors with rapid switching arrangements. For much of the generating cycle the machines are not fully loaded and this should increase the stability margin. Fast acting field forcing may also assist stability together with suitably rapid fault clearing equipment. Further studies will be required and adjustments to the plant arrangements may be desirable.

ALTERNATIVE CONNECTIONS

19. As alternatives to the preferred 400kV connections to shore overhead lines, SF_6 trunking and air insulated busbars have been considered. Overhead lines would be subject to salt water pollution and could therefore be unreliable. SF_6 trunking proved to be considerably more expensive than cable and air insulated busbars would require expensively large ducts to protect them from atmospheric pollution. However it would be wise to consider all such alternatives again before detailed design commenced.

CONCLUSIONS

20. An arrangement using a single transformation from generator to transmission voltage can be combined with a switching station on the barrage to provide an economic method of connecting the generators to the national grid with a satisfactory level of reliability.

21. Conventional controls with automatic start up synchronizing and loading sequences can be used to bring the barrage generators on to the grid without exceeding permissible

loading rates.
22. Cables are at present the most satisfactory method of
connection but other methods should be checked before a final
decision is made.
23. Pumping can be accommodated relatively simply.
24. DC transmission could be used to improve stability and
energy capture but does not appear to be economic.

REFERENCES

1. DEPARTMENT OF ENERGY. Energy paper number 46 1981. Tidal
power from the Severn Estuary Volume II.

ACKNOWLEDGEMENTS

The author wishes to thank the Department of Energy for
permission to publish this paper. He also acknowledges that
the proposed system was developed by his colleagues at
Balfour Beatty Engineering Limited.

8. Transmission system reinforcement for Severn tidal power barrage system

P. BAXTER, BEng, MIEE, *System Design Manager*, G. E. GARDNER, BSc, FIEE, *Head of System Compensation and Analytical Support Group*, C. D. AKERS, BSc, MIEE, *System Design Engineer*, and P. W. BAILLIE, MA, *System Design Engineer, CEGB*

SYNOPSIS. The construction of a very large tidal barrage on the Severn would result in a major redistribution of power flow within the electricity supply network at times of high output from the barrage. It would also affect the way in which other generating plant would be operated to meet system load. A preliminary survey has been made by the CEGB of the potential impact of either the 7.2 GW or 1 GW barrage proposals and requirements for system connection and reinforcement have been identified.

INTRODUCTION

1. Once the high installation cost of a tidal barrage has been justified the low operating costs make it attractive when compared with all thermal sources of generation. Hence there will be a strong incentive to operate the barrage for maximum energy output and to operate the electricity supply system to accept the power whenever it is available. Although it is in principle possible to regulate the barrage output to match system demand, overall system economics are more likely to use the higher fuel cost thermal plant for this role. Consequently the introduction of the barrage will impose additional constraints on the operation of other generating plant. Similarly, there will be a strong incentive to provide sufficient transmission connections and system reinforcement to be able to handle the full barrage capacity under all conditions. However, the intermittent and variable nature of the barrage output results in relatively poor utilisation of any added transmission capacity and economic considerations may dictate some limitation in the capability provided.

2. This paper examines the power transmission requirements for both the very large 7.2 GW barrage and the much more restricted English Stones scheme with maximum capacity of 1 GW. Since both schemes are in electrically similar positions with regard to the main 400 kV transmission network, the different transmission reinforcement requirements are directly indicative of the system impact of the barrage capacity.

SYSTEM IMPACT

3.   The large barrage with an installed capacity of 7.2 GW
represents an injection of power at one location equivalent
to nearly 4 of the largest thermal stations operated by the
CEGB. At its peak operating level of 6.86 GW, assuming 95%
availability, it will cause a major distortion to the real
and reactive power flow pattern for the country and to assess
this it is necessary to consider both the nature of the
barrage output and its location. Whereas thermal power
stations can and do adjust their output power to the
requirements of the system, the barrage output is set by the
tidal conditions. Fig. 1 shows the range and temporal
distribution of the tides, each dot representing an
individual tidal level. The power pulses corresponding to
the maximum and minimum tidal ranges show the enormous
variation in output energy. This is further illustrated in
Fig. 2 which shows the order of variation in peak power level
and energy output over the 14 day tidal cycle.

Operation of generating plant

4. In this context it is the larger power levels which are
most significant since acceptance of barrage energy implies
power reduction from other sources. Fig. 1 shows that the
highest tidal ranges tend to occur between 0600 h and 1000 h
and between 1800 and 2200 h, though the actual period of
power generation will be delayed by about 3 hours. This
means that while one group of large power pulses will
correspond to times of high system load, those corresponding
to peak tides between 1800 and 2200 h will occur when system
demand is low. Fig. 3 illustrates the problem that could
arise when the barrage output is high on a typical summer day
for the conditions which could obtain in the mid 1990s. This
is based on the medium growth scenario C as presented to the
Sizewell Inquiry and the generation distribution includes no
new nuclear plant since this gives maximum scope for
accepting the barrage output. Without the barrage the
available nuclear plant would be operating at constant load
and there would be an adequate amount of the more flexible
fossil-fired plant available for load following etc. The
inclusion of a large pulse from the barrage during the early
hours of the morning requires much greater flexibility from
the fossil-fired generation corresponding to part-loading to
60%. This type of plant is not normally loaded to below 50%
for both technical and economic reasons so for the condition
shown on Fig. 3 effective operation should still be possible
though there would be an economic penalty. However the lower
summer weekend load condition, shown by the dotted curve,
could present serious operational problems either requiring
some part-loading of the less flexible nuclear plant or a
limitation on the power which could be accepted from the
barrage. Similarly the introduction of more nuclear
generation will raise the minimum load level at which this
type of selection will occur.

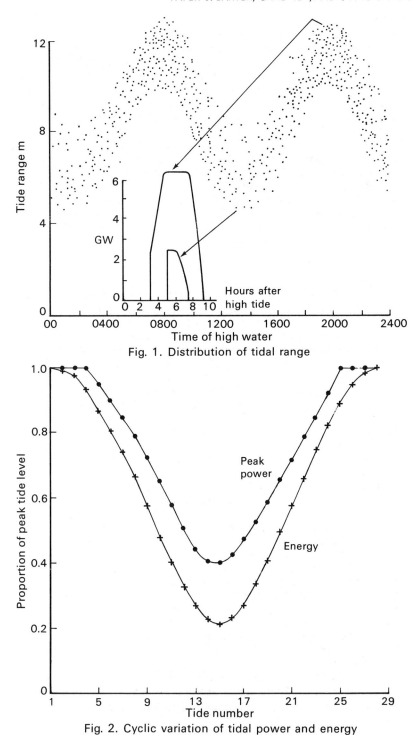

Fig. 1. Distribution of tidal range

Fig. 2. Cyclic variation of tidal power and energy

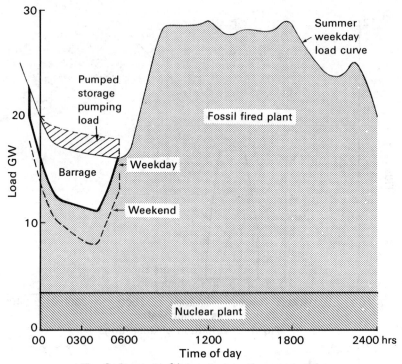

Fig. 3. Impact of barrage on other generation

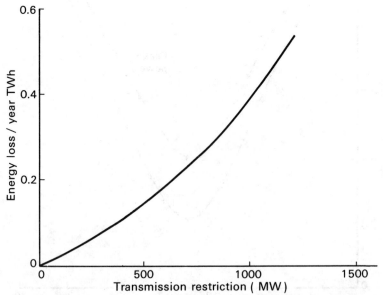

Fig. 4. Maximum energy loss due to limitation of peak power level

5. It is seldom economic to shut down coal fired-generation for less than 6 hours since the start up energy requirement is equivalent to about 1 hour's operation at full load. This compares with the energy losses during part-load operation which are of the order of 15% of the power reduction. These are charges which need to be included in the calculation of the value of the energy obtained from the barrage. A further cost component should be included to cover the increased maintenance and plant deterioration caused by the load cycling but this is less easy to quantify.

6. The efficient operation of the barrage would dictate a rapid build up of power at the start of the generation period in order to maximise the total energy. While at times of high system load this is unlikely to be a serious problem it may be more difficult to accommodate when the load is low or when it is falling rapidly between 2230 h and 2400 h. In general existing generation can be unloaded at rates up to about 5% of the maximum continuous rating (MCR) per minute and this implies that an initial load pick-up of the order of 4 GW over a period of 5 minutes would require 16 GW of responsive generation for the barrage alone before any allowance is made for the system load reduction. Fortunately the initial build up of the barrage output can be spread over a period of up to 20 minutes with little loss of energy. It should also be possible to use the 1.8 GW pumping capacity of Dinorwig to further alleviate the impact on other generation both in terms of rate and extent of the load reduction.

7. At the end of the barrage power pulse the system load must be taken up by other generation. Fortunately even for maximum tidal range this only represents a loading rate of the order of 100 MW/min which would be equivalent to less than 1.5% MCR/min for the generation required to accept the load. However where this coincides with the build up of the morning load which starts at about 0600 h the amount of thermal generation which must be brought into operation over a relatively short period is further increased.

8. Unplanned loss of barrage output. The multiplicity of generators required for the complete tidal scheme means that loss of a single generator would have an insignificant effect on the supply system. However, it is important that faults in the power collection system should not result in a very large loss of power. The supply system is operated so that it can withstand the loss of the largest unit which can be disconnected as a result of a single outage without causing load disconnection by operation of underfrequency relays. At present the loss incident limit is 1 GW as represented by an outage of the double circuit line connecting Wylfa Power Station into the system or the loss of 1 bipole of the Cross Channel HVDC Link. To secure the system against this loss a certain amount of generation is operated at less than maximum output and some of the Dinorwig pumped storage generating capacity is kept spinning in air so as to be capable of rapid

loading. Provided the maximum loss incident level attributable to the barrage does not exceed 1 GW no additional spinning reserve would be required, but for a design which could result in higher loss levels the cost of the additional reserve capacity would have to be carried by the barrage. This might be provided from additional machines at Dinorwig operating in air but if this prevented their use for merit order generation there would be a cost penalty which could be of the order of £12/MWh based on present day fuel costs. This is an average value for the whole year. It represents the difference between the merit order cost of generation from pumped storage and the operation of the marginal plant including the start-up and standby costs incurred in meeting the load which the pumped storage plant would otherwise have provided.

9. Barrage operation in pumping mode. The inclusion of pumping to increase the energy obtainable from the barrage is likely to be of greatest value when the natural tidal range is low. It is only likely to be of interest when the value to the system of the tidal energy is high. Hence it could be of value when the pumping can be done at times of light load prior to the build up of the morning load. The additional generation providing the pumping power will need to be off-loaded for the period between pumping and the increase of the morning load. This will entail some additional no-load heat costs but these need not be large since the machines involved would be those due to take up the first part of the load increase. The pumping power is apparently unlikely to exceed 1 GW and this is not likely to distort significantly the overall pattern of operation of generation plant or the transmission system.

POWER TRANSMISSION

10. The major analysis work of 1984 and 1985 was directed towards exporting 5.34 GW from the barrage, and this work had indicated that the output could be exported through one 400 kV substation on the English side. When the proposals reverted to the original 7.2 GW (192 x 37.5 MW) installed capacity with a required export capacity of 6.84 GW, it was the 5.34 GW scheme which was re-examined, with a view to assessing its capability for handling the higher output, notwithstanding the fact that in the ICE Paper (Ref. 1), CEGB had stipulated that this level of power would have to be split between 400 kV substations on each side of the estuary.

11. CEGB Transmission security standard. The supergrid transmission system has been designed around the need to be able to supply peak winter demand after the loss of any double circuit or 2 single circuits through failure for any reason. This has traditionally been the most onerous condition, but there is also the day-to-day operational

criterion of being able to supply the demand at all times after the loss of a double circuit route, and, applied during the summer maintenance season as a 3 circuit outage, (one on maintenance, followed by a double on fault), with the remaining lines on lower summer ratings, it can be an arduous condition but it is considered to be credible given the average number of double circuit faults on the system (9/year), and the normal maintenance regime of 6 weeks per circuit every 2 years.

12. <u>The proposed connection</u>. The connections now proposed for the Cardiff-Weston barrage are shown on Fig. 5. The barrage power will be brought into a new 400 kV supergrid substation on the English side at Brean Down. This substation will be connected into the supergrid via three L6 lines - to Aust (47 km), to Melksham (64 km), and to Bridgwater (25 km). Part of the Hinkley-Melksham line (48 km), from where it crosses the Aust-Bridgwater line to a new sub-station on this line at Frome will be rebuilt to L6 standard. Additional 400 kV lines which would be needed would be Frome-Nursling (65 km), and Ross-Hereford-Bishops Wood (63 km). A reconnection as a double tee would be required at Dowlais and a new substation with full bussing would be required at the Seven Springs junction. The 275 kV circuits around the west of London, between Iver and Beddington (59 km), would require uprating to 400 kV. Finally, a case could be made for rebuilding the Kemsley-Sellindge line (53 km) to L6 standard, although more work is required to clarify this need. A more detailed exposition of these reinforcements, and the reasons for them, are contained in ref. 2. The problem of voltage control of the transmission system throughout a barrage cycle has not been examined in detail, but an approximate assessment based upon the amount of generation which will be displaced has indicated a need for approximately 2300 MVAr of reactive compensation.

13. For the reasons explained in paragraph 10, these proposed connections meet the CEGB security standards for a barrage output of 5.34 GW without any restriction. After the reversion to a 7.2 GW installed capacity for which the CEGB had originally specified connections on both sides of the Severn estuary, some additional work was carried out, and it was concluded that there is no case for reinforcement in addition to what has already been specified provided that the barrage output can be rapidly reduced in certain specific circumstances. These circumstances would occur during summer nights, only when either a single Brean Down-Aust or a Brean Down-Melksham circuit would be out for maintenance. A trip on fault of the other double circuit would overload the circuit associated with the maintained circuit. The flows would be within the short time overload ratings, and the barrage output would need to be reduced to about 6.2 GW within about 10 minutes to alleviate the problem.

FIG. 5 SUPERGRID NETWORK
SHOWING PROPOSED REINFORCEMENTS
FOR CARDIFF - WESTON SEVERN BARRAGE

14. The cost of the energy which would be lost over a year by a permanent generation restriction is shown on Fig. 4, but in this case the output would not be expected to exceed 6.84 GW (i.e. 95%); about 2/3 of tides would not even produce that output (Fig. 2), and furthermore, many double circuit line outages are restored automatically within minutes, so it can be deduced that any loss of revenue from a brief fault outage would be insignificant and would certainly not justify any additional reinforcement. Finally there may in the event be CEGB operational reasons for limiting barrage output over the summer period.

15. <u>Operation of the transmission system</u>. The CEGB transmission system has been designed around existing sources of generation and main load centres. With much of the efficient coal-fired generation in the midlands, and a major load centre in the south east, the predominant flows have tended to be from the north through the midlands to the south east. The introduction of a 7.2 GW barrage will back off and at times reverse these flows, at the same time imposing a significant west to east flow across the south of the system. Figs. 6a to f show schematically snap shots of the system at three typical demand levels, with and without maximum barrage generation. The thickness of each bar is scaled approximately to the magnitude of the power transfer, and the way in which they change should give some indication of the operational problems of absorbing the barrage output, and will give an indication of why some of the more remote reinforcements are required.

16. An unusual feature of the barrage from the point of view of operating the system is the cyclical nature of the output. All other large generation stations have been commissioned in the expectation that they will contribute continuously to the system, as determined by their positions in the merit order. They are therefore contributing at times to suit the needs of the system. Planning for the barrage on the other hand has necessarily been carried out on the basis that the system must also work without the barrage and all other tentative reinforcements for generation and transmission have to be included for the next 10 or 15 years, whereas if the barrage was base load, one would certainly consider it as an alternative to any generation development contemplated for the south or south west, and would realise consequential capital savings.

Environmental assessment

17. Apart from the impact of the 7.2 GW barrage itself and three new supergrid substations, the current proposals would entail the construction of approximately 260 km of double circuit overhead line, in addition to uprating by rebuilding an additional 110 km. This will generate a significant amount of public antipathy, and the difficulties of obtaining consent for this work against organised opposition will be

143

REGIONAL IMPORTS (−VE) AND EXPORTS (+VE); MW

WITHOUT BARRAGE GENERATION

WITH 'MAX' BARRAGE GENERATION

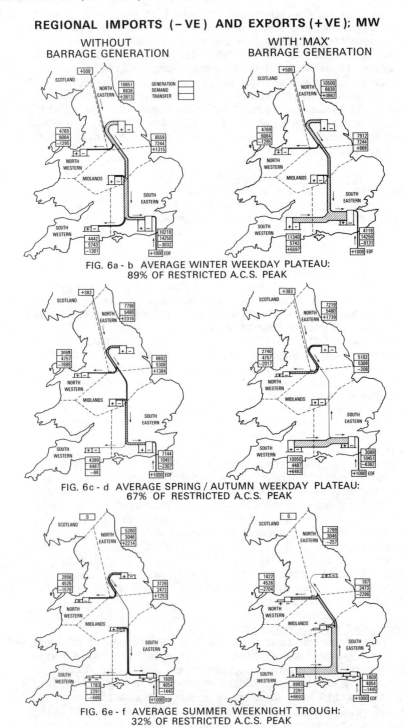

FIG. 6a - b AVERAGE WINTER WEEKDAY PLATEAU:
89% OF RESTRICTED A.C.S. PEAK

FIG. 6c - d AVERAGE SPRING / AUTUMN WEEKDAY PLATEAU:
67% OF RESTRICTED A.C.S. PEAK

FIG. 6e - f AVERAGE SUMMER WEEKNIGHT TROUGH:
32% OF RESTRICTED A.C.S. PEAK

considerable. Given that underground cabling costs about 20 times as much as overhead lines, it would not take the imposition of much undergrounding to undermine the economics of the whole scheme.

18. The STPG report on transmission (ref. 2) has already conceded that the first 1 km of each of the circuits from Brean Down should be cabled. This was not costed in 1984, and will add over £30M to the quoted costs.

Transmission costs

19. The costs currently being quoted for the connection of the barrage (ref. 2) are £392M. These were calculated at March 1984 prices, and include a sum of £25M for capacitive compensation, at locations which can only be determined by additional much more detailed study work. The cable quoted in paragraph 18 will cost an extra £30M, but will also be worth 240 MVAr at Brean Down, i.e. 10% of the estimated necessary compensation.

English Stones Scheme

20. The proposed English Stones scheme has an installed capacity of 972 MW (923 MW at 95%). This barrage would be connected into a new 400 kV substation on the Welsh side of the estuary, which in turn would be connected into the supergrid by turning in one circuit of the Cilfynydd-Melksham 400 kV route. This would require the construction of 3.5 km of L6 line, at a total cost including the substation of about £13M.

21. The transmission system in South Wales has been designed to cater for a significant amount of generation, including the 2000 MW oil-fired station at Pembroke, and the addition of a 972 MW barrage would not be a serious problem. An examination of the network, with the application of the security standards described in paragraph 11, has indicated that the only likely restriction would be caused by the Severn crossing cables in the Cilfynydd-Melksham circuits, which could be uprated to L6 rating by the installation of extra cooling capacity at a cost of about £0.5M.

22. This barrage will require no significant remote system reinforcements.

REFERENCES
1. ARNOLD P.J. Transmission Aspects: Severn Barrage. Proceedings of a symposium organised by the Institution of Civil Engineers, London, 8-9 October 1981.
2. Severn Tidal Power Group: Transmission System and Control, Report Reference: SD/3.8/STPG/BB, March 1986.

ACKNOWLEDGMENT
The authors acknowledge with thanks, permission from the Central Electricity Generating Board to publish this paper.

Discussion on Papers 5–8

MR J. TAYLOR, Merz and McLellan
To overcome the evident differences in approach by the European
turbine companies giving advice, will a design consortium be
established and can British water turbine design experience be
utilized?

What conditions apply to the generating sets when acting as
sluices and what special design measures will be required? For
example, would they operate at a controlled runaway speed
condition, or would they be synchronized with each other?

No doubt the 'further consideration of operating experience'
included the Rance scheme. How were the problems which were
encountered on that scheme reconciled to justify 95%
availability? Were they, for example, simply regarded as
teething troubles for long life plant?

The paper on the Rance scheme in 1973 entitled 'Six years of
operation of the Rance tidal scheme' supports the 95%
availability value. However, a second paper in 1984 entitled
'Twenty one years after the Rance — a tidal power experience'
gives rise to doubts about the 95% availability value. For
example

(a) quite a lot of renewal and redesign of the runner blades
    seals were necessary
(b) the generators were a major preoccupation: all 24
    generators had suffered air gap reduction and in January
    1975 one of the rotors rubbed the stator
(c) all stator fixing bars had to be replaced involving
    special arrangements
(d) an additional stator frame and core was ordered to
    reduce the outage time
(e) the pinning of stator cores on seven of the generators
    least affected was carried out to avoid further
    deterioration
(f) the piercing of the bulb enclosure to retrieve the
    fixing bars and to pull them back, then the sealing of
    the holes to prevent entry of sea water was carried out:
    4000 holes had to be drilled

(g) electroerosion occurred on each of the 64 poles of 24 generators, i.e. 1536 poles had to be dismantled, the damper bars secured and reassembled and all poles reinsulated in the process

(h) three to three and a half generators per year were refurbished with an outage time of 9 months each. The availability varied over this period from 70% to 94% with an average annual energy loss of 100 GWh for the 7 years needed for repairs.

Is the generator efficiency of 97.5% stated the generating set efficiency (i.e. turbine efficiency x generator efficiency)?

The generating sets will constitute an extrapolation beyond the present experience limit and will therefore be a test bed on which some adverse effects will be encountered. Which of these sets have been recognized as being prone to suffer undue wear and tear, or even failure, in service and what special measures are being considered to eliminate or reduce their influence on plant reliability?

MR G. L. DUFFETT and MR G. B. WARD
There is an open mind on the most appropriate method of organizing the design development of turbine generators, and that will be one of the matters to be dealt with in the next phase of work which is at present being defined. The two British manufacturers are working together and, since the choice at present is based on bulb turbines, co-operation with the two major European companies who have supplied a large proportion of this type of machine throughout the world will continue to form the basis for future investigations.

The turbines will be allowed to rotate in the reverse direction at speeds up to normal full forward speed under the control of the variable vanes and/or control gates depending on the final design chosen. Model tests will be required to check the hydraulic conditions and the bearings will have to be designed to cope with the reverse running. If finally pump/turbines are used there will be no major special design requirements although sluicing hydraulic conditions would again form part of the model testing.

The Rance scheme was included in the assessment of availability. The generators for La Rance and several other large machines have had problems in the past but it was concluded that all this experience formed part of a learning process for the designers and that new designs, which will be in the range of power and speed of existing machines, should be capable of meeting the run-in availability of these machines early in their life. The choice of turbine rotor diameter is linked to this subject in that it is not a large extrapolation from the biggest existing machines and will help to minimize potential new problems.

The value 97.5% is the full load efficiency of the generator

only. During the tidal cycle the turbine operates over a large
part of its hydraulic characteristic. The peak turbine
hydraulic efficiency is likely to be about 92% but in any cycle
the machine could operate between this level and the very low
value when the machines are shut down at the end of the
generating period.

The machines proposed are not an extrapolation beyond present
experience. As stated in the Paper the proposed major
parameters are a diameter of 8.2 m, a speed of 55.5 r.p.m. and
a power of 37.5 MW. Although this combination has not been
used before, there are machines of comparable diameter and
speed and there are machines with power ratings up to 54 MW at
85.7 r.p.m. Detailed design work has not yet been undertaken
but as noted previously the machines proposed are not
considered to present any new design problems.

MME C. BESSIERE, Société d'Etudes et d'Equipements
d'Enterprises

POWER GENERATION STUDIES OF A DOUBLE-RESERVOIR SCHEME ALONG THE
COTENTIN COAST IN FRANCE

In 1980-81, the Société d'Etudes et d'Equipements d'Enterprises
(SEEE) was involved in a programme of research developed under
the management of Electricité de France (EDF) in charge of a
comparison between different types of scheme for the same site,
located along the Cotentin Coast in France. This contribution
presents the realized studies and their main results concerning
the benefits and costs and the possibility of maximum power
generation during the peak electricity supply hours
independently of tidal characteristics.

BASIS OF COMPARISON
The site is located along the Cotentin Coast roughly between
Jersey and Chausey islands. The total reservoir area is the
same for the different types of scheme and equal to 200 km^2.

The power plant is equipped with bulb units each having an
output of 40 MW. The number of units varies between 40 and 54
according to the type of scheme. These units may be used for
pumping.

TYPE OF SCHEME
The comparison was made between six types of scheme (Fig. 1).
The high water reservoir is always close to the shore for
environmental reasons.

INVESTMENT COST ESTIMATE
For the six schemes the investment costs were estimated on
the basis of

(a) technical proposals for the electromechanical equipment
(b) feasibility studies for the civil works.

149

These feasibility studies were based on all the available data concerning the site conditions.

The results, given in Table 1, are at July 1980 prices.

POWER GENERATION

The studies were made on SEEE's computer, using 19 years of tidal conditions and also typical periods of 2 weeks chosen as good representations of the particular site conditions.

The tidal amplitude ranged between roughly 4 m and 12 m with an average around 8 m.

Two types of power production were analysed:

BHU54

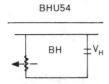

— High water reservoir only
— Installed power 54 units of 40 MW
— Gates section VH = 15 700 m^2

BEL52

— High and low reservoirs (Belidor schemes)
— Installed power 52 units of 40 MW
— Gates sections VH = 20 500 m^2
                    VB = 18 500 m^2

PM40

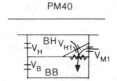

— High and low water reservoirs
— Installed power 40 units of 40 MW
  with multiple inlets (high
  reservoir and sea
— Gates section 42 900 m^2

PM20   RM20

— High and low water reservoirs
— Installed power
  20 units of 40 MW with multiple inlets
  (high reservoir and sea)
  20 units of 40 MW with multiple outlets
  (low reservoir and sea)
— Gates section 44 900 m^2

PRM40

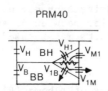

— High and low water reservoirs
— Installed power 40 units of 40 MW
  with multiple outlets and inlets
— Gates section 50 200–61 800 m^2

PM40   PP20

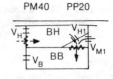

— High and low water reservoirs
— Installed power
  turbines = 40 units of 40 MW
  pumps = 20 units of 30 MW
— Gates section 42 900 m^2

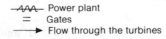

–ᴧᴧᴧ– Power plant
=   Gates
——➤   Flow through the turbines

Fig. 1

Table 1

Scheme	Investment cost I: FFr 1000 million	Maximum yearly production E: x 10^9 kWh	I/E: FFr/kWh
BHU 54	9.13	5.59	1.63
BEL 52	10.54	5.58	1.89
PM 40	9.14	5.57	1.64
PM20-RM20	9.24	5.55	1.66
PRM 40	9.33-9.80	5.95	1.57
PM40-PP20	10.79	6.10	1.77

(a) the maximum yearly energy production
(b) production following demand with pumping during week-ends and nights and full power for 6-8 hours per day whatever the tidal conditions.

The results for the maximum yearly production are given in Table 1 as the ratio investment cost divided by the yearly production (I/E). Table 1 shows that the double-reservoir solutions other than BELIDOR can ensure for a similar investment cost a maximum energy production that is higher than for a simple reservoir scheme. They have another advantage in the flexibility of production according to the electricity consumption.

Even during a week when the average tidal amplitude is 5 m, schemes PRM 40 and PM40-PP20 can ensure the maximum power of 1600 MW for 6 hours per day from Monday to Friday (08.00 h to 12.00 h and 18.00 h to 20.00 h) and 4 hours on Saturday mornings.

The yearly power generation under these conditions is reduced to 2.6 x 10^9 kWh and 3.3 x 10^9 kWh respectively for schemes PRM 40 and PM40-PP20.

The two other double-reservoir schemes also allow flexibility in power generation, but the results are slightly lower: the guaranteed power for 6 hours per day is only 1500 MW for an installed power of 1600 MW, and the yearly production is reduced to 2.6 x 10^9 kWh and 2.3 x 10^9 kWh.

CONCLUSIONS

The results obtained by the SEEE were confirmed a few months later by EDF's analysis. Unfortunately, EDF's nuclear programme was reduced at that time and the envisaged tidal plant allowing energy transfer was delayed.

For all the sites where tidal power plants are envisaged, solutions with high and low water reservoirs with multiple inlets (sea and high water reservoir) and multiple outlets (low water reservoir and sea) can allow flexible power generation, the maximum energy being a few per cent higher than in a single-reservoir plant. This result would be obtained for an investment cost that is similar to that for a single-reservoir plant.

MR M. BALLS, Salford Civil Engineering Limited
With regard to the addition of sea-to-basin pumping operation into the ebb generation mode of operation, Mr Duffett referred extensively to results of energy calculations performed by Salford Civil Engineering Limited (SCEL) for the _evern Tidal Power Group and in doing so introduced some of the parameters the effects of which the Salford flat water surface model is designed to test. This model does not model the dynamic response of the estuary but it does aim to model the turbine and generator performance fully and to incorporate all the constraints on operation, such as the cavitation limits and generator capacity limits. The simplification of the model allows many variables to be tested independently before selection of the more promising combinations for an evaluation on the more complex models.

With respect to pumping to augment the energy yield of ebb generation operation, there is conflicting evidence on the benefit of pumping. Firstly Dr Keiller indicated that the Binnie one-dimensional model shows a minimal net energy gain, whereas Dr Nichols suggests that the Reading one-dimensional model shows a significant energy gain. SCEL's results may help to put these claims into perspective. SCEL's calculations have shown that the benefits of pumping are sensitive to the turbine and pump performance characteristics used and the relative efficiencies of these components. It has been found that the better the turbine performance the poorer the energy gain from pumping.

This highlights another potential development since recently improved pump characteristics have been tabled by both European manufacturers, but correspondingly improved turbine curves have not. Before pumping is firmly adopted for the Severn or Mersey barrages it would be advantageous to obtain state of the art performance curves for turbines and pumps and to make a thorough evaluation of the benefits (and disbenefits).

MR G. L. DUFFETT and MR G. B. WARD
We concur with Mr Balls' comments. As stated in the Paper the potential for pumping needs much further investigation using a more sophisticated programme. A complete state of the art pump/turbine characteristic has not yet been formulated. Separate characteristics have been used with the assurance from the designers that they can design a machine which will combine them.

DR H. R. SHARMA, Central Electricity Authority, Government of India
For the energy computation and optimization studies,

Electricité de France suggest a seven-day tidal cycle, and Fundy tidal project studies recommend 50-60 tidal cycles. How many tidal cycles have been considered for energy computation and optimization in the present case? Is the recommendation based on a single tidal cycle for different ranges?

The operating regime finally selected is ebb generation with flood pumping. In this respect the following points need to be clarified:

(a) what is the extra energy generated compared with ebb generation alone?
(b) how does the cost of energy compare in the two cases?

Has the alternative of using Straflo turbines instead of bulb turbines been studied in detail? If so, what is the energy cost in the two cases?

What is the design head for the proposed Severn Cardiff-Weston barrage scheme? What criteria have been adopted to arrive at the proposed design head? A high design head will have low efficiency at smaller tidal ranges whereas a lower design head will increase the cost of the turbine for a given capacity.

MR G. L. DUFFETT and MR G. B. WARD
Most of the work has been done using the Salford Civil Engineering Limited model which effectively performs calculations for a single tide and then sums the answers for any particular year's variations in range. Any new more accurate complex model should incorporate a series of tides. It is not certain how many should be in the series and perhaps model development should investigate this. A maximum practicable level would seem to be a lunar cycle.

Flood pumping needs to be investigated in more depth, so a complete answer to questions concerning this cannot be given.

(a) There is a net energy gain from pumping provided that the operation is arranged so that pumping takes place towards the end of the flood tide when the head is relatively low. This produces a significant increase in head for generation during the ebb tide that immediately follows and, provided that the operating periods in both pumping and generating modes are sensibly optimized, there is a significant net energy gain. Estimates are provided in table 1 of the Paper, where the critical importance of the relative characteristics of the impeller (as a pump and as a turbine), which was mentioned by Mr Balls, is clearly shown.
(b) In a connected electrical supply system, the value of electrical power varies with the time of day and with the season. The operator can take advantage of this to discriminate between tides, choosing on the whole to

> pump at times of low value and to generate at times
> of high value, and he can arrive at a different
> optimization from that for energy yield on its own,
> leading to a greater net economic benefit. Preliminary
> values are presented in table 2 of the Paper.

The use of Straflo turbines has not been studied because much
more experience is required to be sure that large Straflos are
an acceptable alternative. A further point is that if pumping
is proven to be advantageous then variable rotor blade angle
machines would be required; this is less easy to engineer on a
rim generator turbine than on a bulb turbine: continuous double
regulation is not essential, since the switch from one mode to
the other can be performed off load.

The design head is virtually a notional value for a tidal
machine. The initial aim of calculations is to maximize the
energy output during any tide series by varying the time of
starting generation and by obtaining the optimum path across
the turbine hydraulic characteristic as the head varies noting
the cavitation criteria and any other limits such as generator
rating value which is being investigated. These are not simple
calculations. The ultimate criterion is the optimum economic
benefit.

PROFESSOR E. M. WILSON, Salford Civil Engineering Limited
How are the 1600 t lifts of complete bulb turbines to be
accomplished? Is a lift barge always to be available for
maintenance as well as for initial installation?

MR G. L. DUFFETT and MR G. B. WARD
A lift barge or a slide barge is proposed only for initial
installation. The layout of the turbine caisson and bulb is
proposed such that the items expected to require outside
maintenance can be removed individually and only for complete
replacement at some far time in the future would whole units be
removed.

MR J. B. SCIVIER, Binnie & Partners
When the original studies were carried out on the Severn
barrage the turbine performance data used were based on what
was available from manufacturers. This involved an
extrapolation of existing model test data. The data used were
therefore not necessarily the best that could be obtained.
Would it be sensible to arrange a design competition? That is
to say at a very early stage turbine designers would be given
all the tidal data over one year and they would be invited to
make and test model turbines to give maximum annual energy
output. Eventually all the models would be tested on a single
test stand so that test conditions and measurement accuracies
would be identical. The winner would be the model which

produced the maximum annual energy output. Thereafter manufacturers would be invited to tender for the mechanical design and manufacture of the winning hydraulic design. This would have the advantage of enabling civil design work to be started at an early stage since depths, submergence, draft tube shape and upstream entry would be fixed.

On another subject, mention has been made of the extensive repairs and modifications required to the Rance generators. Was this because they also operated as motors when in the pumping mode, with consequent high acceleration torques?

MR G. L. DUFFETT and MR G. B. WARD
As noted in answering the first question from Mr Taylor the subject of organizing design is a matter currently being reviewed by the Severn Tidal Power Group. However, it would be a mistake to suppose that impeller performance was the overriding consideration in selecting turbine generator technology. A design competition is by no means the only way forward, but if it were envisaged other factors associated with the engineering, manufacture, operational features and reliability of the sets would have to be taken into account.

Part of the problems of the Rance generator/motors was associated with the methods adopted to start up in the pumping mode but changes in methods of operation have been developed to obviate these problems. If pumping is adopted on any new scheme then the pump start-up will require careful study, but solutions to potential problems are available.

PAPER 6

MR J. TAYLOR, Merz and McLellan
It has long been recognized that bulb turbine generators may lose synchronism with the supply network during a fault. The principal reasons are the low inertia (usually about 1 MWs/MVA or less), the small size of the generator which imposes limitations on rotor size and the penstock water time constants. In the case of Arroyitos, a 120 MW 'run of the river' scheme, comprising four 30 MW sets, in Argentina, it was accepted that synchronism would be lost if a supply network fault occurred. The loss of 120 MW from the relatively small system to which it was connected was covered by comprehensive load shedding schemes. However, the potential loss of 7200 MW for a system even as large as the Central Electricity Generating Board's network would not be acceptable. Perhaps a new level of engineering is required for such a scheme incorporating thyristor-controlled rather than circuit-breaker-controlled braking resistors. Security of generation is of paramount importance and may go against operating the machines synchronously even though the cost may be higher for variable speed operation, e.g. through a zero-length direct current link.

MR W. P. WILLIAMS
The electrical design of the barrage generation in relation to
the Central Electricity Generating Board's system must preclude
the possibility of loss of synchronism due to a credible fault
on the system. The aim is to achieve this by optimization of
the mechanical and electrical parameters of the turbine
generator sets and their control systems. As mentioned in the
Paper, braking resistors are a possible back stop. The use of
thyristor-switched braking resistors would have great
advantages particularly with regard to speed of operation and
lack of mechanical linkages requiring maintenance. This does
not require a 'new level of engineering'. The application of
power thyristors in both high voltage direct current converters
and alternating current static var compensators is well
established. This particular application should be a
relatively easy duty.

MR G. E. GARDNER, MR P. BAXTER, MR C. D. AKERS and MR P. W.
BAILLIE
With reference to the use of braking resistors for maintaining
stability, although thyristor control would provide greater
precision in the application of the braking force it is
debatable whether it is necessary or cost effective. The
Central Electricity Generating Board will be carrying out tests
on the use of braking resistors for maintaining the stability
of one of their large alternators under severe system fault
conditions. For this application calculations suggest that
there should be adequate control using circuit-breaker
switching and this will be used in these tests. For the
inertia values that are appropriate to a large turboalternator,
shunt-connected braking resistors should be adequate so there
would be no operating penalty with thyristor control. However,
calculations made at the time of the Bondi study suggested that
for the low inertia of the bulb-type machine proposed for the
barrage there would be difficulty in maintaining a sufficient
terminal voltage for shunt resistors to be effective. The
Board's studies showed that series-connected braking resistors
would be required so that in normal operation the switch is
closed to short the resistor. In this situation a thyristor
control system would introduce additional losses during normal
machine operation unless a separate high speed shorting switch
is also provided.

PAPER 8

MR G. E. GARDNER
Paper 8 gives details of the way in which the Severn barrage
would affect the operation of the other generating plant on the
system and the electrical transmission reinforcement required.
The low operating cost of the tidal power system means that it
would displace all other generation and with an installed
capacity of 7.2 GW it would represent of the order of 12% of
the total generating capacity and provide of the order of 5% of

the annual demand. The variable nature of the tidal output
means that other generating plant output must be adjusted to
match the resultant variation in demand. This incurs
additional operating costs which need to be quantified in
estimating the value of the barrage energy. The Paper details
some of these constraints but does not give an overall estimate
of the economic penalty. Although the maximum penalty
associated with part loading generation is of the order of 15%
of the energy supplied it should be possible by careful
management taking advantage of the predictability of the
barrage output to reduce the value. However, the penalty is
unlikely to be less than about 5% of the total energy supplied
by the barrage.

The pulse nature of the barrage output means that there will
be significant periods with zero generation although the
transmission circuits have to be rated for the full output and
the supply system must have sufficient additional generating
plant to make up for the whole barrage capacity. Fortunately
statistics show that even in this situation some part of the
total barrage capacity can be credited as contributing to the
total system capacity. However, for such a large installation
the capacity credit is unlikely to be more than 15% of the
installed capacity.

In examining the requirements for transmission connections
for the 7.2 GW barrage the studies showed the need for an
additional 260 km of overhead line. In the cost estimates an
allowance was made for a short cable length near the barrage
for environmental reasons. However, much of the additional
overhead line routes would cross environmentally sensitive
areas and it is possible that there could be considerable
opposition requiring additional underground cabling. In this
context the costs for cable connections are about 20 times
those for overhead lines and have a significant impact on the
overall economics of the barrage.

MR J. TAYLOR, Merz and McLellan
The incorporation of the Severn barrage into the Central
Electricity Generating Board's (CEGB's) system should include
planning of future thermal stations and pumped storage schemes.

If the percentage of nuclear power generation is increased on
the CEGB network then the inflexibility of nuclear generation
will make it difficult to accept pulses of energy into the CEGB
network.

However, the ability to operate two shifts on even large
coal-fired power-stations like DRAX during the miners' strike
showed that two-shift operation can be achieved with large
turbogenerating sets and future stations may be designed for
two-shift operation.

The CEGB has only considered the pumped storage available to
itself, but other major schemes already exist or are planned,
i.e. Cruachan and Foyers and Craigroyston in Scotland. An
overall UK solution to energy storage is required rather than a
polarized north/south approach.

DISCUSSION ON PAPERS 5–8

MR G. E. GARDNER, MR P. BAXTER, MR C. D. AKERS and MR P. W. BAILLIE

We agree that the incorporation of the Severn barrage into the supply system would affect planning decisions for other plant. In this context the capacity credit which can be assigned to the barrage is important in that it represents the amount of other generation which would not need to be built. For a large installation such as proposed for the Cardiff-Weston line it is unlikely that more than 15% of the total installed capacity could be allowed as displacing other generation construction. With regard to pumped storage plant, the existing capacity is sufficient for reserve and frequency regulation requirements. Any further pumped storage would have to be economic for merit-order generation and needs to be justified. The use of pumped energy storage in the pumping mode can also avoid some of the losses associated with part load or double shifting of thermal plant. However, it has to be established that there will be an economic benefit such that the effective cost of the generation used to provide the pumping energy divided by the total pumping and generating efficiency of the storage is less than the cost of the generation which it could displace later in the daily load cycle. When the energy from pumping comes from coal and it is coal-fired generation which will be displaced later in the day it may not be possible to justify the use of pumped storage in this role. If there is more nuclear plant on the system such that this provides the energy for pumping while it is fossil fuel generation which is displaced at peak, the economics are more favourable.

With reference to the comment on the inflexibility of nuclear power it should be pointed out that the pressurized water reactor design of nuclear plant does allow part loading but its generation costs are such that this would only be done when all the economic part loading and double-shifting capability on fossil-fired plant had already been utilized. Double-shifting of both oil- and coal-fired power-stations is now an established technique which is regularly used during the normal operation of the system wherever it is economic. However, as pointed out in the Paper, the energy requirements for restarting after an overnight shut-down is equivalent to one hour's operation at full load. This means that it is uneconomic for the station to be shut down for any period less than six hours. To optimize on system operation to accept the tidal power the aim would be to shut down as much thermal capacity as will not be required for at least six hours and to use part loading for the rest.

Energy storage should be treated nationally rather than adopting a polarized north/south approach. However, to be of value the stored energy has to be economic and has to be in the right place. Consideration must be given to constraints and losses introduced by transmission. Effective used by the Central Electricity Generating Board of more pumped storage in Scotland would require transmission capability and would involve bulk transmission over a long distance to provide support for the barrage.

158

9. The economics and possible financing of the two Severn barrages

J. G. CARR, BA, PhD, *Assistant Director, County Group Limited*

SYNOPSIS   This   paper   considers   the   economics   of   two
possible Severn Barrages, one built on an alignment from
Cardiff   to   Weston-Super-Mare,   and   the   other   on   an
alignment   at   the   English   Stones.   Two   parameters   are
chosen to measure the economics. The first on the real
pre-tax internal rate of return, while the second is the
cost of a unit of electricity at a real discount rate of
5%.   The   implications   of   the   economics   for   the   prospects
of financing either Barrage as a private sector project
are then discussed.

ECONOMIC ANALYSIS
Scope
1.   This paper first examines the economic performance
of the two Barrages. The performance is assessed by means
of two measures. The first is the real pre-tax internal
rate of return (IRR) - in other words that discount rate
which results in the net present value of costs and
benefits being zero - while the second is the cost, in
pence/kWh, at a discount rate of 5%. The basic approach
has been to analyse the economics against the parameters
for the capital cost, energy capture, and other relevant
variables which are set out in other papers, and against a
number of Scenarios for the value of the energy. A number
of sensitivities are also considered.

Input Parameters
General
2.   The basic parameters which are involved in this
assessment of the economics of the Barrages are the
following:-
    (i)   Capital costs and their timing;
    (ii)   Operating and Maintenance Costs; and
    (iii) Value of the Energy and Power Capacity.
3.   While (i) and (ii) are to a large extent governed by
factors internal to the Barrages, the reverse is largely
the case with (iii), where factors affecting the overall
power system are involved, including:-
    (a)   Timing constraints of Barrage output;

(b)   Tariff arrangements;
(c)   Offload heat losses;
(d)   Transmission reinforcement costs; and
(e)   Transmission losses.

Capital Costs and Timing

4.   The capital cost estimates are discussed in detail in other papers. Tables 1 and 2 summarise the breakdown of the capital costs, expressed in January 1984 prices, by area and year for the Cardiff-Weston and English Stones Barrages (hereinafter referred to as the CW Barrage and the ES Barrage respectively).

Operating and Maintenance Costs

5.   The study by the Severn Barrage Committee (Reference 1) estimated operating and maintenance costs to be at most 1% of the capital cost of plant and 0.75% of the capital cost of the structure. For large hydro electric power plants (over 500MW) annual operating and maintenance costs can be as low as £2 per kilowatt (1984 prices) for manually controlled projects and £1 per kilowatt for remote controlled projects.

6.   Although the Severn Barrage is planned to be remotely controlled, operating and maintenance costs will be higher than for a large hydro plant due to the more severe maritime environment and the diurnal cycle of operation. The annual operating and maintenance cost has been taken as 0.5% of the total capital cost for the purpose of the present analysis. This is equivalent to about £4 per kilowatt. It seems likely that this figure may be reduced further as the project details are developed and an estimate of actual operating and maintenance costs can be built up.

7.   In addition, a further 0.5% of the capital costs has been charged against annual revenue. This covers an allowance for ancillary works and other external charges, such as rates and royalty to the Crown Estate Commissioner. It could be argued that, in national terms, these latter costs merely represent a redistribution, and should not be taken into account in assessing the economics of the Barrage. However, they are costs that will have to be met by the Barrage owning company, and it has been judged more appropriate to include them. Paragraph 38 below, which analyses the sensitivity of the Project to changes in the level of operating and maintenance costs, includes a consideration of the effect of not including this allowance for external charges.

Value of Energy

(a)   Timing constraints of Barrage output
8.   Other papers report in detail work on Energy Capture Evaluation. For the purposes of the economic analysis,

Table 1. Cardiff Weston barrage: cash flow schedule of costs (£M at 1984 prices)

Year	-6	-5	-4	-3	-2	-1	1	2	3	4	5	6	7	8	9	10	Total Cost	Total Cost
Capital Costs																		
Civil Works																		
Construction Yards							140	141	48								329	
Caisson Construction								229	265	266	265	209	29				1263	
Dredging							12	30	31	31	30	19					153	
Caisson Installation									70	117	144	108	80	29			548	
Foundations/Grout									10	24	41	52	38	27			192	
Embankments									62	63	63	63	30				281	
Breakwaters									10	24	24						58	
Locks (small)								15	33	33							81	
																	2905	
Contingencies 20%							30	83	104	114	122	88	35				581	3486
Turbine-Generator Works																		
Capital Plant							40	60	34								134	
Manufacture									141	141	141	141	141	141	141	37	1024	
Transport to Site									1	2	2	3	3	3	3	1	18	
Install & Commission									2	4	4	4	4	4	4	2	28	
																	1204	
Contingencies *																		1204
Transmission & Control Works																		
Manufacture & Delivery										24	37	44	57	37	28		227	
Install & Commission												16	29	42	48	24	159	
																	386	
Contingencies 10%										2	4	6	9	8	8	2	39	425
Sub-Total							222	587	807	862	921	739	444	235	232	66		5115
Feasibility Studies	1	4	4	6														15
Management, Engineering				2	32	39	41	36	34	34	28	25	18	4	3	1		297
Non-energy costs											20	30	30					80
Parliamentary and Environment	2	7	10	8	6	3												36
Total	3	11	14	16	38	42	263	623	841	896	969	794	492	239	235	67		5543

* Included in estimated costs

Table 2. English Stones barrage: cash flow schedule of costs (£M at 1984 prices)

Year	-8	-7	-6	-5	-4	-3	-2	-1	1	2	3	4	5	6	7	Total Cost	Total Cost
Capital Costs																	
Civil Works																	
Construction Yards									38	38						76	
Caisson Construction										67	90	58				215	
Dredging									5	15	10					30	
Caisson Installation											62	60	23	4		149	
Embankment										10	17	28	27	11		93	
Locks (small)									4	9	5					18	
																581	
Contingencies 20%									9	28	37	29	10	3		116	697
Turbine-Generator Works																	
Capital Plant									20	30	17					67	
Manufacture											45	45	45	26		161	
Transport to Site												1	2	2		5	
Install & Commission												1	2	2	1	6	
Contingencies *																–	239
Transmission & Control Works																	
Manufacture & Delivery										4	4	8	9	5	2	32	
Install & Commission											1	4	9	6	4	24	
																56	
Contingencies 10%											1	1	2	1	1	6	62
Sub-Total									76	201	289	235	129	60	8		998
Feasibility Studies	0.5	0.5															12
Management, Engineering			2	3	3	3	8	11									65
Non-energy costs			1	4	7	7	5	4									47
Parliamentary																	28
Environment																	
Total	0.5	0.5	3	7	10	12	13	15	85	210	297	255	151	83	8		1150

* Included in estimated costs

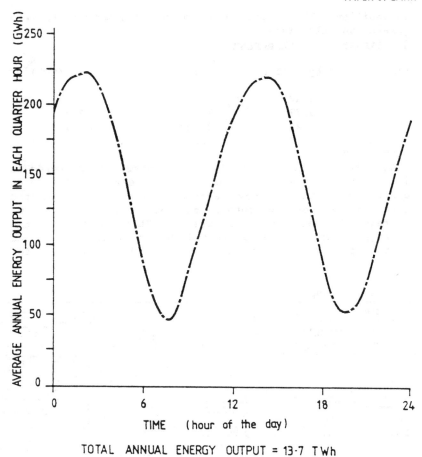

TOTAL ANNUAL ENERGY OUTPUT = 13·7 TWh

Fig. 1. Average energy output per quarter hour

outputs of 13.7 TWh per annum for the CW Barrage, and 2.7 TWh per annum for the ES Barrage before any increase as a result of pumping, have been taken which represents the energy that would be generated in a year of average tides.

9.   The + 7% - 6% variation of the tides which takes place on an approximate 19 year cycle is not taken into account in this economic analysis.

10.   The build-up to this level of output is assumed to be pro-rata with the commissioning of the turbines (see the paper of Gordon Duffet and Bernard Ward). This assumption balances out the fact that fewer turbines could be expected to produce proportionately more energy, but that, during this commissioning phase, they are likely to operate at a lower efficiency or with a lower

reliability. The actual energy output assumption, for ebb generation only, is:-

CW Barrage		ES Barrage	
Year	Output (TWh)	Year	Output (TWh)
8	4.42	6	1.73
9	10.77	7	2.65
10	13.24	8 and	2.70
11 and thereafter	13.70	thereafter	

11. The time of day at which the energy is delivered is important to the economics. The breakdown for each quarter hour of the day, averaged over a year, of the total annual energy output of 13.7 TWh (for the CW Barrage) is set out in Figure 1. This shows that the average annual energy generated during a quarter of an hour period, averaged over the year, for

Table 3. Bulk supply tariff

CENTRAL ELECTRICITY GENERATING BOARD

BULK SUPPLY TARIFF 1984/85   TARIFF FOR BULK SUPPLIES TO AREA BOARDS

Fixed by Central Electricity Generating Board (CEGB) pursuant to Section 37(1) of the Electricity Act 1947

Each Area Electricity Board (Area Board) in England and Wales shall pay CEGB for electricity supplied in the year ending 31 March 1985 in accordance with the following charges, rates, and adjustments

* * *

CHARGES
SYSTEM SERVICE CHARGE

1. Each Area Board shall pay the charge indicated in the following schedule, related to costs and expenses incurred in respect of the bulk supply points, and other services

	£m
London	48 395
South Eastern	40 700
Southern	60 399
South Western	27 634
Eastern	66 458
East Midlands	50 966
Midlands	55 971
South Wales	26 911
Merseyside and North Wales	37 465
Yorkshire	58 571
North Eastern	33 680
North Western	49 701

CAPACITY CHARGES

2. For the purpose of this BST kW means twice the number of kWh measured over 30 consecutive minutes starting either on, or thirty minutes after, the hour.

3. The Capacity Charges set out in paragraphs 4 and 5 relate to kW taken during half hours when System Demand attains the respective levels specified in those paragraphs. System Demand means the kW sent out from Area Board stations plus the kW acquired by CEGB from other sources minus the kW supplied by CEGB outside England and Wales during any half hour.

BASIC CAPACITY CHARGE

4. The basic capacity charge shall be £31 for each kW taken by the Area Board on average at times of Basic Demand. Basic Demand means the average System Demand when the latter is within the band 90% to 91% of Chargeable Peak System Demand. Chargeable Peak System Demand means the average of System Demand prevailing during the following half hours:
(a) the half hour of the highest System Demand;
(b) the half hour of the highest System Demand occurring other than on the day identified under (a) above or within ten days thereof;
(c) the half hour of the highest System Demand occurring other than on either of the days identified under (a) or (b) above or within 10 days of those days.

PEAKING CAPACITY CHARGE OR REBATE

5. The peaking capacity charge (or rebate) shall be £24 for each kW by which the average kW taken by the Area Board at times of Chargeable Peak System Demand exceeds (or falls short of) average kW take at times of Basic Demand.

RATES
UNIT RATES

	RATE p/kWh		
6. Times Days and half hours applicable each day	Summer Period 29 May to 30 September inclusive	Periods other than summer 1 April to 28 May and 1 October to 31 March, inclusive	
WEEKDAYS			
2400 - 0100	1 59		1 76
0100 - 0400	1 17		1 34
0400 - 0600	1 17		1 34
0600 - 0800	1 92		1 92
0800 - 1300	3 31		3 29
1300 - 1600	1 47		2 74
1600 - 1800	2 76		2 76
1800 - 2100	3 52		3 76
2100 - 2400	1 55		1 76

SATURDAYS, SUNDAYS & PUBLIC HOLIDAYS

2400 - 0100	1 61		1 72
0100 - 0300	1 13		1 24
0300 - 0700	1 13		1 17
0700 - 0800	1 47		1 39
0800 - 1330	3 33		3 25
1330 - 1400	2 32		2 04
1400 - 1430	2 04		2 04
1430 - 1700	2 04		2 04
1700 - 2400	2 16		2 32

In addition to the above rates a peak surcharge rate of 2.28 p/kWh applies in the half hour of highest system Demand in the period 0830-2330 and in each immediately adjacent half hour on each day except on weekdays in the Summer Period.

FUEL PRICE ADJUSTMENT

7. All the above rates, except the peak surcharge rate, shall be increased or reduced in their application in each month provided in each month of the year by 0.01p for each £0 25 by which the national fuel price per tonne in the relevant month (rounded to the nearest £0 25) differs from £45.

8. "National fuel price per tonne" means the replacement value of fuels consumed in the relevant month multiplied by 25 and divided by the net heat content of such fuel in gigajoules.

9. "Replacement value of fuel consumed in the relevant month" means the sum of the product for each Generating Board station of the net heat content in gigajoules of coal, coke, oil and gaseous fuels consumed in the relevant month and the average delivered price per gigajoule of fuels of a like kind delivered to the station in that month, or in the month when last there were deliveries.

10. The national fuel price per tonne shall be estimated by the Generating Board in the relevant month and corrected if necessary to take account of any differences between actual and estimated value not already taken into account for any month previous to the relevant month.

* * *

ADJUSTMENTS
LOAD MANAGEMENT ADJUSTMENT

11. Where an Area Board adopts Load Management, the total of the sums payable by the Area Board pursuant to paragraphs 1 to 10 above shall be adjusted as set out below. Load Management means the reduction, on a Notice issued by CEGB to the Area Boards, of the kW take by the Consumers registered by Area Boards in accordance with paragraphs 15 and 19. There are three Categories of Load Management, A, B and C as follows:

CATEGORIES A AND B

12. Category A Load Management is that in respect of which CEGB has by 1700 hours issued a Category A Notice that Load Management is required on specified hours the following day. Where such a Notice calls for Load Management to be implemented after 1300 hours it may be cancelled by CEGB no later than 0900 hours on that day.

13. Category B Load Management is that for which CEGB has by 0900 hours issued a Category B Notice that Load Management is required at specified times not earlier than 1600 hours on the day on which the Notice is given.

14. Category A or B Notices shall be confined within 1 October 1984 to 31 March 1985 and shall not be issued in respect of Saturdays, Sundays or Public Holidays. The aggregate number of hours for each Category for which Notices may be issued shall not exceed 50 except that a period for which a Category A Notice is cancelled shall count as only half of the period for which the Notice was issued.

15. The adjustments referred to in paragraph 11 shall be in respect of each Consumer
(a) who has by 31 March 1984 been registered by the Area Board with CEGB as likely to reduce his load significantly in response to Category A or B Load Management Notices as appropriate;
(b) whose take of kW during Category A or B Load Management periods and at times of Chargeable Peak System Demand is measured and certified by the Area Board;
(c) in respect of whom the Area Board indicates to CEGB by 31 March 1984 the reduction in the take of kW expected in response to Category A or Category B Notices.

16. The above adjustments shall be a rebate (charge) of £24 times the number of kW by which:
(i) the average kW taken by the Area Board's Category A or Category B Consumers at times of Chargeable Peak System Demand exceeds (falls short of);
(ii) the average kW taken by those Consumers in Category A or Category B Notice periods as appropriate.

CATEGORY C

17. Category C Load Management is that in respect of which CEGB has issued a Category C Notice requesting load reduction no sooner than fifteen minutes after receipt of the Notice

18. The aggregate number of hours for which such Notices may be issued shall not exceed two on any one day or 273 in the year

19. The adjustments referred to in paragraph 11 shall be in the case of Category C Load Management be in respect of each Consumer:
(a) who is also a Category A or B consumer;
(b) who has by 31 March 1984 been registered by the Area Board with CEGB as likely to reduce his load in response to Category C Load Management Notices by no less than 2MW in normal circumstances;
(c) whose take of kW during times of Basic Capacity and Peak Surcharge is measured and certified by the Area Board;
(d) in respect of whom the Area Board indicates to CEGB by 31 March 1984 the reduction in kW expected in response to Category C Notices.

20. The above adjustments shall be:
(a) a rebate (charge) of £7 times the number of kW by which:
(i) the average kW taken by the Area Board's Category C Consumers at times of Basic Demand exceeds (falls short of);
(ii) the average kW taken by these Consumers in Category C Notice periods
(b) a rebate of 2 28p for each kWh taken by the Area Board's Category C consumers at times of Peak Surcharge;
(c) a surcharge of 3.42p for each kWh taken by those consumers in Category C Notice periods

CONTRACTED CONSUMER ADJUSTMENTS

21. The total of the sums payable by the Area Board under paragraphs 1 to 20 above shall, in respect of the loads of Contracted Consumers, (as defined in the Bulk Supply Tariff published by the CEGB for the year 1982/83), be subject to adjustment as provided in the Bulk Supply Tariff for the year 1982/81, and on the basis of the tariff structure existing in that year

TARIFF FOR USE OF THE SYSTEM - 1 APRIL 1984 to 31 MARCH 1985
Fixed by the Central Electricity Generating Board (CEGB) pursuant to Section 8(1) of the Energy Act 1983 for the year ending 31 March 1985

1. Private producers of electricity supplying their own consumers via the CEGB grid system shall pay in respect of each private consumer so supplied:
(i) 43.9 pence per month, per kilowatt of Maximum Power Required, and,
(ii) 0.21 pence per month, per kilometre of the Distance Conveyed multiplied by the Maximum of Maximum Power Required.

2. In the above:
(a) Maximum Power Required shall mean the relevant demand in kilowatts of each private consumer as determined by the Area Board Concerned

(b) Distance Conveyed shall mean the distance in kilometres of the shortest practical grid connection existing between the private generator and the consumer.

3. Where the arrangement for the use of the system is made by and through an Area Board, that Board shall be responsible for payment of these charges on behalf of the private producer

4. The CEGB reserves the right to replace this Tariff with Special Terms where unusually large loads are connected or where technical difficulties are foreseen

a specific quarter of an hour period can vary between about 225GWh (at around 0230 hours and again some 12 1/2 hours later), and about 50GWh (at around 0730 and again some 12 1/2 hours later). The average energy generated in a quarter of an hour period, average over the year and the day is 142.7GWh. An analogous pattern of output pertains for the ES Barrage.

(b) Tariff Arrangements

12. If the Barrage is to be constructed and owned in the private sector, the value of the electricity generated by the Barrage is the price that would be paid for it by a purchaser under an appropriate sales and purchase contract. With the present structure of the Electricity Supply Industry, the quantity of electricity generated is such that the only possible purchaser is the CEGB. A number of discussions were held with the CEGB as to the contents of such a contract, the terms of which would be of critical importance to financing. These discussions are described in detail in paragraphs 54-59 below.

13. For the purposes of economic analysis, the economics of the Barrage were assessed against 13 scenarios for the value of the electricity.

14. The first scenario is to value the electricity as against the unit rates for electricity in the 1984/85 Bulk Supply Tariff (BST). For ease of reference, a copy of this is attached at Table 3. Two further scenarios have been considered related to the BST. In the first scenario, the BST rates increase in real terms at 2% per annum, and in the second the BST rates decrease in real terms at 1% per annum. These three scenarios are hereinafter referred to as BST, BST +2 and BST -1 respectively.

15. The remaining ten scenarios relate to the five economic scenarios, Scenarios A-E, against which the CEGB have assessed the economic arguments for constructing the Sizewell B power station (for each scenario, two cases are considered, depending upon the nuclear background). (See Reference 2 for a detailed description of these scenarios).

16. Table 4 shows, for the CW Barrage, the revenue generated in a year for each tariff period set out in the BST. This table has been calculated on the basis of the unit rates set out in Table 3, and assumes that the Peak Surcharge Rate applies between 16.30 and 18.00 in those periods when it applies. The unit rates of the BST vary according to how much the national fuel price varies from £45/tonne (see Paragraph 7 of Table 3). Discussions with the CEGB have led to the view that an appropriate assumption for the national fuel price is £49.25/tonne. This increases the revenue to £322.58M per annum. This corresponds to an average tariff of 2.35p/kWh.

17. It is noteworthy that a simple weighted average of the BST rates is 2.44p/kWh. This shows that the timing of the electricity generated by the Barrage in effect reduces its value. A similar phenomenon pertains for the ES Barrage. An

Table 4. Revenue components

Time	Summer Weekday	Winter Weekday	Summer Weekend	Winter Weekend
(GMT)	£M	£M	£M	£M
0.00	1.130	3.151	0.471	1.397
0.30	1.196	3.334	0.499	1.478
1.00	1.233	2.659	0.516	1.103
1.30	1.247	2.673	0.520	1.111
2.00	1.259	2.703	0.525	1.122
2.30	1.233	2.648	0.514	1.099
3.00	1.172	2.516	0.489	0.985
3.30	1.099	2.360	0.453	0.924
4.00	0.995	1.977	0.415	0.836
4.30	0.862	1.713	0.360	0.725
5.00	1.172	1.412	0.296	0.598
5.30	0.930	1.120	0.235	0.474
6.00	0.729	1.367	0.240	0.372
6.30	0.576	1.080	0.189	0.294
7.00	0.814	0.884	0.234	0.286
7.30	0.751	0.817	0.216	0.264
8.00	0.855	1.319	0.246	0.486
8.30	1.101	1.699	0.317	0.625
9.00	1.390	2.145	0.400	0.790
9.30	1.687	2.602	0.486	0.958
10.00	2.013	3.106	0.580	1.144
10.30	2.357	3.637	0.679	1.339
11.00	2.689	4.149	0.774	1.528
11.30	2.977	4.593	0.857	1.692
12.00	2.355	4.899	0.914	1.804
12.30	2.482	5.162	0.963	1.901
13.00	2.583	5.373	0.921	1.979
13.30	2.612	5.433	0.932	1.814
14.00	2.631	5.473	0.938	1.827
14.30	2.593	5.393	0.925	1.801
15.00	2.746	5.149	0.883	1.719
15.30	2.560	4.799	1.743	1.602
16.00	2.292	4.298	1.604	1.435
16.30	1.983	6.813	1.338	2.739
17.00	1.404	5.672	0.562	2.230
17.30	1.129	4.563	0.452	1.834
18.00	0.894	1.971	0.353	0.716
18.30	0.715	1.576	0.286	0.573
19.00	0.619	1.365	0.248	0.496
19.30	0.600	1.322	0.240	0.480
20.00	0.643	1.417	0.257	0.515
20.30	0.765	1.686	0.306	0.613
21.00	0.949	1.819	0.330	0.751
21.30	1.174	2.243	0.470	0.940
22.00	1.415	2.710	0.566	1.134
22.30	1.651	3.163	0.661	1.323
23.00	1.275	3.532	0.553	1.498
23.30	1.421	3.788	0.621	1.668

Total revenue is £M 299.283

Total energy available each year is 13.701 TWh

analysis of the possibility of retiming the generation so as to optimise the value of electricity generated, measured against the BST, as opposed to maximising energy generation, showed that such a course of action would only increase revenues by about 1%.

18. As far as the Sizewell scenarios are concerned, we have assessed the value as being the system marginal generation costs that are implied by the various scenarios, before account is taken of the losses described in Paragraphs 22-27 below. For each economic scenario, consideration has been given for two sets of system marginal generation costs - against a high and low nuclear background respectively. These figures were provided by the CEGB, and are set out in Table 5.

19. These system marginal generation costs have been translated into an annual tariff, taking into account the timing of generation throughout the day, as set out in Table 6. For the calculation, it was necessary to assume specific calendar years for the Barrage construction and operation. For these purposes, we have assumed the first year of full generation of the CW Barrage - year 11 - to be the year 2001, and the first year of full operation of the ES Barrage - year 8 - to be the year 2000.

20. It should be noted that the tariff for the low nuclear background case is essentially unaltered, provided that the nuclear background is such that the electricity generated by the Barrage is not displacing nuclear plant. This is likely to be the case certainly with a medium nuclear background (40% installed capacity).

(c) Pumping

21. The paper (number 5) by Gordon Duffet and Bernard Ward describes the benefits, in both energy and value terms, of pumping. In summary, pumping was likely to increase the energy output by some 5%, and the value of the energy output by some 7%. In the economic analysis, this 7% increase has been applied to all the revenue figures that follow from the assumptions on output and tariff described above.

(d) Offload Heat Losses

22. Following the introduction of tidal generation onto the electricity supply system, it will be necessary to have spinning reserve and part load operation of coal fired generation plant to maintain the required degree of system security. Spinning reserve requires a coal burn which is totally non-productive; part loan operation causes coal to be burnt at reduced efficiency.

23. The revenue figures implied by the scenarios for the value of electricity described above have been reduced by 5% to take account of these off load heat losses.

Table 5

System Marginal Generation Costs

£/MWh - 31/3/84 Price Level

Scenario Background		A High Nuc	A Low Nuc	B High Nuc	B Low Nuc	C High Nuc	C Low Nuc	D High Nuc	D Low Nuc	E High Nuc	E Low Nuc
YEAR 1995	D	46		54		37		28		27	
	N	40		41		30		24		25	
2000	D	44	44	53	53	45	46	29	35	28	31
	N	31	31	40	45	31	35	25	31	26	29
2010	D	50	54	45	49	36	46	35	44	28	35
	N	10	43	10	44	19	35'	31	35	28	30
2020	D	(9)57	60	(9)44	48	(9)38	43	(9)38	43	(9)31	32
	N	(15)11	59	(15)11	46'	(15)11	42	(15)11	42	(15)11	32
2030	D	(6)66	71	(6)45	50	(6)45	45	(6)44	48	(6)35	35
	N	(18)11	66	(18)11	48	(18)11	46	(18)11	47	(18)11	35

D and N refer to Day and Night Generation respectively
Day 16h. Night 8h except where shown ( )

System Capacity Credit Values

£/kW pa - 31/3/84 Price Level

YEAR	Plant Displaced Conventional Coal Fired	Plant Displaced Nuclear *
1995	0	-60
2000	0	-60
2010	30	-60
2020	40	-80
2030	40	-90

* Effective capacity credit appropriate to economic appraisals where nuclear capacity is to be displaced

Table 6. Annual tariffs (p/kWh)

Year	1995	1996	1997	1998	1999	2000	2001	2002	2003	2004	2005	2006
High Nuclear Background												
Scenario A	4.39	4.30	4.21	4.12	4.03	3.95	3.91	3.88	3.84	3.81	3.77	3.74
B	4.93	4.93	4.91	4.89	4.87	4.85	4.69	4.53	4.37	4.22	4.06	3.90
C	3.46	3.57	3.68	3.79	3.90	4.01	3.91	3.81	3.71	3.61	3.51	3.41
D	2.66	2.68	2.70	2.72	2.74	2.76	2.82	2.88	2.94	3.00	3.06	3.12
E	2.63	2.65	2.67	2.69	2.71	2.73	2.74	2.74	2.75	2.76	2.77	2.77
Low Nuclear Background												
Scenario A	4.39	4.30	4.21	4.12	4.03	3.95	4.05	4.16	4.27	4.37	4.48	4.59
B	4.95	4.96	4.98	4.99	5.01	5.02	4.99	4.96	4.93	4.90	4.87	4.84
C	3.46	3.61	3.76	3.91	4.06	4.22	4.22	4.22	4.22	4.22	4.22	4.22
D	2.66	2.80	2.94	3.08	3.22	3.36	3.43	3.51	3.58	3.65	3.72	3.80
E	2.63	2.71	2.79	2.87	2.95	3.03	3.06	3.09	3.12	3.15	3.18	3.21

Year	2007	2008	2009	2010	2011	2012	2013	2014	2015	2016	2017	2018
High Nuclear Background												
Scenario A	3.70	3.67	3.63	3.60	3.56	3.51	3.47	3.42	3.38	3.33	3.29	3.24
B	3.75	3.59	3.43	3.28	3.21	3.13	3.06	2.99	2.92	2.85	2.78	2.71
C	3.31	3.21	3.11	3.01	2.94	2.87	2.80	2.73	2.66	2.59	2.52	2.45
D	3.18	3.24	3.30	3.36	3.25	3.15	3.04	2.94	2.83	2.73	2.62	2.52
E	2.78	2.79	2.79	2.80	2.72	2.64	2.56	2.48	2.40	2.32	2.24	2.15
Low Nuclear Background												
Scenario A	4.69	4.80	4.91	5.02	5.11	5.21	5.30	5.40	5.49	5.59	5.68	5.78
B	4.81	4.78	4.75	4.73	4.73	4.73	4.73	4.73	4.73	4.73	4.73	4.73
C	4.22	4.22	4.22	4.22	4.22	4.23	4.23	4.24	4.24	4.25	4.25	4.26
D	3.87	3.94	4.01	4.09	4.10	4.12	4.14	4.16	4.18	4.19	4.21	4.23
E	3.24	3.27	3.30	3.33	3.31	3.30	3.29	3.28	3.26	3.25	3.24	3.23

Year	2019	2020	2021	2022	2023	2024	2025	2026	2027	2028	2029	2030
High Nuclear Background												
Scenario A	3.20	3.15	3.10	3.05	2.99	2.94	2.88	2.83	2.77	2.72	2.66	2.61
B	2.64	2.57	2.52	2.47	2.41	2.36	2.30	2.25	2.20	2.14	2.09	2.03
C	2.38	2.31	2.28	2.25	2.22	2.20	2.17	2.14	2.12	2.09	2.06	2.03
D	2.41	2.31	2.28	2.25	2.22	2.19	2.16	2.13	2.10	2.07	2.04	2.01
E	2.07	1.99	1.97	1.95	1.92	1.90	1.88	1.85	1.83	1.81	1.78	1.76
Low Nuclear Background												
Scenario A	5.87	5.97	6.06	6.16	6.25	6.35	6.45	6.54	6.64	6.73	6.83	6.93
B	4.73	4.73	4.75	4.77	4.79	4.81	4.83	4.85	4.87	4.89	4.91	4.93
C	4.26	4.27	4.29	4.32	4.35	4.37	4.40	4.43	4.45	4.48	4.51	4.53
D	4.25	4.27	4.32	4.37	4.42	4.47	4.52	4.57	4.62	4.67	4.72	4.77
E	3.21	3.20	3.23	3.26	3.29	3.32	3.35	3.38	3.41	3.44	3.47	3.50

(e) Transmission Reinforcement

24. There is a need for expenditure of some £392M on transmission reinforcement for the CW Barrage. The calculations in this paper are based on the assumption that the CEGB themselves bear the capital cost, and are remunerated out of Barrage revenues. This remuneration has been taken at £27.5M per annum for the first 30 years of full operation of the Barrage.e This figure gives a 5% real pre-tax internal rate of return to the CEGB on the basis of a phasing of the capital expenditure as follows:-

Year (in relation to Barrage Construction)	Cost (£M)
8	135
9	205
10	52

25. The requirement for transmission reinforcement for the ES Barrage is less. Hence, the analogous figures are £12M on the reinforcement, remunerated at £0.85M per annum for the first 30 years of Barrage operation.

(f) Transmission Losses

26. The revenue figures implied by the Scenarios for the value of the electricity described above have been reduced by 2% and 2.4% to take account of transmission losses for the CW and ES Barrages respectively. These figures have been arrived at after consideration of the losses in the CEGB transmission system over the 10 years from 1976 to 1985. Average losses at 2.4% of the units available are consumed on the system.

27. The overhead line associated with the CW Barrage will be lightly loaded at times due to variations in the tidal cycle, therefore the losses will be less than this average system figure. It is considered that 2.0% is a reasonable estimate at this stage. For the ES Barrage, where no extensive transmission reinforcement is necessary, the appropriate figure is the average figure of 2.4%.

Siltation

28. The paper (number 11) by Dr. Kirby explains that there is a potential problem of siltation for the ES Barrage. In view of the uncertainties surrounding this problem, the economic implications of siltation have been considered only by means of a sensitivity analysis. This is done in Paragragh 40 below, which considers the implications of energy output being halved over a range of periods from 10 to 100 years due to siltation.

Value of Power Capacity

29. The firm power contribution of the Barrage is determined by the system as a whole. The STPG has held discussions with the CEGB on this, which have concluded that the original

assumption in Reference 1, that the firm power contribution or the CW Barrage is 1.1. GW, remains valid. This figure represents around 15% of the installed power of 7200MW. Discussions with the CEGB suggested that, for the ES Barrage, with a smaller total installed capacity, a figure of 25% would be more appropriate. A figure of 250MW was therefore chosen.

30. These firm power contributions are assumed to apply once the Barrage is in full operation. As far as the value of these firm power contributions are concerned, in the BST cases, a figure of £10/kW has been taken. In the Sizewell economic scenario cases, the basic assumptions are set out in Table 5. For the low nuclear background cases, it has been assumed that the electricity generated by the Barrage displaces coal-fired capacity all the time. For the high nuclear background, it has been assumed that it displaces nuclear generated electricity during the periods of low system marginal cost.

Results of Economic Analysis

31. The IRR for the thirteen selected scenarios for future tariff levels of the Barrage with the other assumptions as set out above, are as follows:-

		IRR %
Scenario	CW Barrage	ES Barrage
BST	3.90	4.08
BST +2	7.24	7.47
BST -1	2.07	2.27

Sizewell High Nuclear Background

	CW Barrage	ES Barrage
A	5.45	5.74
B	5.32	5.71
C	4.37	4.63
D	3.81	3.95
E	3.00	3.11

Low Nuclear Background

	CW Barrage	ES Barrage
A	8.36	8.73
B	8.28	8.77
C	7.38	7.75
D	6.91	7.19
E	5.77	6.04

32. Detailed cash flow analysis up to the years 95 and 93 for the CW and ES Barrages are at Table 7 and 8 respectively.

33. The cost, at a discount rate of 5%, is 3.00p/kWh for the CW Barrage, and 2.89p/kWh for the ES Barrage. These figures are, in effect, those tariffs, constant throughout the life of the Barrage, which would generate a real rate of return of 5%. No account has been taken in this calculation of any value of

Table 7. Cardiff Weston barrage: cash flow analysis (unescalated)

Year	Capital Costs (£bn)	Capital Costs cum. (£bn)	Energy Generated per year (TWh)	Firm Power (GW)	Tariff (p/kWh) (after allowances for off-load heat & transmission losses, uplifted for pumping)	Firm Power Rate (£/kW)	Tariff Revenue (£bn)	Firm Power Revenue (£bn)	Total Revenue (£bn)	CEGB Trans. Payment (£bn)	Operating Costs (£bn)	Net Cash Flow (pre Finance) (£bn)	Net Cash Flow (pre Finance) cum. (£bn)
-6	0.003	0.003					0.00	0.00	0.00		0.00	(0.00)	(0.00)
-5	0.011	0.014					0.00	0.00	0.00		0.00	(0.01)	(0.01)
-4	0.014	0.028					0.00	0.00	0.00		0.00	(0.01)	(0.03)
-3	0.016	0.044					0.00	0.00	0.00		0.00	(0.02)	(0.04)
-2	0.038	0.082					0.00	0.00	0.00		0.00	(0.04)	(0.08)
-1	0.042	0.124					0.00	0.00	0.00		0.00	(0.04)	(0.12)
1	0.263	0.387					0.00	0.00	0.00		0.00	(0.26)	(0.39)
2	0.623	1.010					0.00	0.00	0.00		0.00	(0.62)	(1.01)
3	0.841	1.851					0.00	0.00	0.00		0.00	(0.84)	(1.85)
4	0.896	2.747					0.00	0.00	0.00		0.00	(0.90)	(2.75)
5	0.969	3.716					0.00	0.00	0.00		0.00	(0.97)	(3.72)
6	0.794	4.510					0.00	0.00	0.00		0.00	(0.79)	(4.51)
7	0.492	5.002					0.00	0.00	0.00		0.00	(0.49)	(5.00)
8	0.239	5.241	4.42		3.90		0.17	0.00	0.17		0.02	(0.09)	(5.09)
9	0.235	5.476	10.77		4.05		0.44	0.00	0.44		0.04	0.16	(4.93)
10	0.067	5.543	13.24		4.20		0.56	0.00	0.56		0.05	0.43	(4.50)
11		5.543	13.70	1.10	4.20	3.00	0.58	0.00	0.58	0.03	0.06	0.49	(4.00)
12		5.543	13.70	1.10	4.20	6.00	0.58	0.01	0.58	0.03	0.06	0.50	(3.50)
13		5.543	13.70	1.10	4.20	9.00	0.58	0.01	0.59	0.03	0.06	0.50	(3.00)
14		5.543	13.70	1.10	4.20	12.00	0.58	0.01	0.59	0.03	0.06	0.50	(2.50)
15		5.543	13.70	1.10	4.20	15.00	0.58	0.02	0.59	0.03	0.06	0.51	(1.99)
16		5.543	13.70	1.10	4.20	18.00	0.58	0.02	0.60	0.03	0.06	0.51	(1.48)
17		5.543	13.70	1.10	4.20	21.00	0.58	0.02	0.60	0.03	0.06	0.51	(0.97)
18		5.543	13.70	1.10	4.20	24.00	0.58	0.03	0.60	0.03	0.06	0.52	(0.45)
19		5.543	13.70	1.10	4.20	27.00	0.58	0.03	0.60	0.03	0.06	0.52	0.07
20		5.543	13.70	1.10	4.20	30.00	0.58	0.03	0.61	0.03	0.06	0.52	0.60
21		5.543	13.70	1.10	4.20	31.00	0.58	0.03	0.61	0.03	0.06	0.53	1.12
22		5.543	13.70	1.10	4.20	32.00	0.58	0.04	0.61	0.03	0.06	0.53	1.65
23		5.543	13.70	1.10	4.21	33.00	0.58	0.04	0.61	0.03	0.06	0.53	2.18
24		5.543	13.70	1.10	4.21	34.00	0.58	0.04	0.61	0.03	0.06	0.53	2.71
25		5.543	13.70	1.10	4.22	35.00	0.58	0.04	0.62	0.03	0.06	0.53	3.24
26		5.543	13.70	1.10	4.22	36.00	0.58	0.04	0.62	0.03	0.06	0.53	3.78
27		5.543	13.70	1.10	4.23	37.00	0.58	0.04	0.62	0.03	0.06	0.54	4.31
28		5.543	13.70	1.10	4.23	38.00	0.58	0.04	0.62	0.03	0.06	0.54	4.85
29		5.543	13.70	1.10	4.24	39.00	0.58	0.04	0.62	0.03	0.06	0.54	5.39
30		5.543	13.70	1.10	4.24	40.00	0.58	0.04	0.63	0.03	0.06	0.54	5.93
31		5.543	13.70	1.10	4.25	40.00	0.58	0.04	0.63	0.03	0.06	0.54	6.47
32		5.543	13.70	1.10	4.28	40.00	0.59	0.04	0.63	0.03	0.06	0.55	7.02
33		5.543	13.70	1.10	4.30	40.00	0.59	0.04	0.63	0.03	0.06	0.55	7.57
34		5.543	13.70	1.10	4.33	40.00	0.59	0.04	0.64	0.03	0.06	0.55	8.12
35		5.543	13.70	1.10	4.36	40.00	0.60	0.04	0.64	0.03	0.06	0.56	8.68
36		5.543	13.70	1.10	4.38	40.00	0.60	0.04	0.64	0.03	0.06	0.56	9.24
37		5.543	13.70	1.10	4.41	40.00	0.60	0.04	0.65	0.03	0.06	0.56	9.81
38		5.543	13.70	1.10	4.44	40.00	0.61	0.04	0.65	0.03	0.06	0.57	10.37
39		5.543	13.70	1.10	4.46	40.00	0.61	0.04	0.66	0.03	0.06	0.57	10.94
40		5.543	13.70	1.10	4.49	40.00	0.62	0.04	0.66	0.03	0.06	0.58	11.51
41		5.543	13.70	1.10	4.52	40.00	0.62	0.04	0.66	0.03	0.06	0.58	12.09
42		5.543	13.70	1.10	4.52	40.00	0.62	0.04	0.66		0.06	0.61	12.70
43		5.543	13.70	1.10	4.52	40.00	0.62	0.04	0.66		0.06	0.61	13.31
44		5.543	13.70	1.10	4.52	40.00	0.62	0.04	0.66		0.06	0.61	13.91
45		5.543	13.70	1.10	4.52	40.00	0.62	0.04	0.66		0.06	0.61	14.52
46		5.543	13.70	1.10	4.52	40.00	0.62	0.04	0.66		0.06	0.61	15.13

Table 7 (continued)

Year	Capital Costs (£bn)	Capital Costs cum. (£bn)	Energy Generated per year (TWh)	Firm Power (GW)	Tariff (unesc) (p/kWh)	Firm Power Rate (£/kW)	Tariff Revenue (£bn)	Firm Power Revenue (£bn)	Total Revenue (£bn)	CEGB Trans. Payment (£bn)	Operating Costs (£bn)	Net Cash Flow (pre Finance) (£bn)	Net Cash Flow (pre Finance) cum. (£bn)
47		5.543	13.70	1.10	4.52	40.00	0.62	0.04	0.66		0.06	0.61	15.73
48		5.543	13.70	1.10	4.52	40.00	0.62	0.04	0.66		0.06	0.61	16.34
49		5.543	13.70	1.10	4.52	40.00	0.62	0.04	0.66		0.06	0.61	16.94
50		5.543	13.70	1.10	4.52	40.00	0.62	0.04	0.66		0.06	0.61	17.55
51		5.543	13.70	1.10	4.52	40.00	0.62	0.04	0.66		0.06	0.61	18.16
52		5.543	13.70	1.10	4.52	40.00	0.62	0.04	0.66		0.06	0.61	18.76
53		5.543	13.70	1.10	4.52	40.00	0.62	0.04	0.66		0.06	0.61	19.37
54		5.543	13.70	1.10	4.52	40.00	0.62	0.04	0.66		0.06	0.61	19.98
55		5.543	13.70	1.10	4.52	40.00	0.62	0.04	0.66		0.06	0.61	20.58
56		5.543	13.70	1.10	4.52	40.00	0.62	0.04	0.66		0.06	0.61	21.19
57		5.543	13.70	1.10	4.52	40.00	0.62	0.04	0.66		0.06	0.61	21.79
58		5.543	13.70	1.10	4.52	40.00	0.62	0.04	0.66		0.06	0.61	22.40
59		5.543	13.70	1.10	4.52	40.00	0.62	0.04	0.66		0.06	0.61	23.01
60		5.543	13.70	1.10	4.52	40.00	0.62	0.04	0.66		0.06	0.61	23.61
61		5.543	13.70	1.10	4.52	40.00	0.62	0.04	0.66		0.06	0.61	24.22
62		5.543	13.70	1.10	4.52	40.00	0.62	0.04	0.66		0.06	0.61	24.83
63		5.543	13.70	1.10	4.52	40.00	0.62	0.04	0.66		0.06	0.61	25.43
64		5.543	13.70	1.10	4.52	40.00	0.62	0.04	0.66		0.06	0.61	26.04
65		5.543	13.70	1.10	4.52	40.00	0.62	0.04	0.66		0.06	0.61	26.65
66		5.543	13.70	1.10	4.52	40.00	0.62	0.04	0.66		0.06	0.61	27.25
67		5.543	13.70	1.10	4.52	40.00	0.62	0.04	0.66		0.06	0.61	27.86
68		5.543	13.70	1.10	4.52	40.00	0.62	0.04	0.66		0.06	0.61	28.46
69		5.543	13.70	1.10	4.52	40.00	0.62	0.04	0.66		0.06	0.61	29.07
70		5.543	13.70	1.10	4.52	40.00	0.62	0.04	0.66		0.06	0.61	29.68
71		5.543	13.70	1.10	4.52	40.00	0.62	0.04	0.66		0.06	0.61	30.28
72		5.543	13.70	1.10	4.52	40.00	0.62	0.04	0.66		0.06	0.61	30.89
73		5.543	13.70	1.10	4.52	40.00	0.62	0.04	0.66		0.06	0.61	31.50
74		5.543	13.70	1.10	4.52	40.00	0.62	0.04	0.66		0.06	0.61	32.10
75		5.543	13.70	1.10	4.52	40.00	0.62	0.04	0.66		0.06	0.61	32.71
76		5.543	13.70	1.10	4.52	40.00	0.62	0.04	0.66		0.06	0.61	33.31
77		5.543	13.70	1.10	4.52	40.00	0.62	0.04	0.66		0.06	0.61	33.92
78		5.543	13.70	1.10	4.52	40.00	0.62	0.04	0.66		0.06	0.61	34.53
79		5.543	13.70	1.10	4.52	40.00	0.62	0.04	0.66		0.06	0.61	35.13
80		5.543	13.70	1.10	4.52	40.00	0.62	0.04	0.66		0.06	0.61	35.74
81		5.543	13.70	1.10	4.52	40.00	0.62	0.04	0.66		0.06	0.61	36.35
82		5.543	13.70	1.10	4.52	40.00	0.62	0.04	0.66		0.06	0.61	36.95
83		5.543	13.70	1.10	4.52	40.00	0.62	0.04	0.66		0.06	0.61	37.56
84		5.543	13.70	1.10	4.52	40.00	0.62	0.04	0.66		0.06	0.61	38.17
85		5.543	13.70	1.10	4.52	40.00	0.62	0.04	0.66		0.06	0.61	38.77
86		5.543	13.70	1.10	4.52	40.00	0.62	0.04	0.66		0.06	0.61	39.38
87		5.543	13.70	1.10	4.52	40.00	0.62	0.04	0.66		0.06	0.61	39.98
88		5.543	13.70	1.10	4.52	40.00	0.62	0.04	0.66		0.06	0.61	40.59
89		5.543	13.70	1.10	4.52	40.00	0.62	0.04	0.66		0.06	0.61	41.20
90		5.543	13.70	1.10	4.52	40.00	0.62	0.04	0.66		0.06	0.61	41.80
91		5.543	13.70	1.10	4.52	40.00	0.62	0.04	0.66		0.06	0.61	42.41
92		5.543	13.70	1.10	4.52	40.00	0.62	0.04	0.66		0.06	0.61	43.02
93		5.543	13.70	1.10	4.52	40.00	0.62	0.04	0.66		0.06	0.61	43.62
94		5.543	13.70	1.10	4.52	40.00	0.62	0.04	0.66		0.06	0.61	44.23
95		5.543	13.70	1.10	4.52	40.00	0.62	0.04	0.66		0.06	0.61	44.83

Table 8. English Stones barrage: cash flow analysis (unescalated)

Year	Capital Costs	Capital Costs cum.	Energy Generated per year	Firm Power	Tariff (after allowances for off-load heat & transmission losses, uplifted for pumping)	Firm Power Rate	Tariff Revenue	Firm Power Revenue	Total Revenue	CEGB Trans. Payment	Operating Costs	Net Cash Flow (pre Finance)	Net Cash Flow (pre Finance) cum.
	(£bn)	(£bn)	(TWh)	(GW)	(p/kWh)	(£/kW)	(£bn)	(£bn)	(£bn)	(£bn)	(£bn)	(£bn)	(£bn)
-8	0.001	0.001				0.00	0.00	0.00	0.00		0.00	(0.00)	(0.00)
-7	0.001	0.001				0.00	0.00	0.00	0.00		0.00	(0.00)	(0.00)
-6	0.003	0.004				0.00	0.00	0.00	0.00		0.00	(0.00)	(0.00)
-5	0.007	0.011				0.00	0.00	0.00	0.00		0.00	(0.01)	(0.01)
-4	0.010	0.021				0.00	0.00	0.00	0.00		0.00	(0.01)	(0.02)
-3	0.012	0.033				0.00	0.00	0.00	0.00		0.00	(0.01)	(0.03)
-2	0.013	0.046				0.00	0.00	0.00	0.00		0.00	(0.01)	(0.05)
-1	0.015	0.061				0.00	0.00	0.00	0.00		0.00	(0.02)	(0.06)
1	0.085	0.146				0.00	6.00	0.00	0.00		0.00	(0.09)	(0.15)
2	0.210	0.356				0.00	0.00	0.00	0.00		0.00	(0.21)	(0.36)
3	0.297	0.653				0.00	0.00	0.00	0.00		0.00	(0.30)	(0.65)
4	0.255	0.908				0.00	0.00	0.00	0.00		0.00	(0.26)	(0.91)
5	0.151	1.059				0.00	0.00	0.00	0.00		0.00	(0.15)	(1.06)
6	0.083	1.142	1.73			0.00	0.07	0.00	0.07		0.01	(0.02)	(1.08)
7	0.008	1.150	2.65		4.03	0.00	0.11	0.00	0.11	0.00	0.01	0.09	(1.00)
8	0.000	1.150	2.70		4.18	0.00	0.11	0.00	0.11	0.00	0.01	0.10	(0.89)
9		1.150	2.70	0.25	4.18	3.00	0.11	0.00	0.11	0.00	0.01	0.10	(0.79)
10		1.150	2.70	0.25	4.18	6.00	0.11	0.00	0.11	0.00	0.01	0.10	(0.69)
11		1.150	2.70	0.25	4.18	9.00	0.11	0.00	0.12	0.00	0.01	0.10	(0.59)
12		1.150	2.70	0.25	4.18	12.00	0.11	0.00	0.12	0.00	0.01	0.10	(0.49)
13		1.150	2.70	0.25	4.18	15.00	0.11	0.00	0.12	0.00	0.01	0.10	(0.38)
14		1.150	2.70	0.25	4.18	18.00	0.11	0.00	0.12	0.00	0.01	0.11	(0.28)
15		1.150	2.70	0.25	4.18	21.00	0.11	0.01	0.12	0.00	0.01	0.11	(0.17)
16		1.150	2.70	0.25	4.18	24.00	0.11	0.01	0.12	0.00	0.01	0.11	(0.06)
17		1.150	2.70	0.25	4.18	27.00	0.11	0.01	0.12	0.00	0.01	0.11	0.04
18		1.150	2.70	0.25	4.18	30.00	0.11	0.01	0.12	0.00	0.01	0.11	0.15
19		1.150	2.70	0.25	4.18	31.00	0.11	0.01	0.12	0.00	0.01	0.11	0.26
20		1.150	2.70	0.25	4.19	32.00	0.11	0.01	0.12	0.00	0.01	0.11	0.37
21		1.150	2.70	0.25	4.19	33.00	0.11	0.01	0.12	0.00	0.01	0.11	0.48
22		1.150	2.70	0.25	4.20	34.00	0.11	0.01	0.12	0.00	0.01	0.11	0.59
23		1.150	2.70	0.25	4.20	35.00	0.11	0.01	0.12	0.00	0.01	0.11	0.70
24		1.150	2.70	0.25	4.21	36.00	0.11	0.01	0.12	0.00	0.01	0.11	0.81
25		1.150	2.70	0.25	4.21	37.00	0.11	0.01	0.12	0.00	0.01	0.11	0.92
26		1.150	2.70	0.25	4.22	38.00	0.11	0.01	0.12	0.00	0.01	0.11	1.03
27		1.150	2.70	0.25	4.22	39.00	0.11	0.01	0.12	0.00	0.01	0.11	1.14
28		1.150	2.70	0.25	4.23	40.00	0.11	0.01	0.12	0.00	0.01	0.11	1.25
29		1.150	2.70	0.25	4.23	40.00	0.11	0.01	0.12	0.00	0.01	0.11	1.36
30		1.150	2.70	0.25	4.26	40.00	0.11	0.01	0.12	0.00	0.01	0.11	1.48
31		1.150	2.70	0.25	4.28	40.00	0.12	0.01	0.13	0.00	0.01	0.11	1.59
32		1.150	2.70	0.25	4.31	40.00	0.12	0.01	0.13	0.00	0.01	0.11	1.70
33		1.150	2.70	0.25	4.34	40.00	0.12	0.01	0.13	0.00	0.01	0.11	1.82
34		1.150	2.70	0.25	4.37	40.00	0.12	0.01	0.13	0.00	0.01	0.12	1.94
35		1.150	2.70	0.25	4.39	40.00	0.12	0.01	0.13	0.00	0.01	0.12	2.05
36		1.150	2.70	0.25	4.42	40.00	0.12	0.01	0.13	0.00	0.01	0.12	2.17
37		1.150	2.70	0.25	4.45	40.00	0.12	0.01	0.13	0.00	0.01	0.12	2.29
38		1.150	2.70	0.25	4.47	40.00	0.12	0.01	0.13		0.01	0.12	2.41
39		1.150	2.70	0.25	4.50	40.00	0.12	0.01	0.13		0.01	0.12	2.53
40		1.150	2.70	0.25	4.50	40.00	0.12	0.01	0.13		0.01	0.12	2.65
41		1.150	2.70	0.25	4.50	40.00	0.12	0.01	0.13		0.01	0.12	2.77
42		1.150	2.70	0.25	4.50	40.00	0.12	0.01	0.13		0.01	0.12	2.89
43		1.150	2.70	0.25	4.50	40.00	0.12	0.01	0.13		0.01	0.12	3.01
44		1.150	2.70	0.25	4.50	40.00	0.12	0.01	0.13		0.01	0.12	3.13

Table 8 (continued)

Year	Capital Costs	Capital Costs cum.	Energy Generated per year	Firm Power	Tariff (unesc)	Firm Power Rate (esc)	Tariff Revenue	Firm Power Revenue	Total Revenue	CEGB Trans. Payment	Operating Costs	Net Cash Flow (pre Finance)	Net Cash Flow (pre Finance) cum.
	(£bn)	(£bn)	(TWh)	(GW)	(p/kWh)	(£/kW)	(£bn)	(£bn)	(£bn)	(£bn)	(£bn)	(£bn)	(£bn)
45		1.150	2.70	0.25	4.50	40.00	0.12	0.01	0.13		0.01	0.12	3.25
46		1.150	2.70	0.25	4.50	40.00	0.12	0.01	0.13		0.01	0.12	3.37
47		1.150	2.70	0.25	4.50	40.00	0.12	0.01	0.13		0.01	0.12	3.49
48		1.150	2.70	0.25	4.50	40.00	0.12	0.01	0.13		0.01	0.12	3.61
49		1.150	2.70	0.25	4.50	40.00	0.12	0.01	0.13		0.01	0.12	3.73
50		1.150	2.70	0.25	4.50	40.00	0.12	0.01	0.13		0.01	0.12	3.85
51		1.150	2.70	0.25	4.50	40.00	0.12	0.01	0.13		0.01	0.12	3.97
52		1.150	2.70	0.25	4.50	40.00	0.12	0.01	0.13		0.01	0.12	4.09
53		1.150	2.70	0.25	4.50	40.00	0.12	0.01	0.13		0.01	0.12	4.21
54		1.150	2.70	0.25	4.50	40.00	0.12	0.01	0.13		0.01	0.12	4.33
55		1.150	2.70	0.25	4.50	40.00	0.12	0.01	0.13		0.01	0.12	4.45
56		1.150	2.70	0.25	4.50	40.00	0.12	0.01	0.13		0.01	0.12	4.57
57		1.150	2.70	0.25	4.50	40.00	0.12	0.01	0.13		0.01	0.12	4.69
58		1.150	2.70	0.25	4.50	40.00	0.12	0.01	0.13		0.01	0.12	4.81
59		1.150	2.70	0.25	4.50	40.00	0.12	0.01	0.13		0.01	0.12	4.93
60		1.150	2.70	0.25	4.50	40.00	0.12	0.01	0.13		0.01	0.12	5.05
61		1.150	2.70	0.25	4.50	40.00	0.12	0.01	0.13		0.01	0.12	5.17
62		1.150	2.70	0.25	4.50	40.00	0.12	0.01	0.13		0.01	0.12	5.29
63		1.150	2.70	0.25	4.50	40.00	0.12	0.01	0.13		0.01	0.12	5.41
64		1.150	2.70	0.25	4.50	40.00	0.12	0.01	0.13		0.01	0.12	5.53
65		1.150	2.70	0.25	4.50	40.00	0.12	0.01	0.13		0.01	0.12	5.65
66		1.150	2.70	0.25	4.50	40.00	0.12	0.01	0.13		0.01	0.12	5.77
67		1.150	2.70	0.25	4.50	40.00	0.12	0.01	0.13		0.01	0.12	5.89
68		1.150	2.70	0.25	4.50	40.00	0.12	0.01	0.13		0.01	0.12	6.01
69		1.150	2.70	0.25	4.50	40.00	0.12	0.01	0.13		0.01	0.12	6.13
70		1.150	2.70	0.25	4.50	40.00	0.12	0.01	0.13		0.01	0.12	6.25
71		1.150	2.70	0.25	4.50	40.00	0.12	0.01	0.13		0.01	0.12	6.37
72		1.150	2.70	0.25	4.50	40.00	0.12	0.01	0.13		0.01	0.12	6.49
73		1.150	2.70	0.25	4.50	40.00	0.12	0.01	0.13		0.01	0.12	6.61
74		1.150	2.70	0.25	4.50	40.00	0.12	0.01	0.13		0.01	0.12	6.73
75		1.150	2.70	0.25	4.50	40.00	0.12	0.01	0.13		0.01	0.12	6.85
76		1.150	2.70	0.25	4.50	40.00	0.12	0.01	0.13		0.01	0.12	6.97
77		1.150	2.70	0.25	4.50	40.00	0.12	0.01	0.13		0.01	0.12	7.09
78		1.150	2.70	0.25	4.50	40.00	0.12	0.01	0.13		0.01	0.12	7.21
79		1.150	2.70	0.25	4.50	40.00	0.12	0.01	0.13		0.01	0.12	7.33
80		1.150	2.70	0.25	4.50	40.00	0.12	0.01	0.13		0.01	0.12	7.45
81		1.150	2.70	0.25	4.50	40.00	0.12	0.01	0.13		0.01	0.12	7.57
82		1.150	2.70	0.25	4.50	40.00	0.12	0.01	0.13		0.01	0.12	7.69
83		1.150	2.70	0.25	4.50	40.00	0.12	0.01	0.13		0.01	0.12	7.81
84		1.150	2.70	0.25	4.50	40.00	0.12	0.01	0.13		0.01	0.12	7.93
85		1.150	2.70	0.25	4.50	40.00	0.12	0.01	0.13		0.01	0.12	8.05
86		1.150	2.70	0.25	4.50	40.00	0.12	0.01	0.13		0.01	0.12	8.17
87		1.150	2.70	0.25	4.50	40.00	0.12	0.01	0.13		0.01	0.12	8.29
88		1.150	2.70	0.25	4.50	40.00	0.12	0.01	0.13		0.01	0.12	8.41
89		1.150	2.70	0.25	4.50	40.00	0.12	0.01	0.13		0.01	0.12	8.53
90		1.150	2.70	0.25	4.50	40.00	0.12	0.01	0.13		0.01	0.12	8.65
91		1.150	2.70	0.25	4.50	40.00	0.12	0.01	0.13		0.01	0.12	8.77
92		1.150	2.70	0.25	4.50	40.00	0.12	0.01	0.13		0.01	0.12	8.89
93		1.150	2.70	0.25	4.50	40.00	0.12	0.01	0.13		0.01	0.12	9.01

firm power. This figure should be compared with the costs of other forms of generation. The CEGB's figures (see Reference 3) are:-

	p/kWh
Sizewell B PWR	2.94
AGR	3.67
Coal-fired station	4.29

The figure for the cost of generation from the Barrage is independent of the assumption about energy prices. The cost of generation from nuclear power station is not greatly affected by the assumption about energy prices, as the main element of the cost of nuclear generation is the capital cost of the plant. The converse is the case for coal-fired power stations.

Sensitivities

General
34. This section considers a number of sensitivities. In each case, the sensitivity is measured against the assumptions set out above against two assumptions for the value of electricity - the BST case, and the Sizewell Scenario C (low nuclear background) case.

Energy Capture

35. Change %	CW Barrage			ES Barrage		
	IRR %		COST	IRR %		COST
	Scenario C	BST	p/kWh	Scenario C	BST	p/kWh
- 20	6.07	2.90	3.75	6.28	3.02	3.61
- 10	6.74	3.42	3.33	7.03	3.57	3.21
0	7.38	3.90	3.00	7.75	4.08	2.89
+ 10	7.99	4.36	2.73	8.44	4.58	2.62
+ 20	8.58	4.80	2.50	9.12	5.05	2.41

It should be noted that, for the rates of return, this is equivalent to a similar sensitivity analysis with the variation being on the tariff.

Capital Cost

36. Change % (assumes identical phasing of expenditure)	CW Barrage			ES Barrage		
	IRR %		COST	IRR %		COST
	Scenario C	BST	p/kWh	Scenario C	BST	p/kWh
- 20	8.73	4.80	2.51	9.27	5.07	2.40
- 10	7.99	4.31	2.76	8.44	4.53	2.65
0	7.38	3.90	3.00	7.75	4.08	2.89
+ 10	6.85	3.55	3.25	7.17	3.71	3.13
+ 20	6.39	3.25	3.49	6.67	3.38	3.38
+ 50	5.33	2.53	4.22	5.50	2.61	4.11

Firm Power Value

37. Change %	CW Barrage IRR %		ES Barrage IRR %		
	Scenario C	BST	Scenario C	BST	
					(the
- 20	7.32	3.87	7.69	4.05	cost
- 10	7.35	3.88	7.72	4.07	in
0	7.38	3.90	7.75	4.08	in-
+ 10	7.40	3.91	7.78	4.10	depen-
+ 20	7.43	3.93	7.81	4.12	dent
					of
					firm
					power
					value)

Operating Costs

38. Change %	CW Barrage IRR %		COST	ES Barrage IRR %		COST
	Scenario C	BST	p/kWh	Scenario C	BST	p/kWh
- 50	7.68	4.31	2.79	8.09	4.52	2.68
- 20	7.50	4.06	2.92	7.89	4.26	2.80
- 10	7.44	3.98	2.96	7.82	4.17	2.85
0	7.38	3.90	3.00	7.75	4.08	2.89
+ 10	7.31	3.82	3.04	7.68	4.00	2.93
+ 20	7.25	3.73	3.08	7.61	3.91	2.98

The 50% reduction case shows the effect of not taking into consideration allowance for external charges, such as rates (see Paragraph 7).

Timing of Capital Costs

39. Time Change	CW Barrage IRR %		COST	ES Barrage IRR %		
COST						
(no change in						
overall cost)	Scenario C	BST	p/kWh	Scenario C	BST	p/kWh
One year early	7.74	4.05	2.88	8.14	4.23	2.77
On time	7.38	3.90	3.00	7.75	4.08	2.89
One year late	7.02	3.75	3.13	7.36	3.92	3.02

While the effect in IRR terms of a change of timing is small, the effect on the overall cost, including escalation and interest during construction, where this is relevant, is significant.

Siltation (ES Barrage)

40. Period over which Annual Energy Halved (Years)	IRR %		Cost
	Scenario C	BST	p/kWh
10	3.65	0.25	5.27
20	5.34	1.64	4.22
30	6.05	2.33	3.82
40	6.43	2.71	3.61
50	6.67	2.56	3.48
60	6.84	3.13	3.39
70	6.96	3.25	3.32
80	7.05	3.34	3.27
90	7.12	3.42	3.23
100	7.18	3.48	3.20
Base Case (ie. period of infinity)	7.75	4.08	2.89

Conclusions on the Economics

41.   The main point to emerge is that the most critical single parameter in determining the return offered by either Barrage is how the value of the electricity it generates varies over time. Capital costs and the absolute level of output or the tariff are comparatively less important. Other factors, such as the firm power contribution of the Barrage, or the level of operating costs, are of minor significance. For example, a 2% per annum real increase in the tariff level from the BST rates nearly doubles the rate of return as opposed to the case where the tariffs remain constant in real terms, whereas a once and for all increase in the level of the BST rates by 20% would only increase the return by less than 1%. Equally, a 20% reduction in the capital cost has a similar effect.

42.   The value of the electricity does not affect the cost of generation. Other than this, the same considerations apply in determining the cost of generation, with capital costs and the level of output being more important than other factors. It is also noteworthy that, before any siltation effect is taken into account, the economics are marginally better for the ES Barrage than for the larger Barrage, even though the proportionate reduction in capital costs (reduced by a factor of 4.82) is lower than for the energy output (reduced by a factor of 5.07). This is more than compensated for by the shorter construction programme.

Financing

Scope

43.   The economics of the CW and ES Barrages have been examined. We now turn to consider the implications of the

economic analysis for the possible financing of either Barrage in the private sector.

44. A potential private sector investor would be concerned with the return he is likely to get from investing in the project, and the risk he faces in making the investment. The position would be very different if the project were being contemplated as a public sector project. In that case, the critical factor would be the relative performance of the Barrage in generating electricity, as compared with other forms of generation such as nuclear or coal fired capacity. For the private sector it is the merits of the project on a stand alone basis that matter, not its relative performance.

45. This gives rise to a major problem in that the economic performance of the Barrage is in many respects actually determined by other forms of generation. The electricity generated by the barrage would be sold to the CEGB, and the price it would pay is largely determined by its own generation costs. The CEGB has to meet financial criteria given to it as a public sector body. It is noteworthy that, as shown in the economic analysis, the cost of generation from either Barrage would bear comparison with other forms of generation; and both Barrages have Internal Rates of Return not dissimilar to other forms of public sector investment.

46. However, as suggested below, this return is unlikely to be acceptable to potential private sector investors.

Financing Considerations

47. Any financing scheme is likely in principle to involve a mixture of equity related finance and debt related finance. The main concern of a potential lender is that there should be little risk of interest payments and capital repayments of the debt not being made. For an equity investor, the two basic parameters would be:

(i)   the return available; and
(ii) the risk.

Debt Finance

48. The economic calculations performed correspond, in particular, to two different conditions for the sale of electricity:-

(i)   a price related to the Bulk Supply Tariff (BST); and

(ii) a price related to Scenario C as defined in the CEGB's evidence to Sizewell i.e. one that assumes that the marginal power that the Barrage is in due course substituting will correspond to the implantation and energy price spectrum represented by Scenario C (see Paragraphs 55-59 below).

49. The cash flow analysis in Table 9 shows that, in the

179

Table 9. Cardiff Weston scheme: escalated (at 6% p.a.) cash flow

Year	Capital Costs (undisc) (£m)	Capital Costs (esc) (£m)	Capital Costs cum. (£m)	Energy Used per year (£m)	Firm Power (£m)	Firm Power (GW)	Tariff (undisc) (p/kWh)	Tariff (esc) (p/kWh)	Firm Power Revenue (£/kW)	Tariff Revenue (£m)	Firm Power Revenue (£m)	Total Revenue (£m)	CEGB Trans. Payment (undisc) (£m)	CEGB Trans. Payment (esc) (£m)	Operating Costs (undisc) (£m)	Operating Costs (esc) (£m)	Net Cash Flow (pre Finance) (£m)	Net Cash Flow (pre finance) cum. (£m)	Drawdown (£m)	DEBT CALCULATIONS [Interest Rate 12%] Interest (£m)	Repayment (£m)	Balance Outstanding (£m)
-7	0.003	0.003	0.003							0.00	0.00	0.00			0.00	0.00	(0.00)	(0.00)	0.01	0.00	0.00	0.00
-7	0.011	0.014	0.012							0.00	0.00	0.00			0.00	0.00	(0.02)	(0.02)	0.01	0.00	0.00	0.02
-6	0.016	0.028	0.015							0.00	0.00	0.00			0.00	0.00	(0.03)	(0.03)	0.02	0.00	0.00	0.04
-5	0.036	0.044	0.031							0.00	0.00	0.00			0.00	0.00	(0.05)	(0.05)	0.03	0.01	0.00	0.06
-4	0.038	0.062	0.051							0.00	0.00	0.00			0.00	0.00	(0.05)	(0.10)	0.06	0.01	0.00	0.12
-3	0.042	0.124	0.100							0.00	0.00	0.00			0.00	0.00	(0.06)	(0.16)	0.08	0.02	0.00	0.26
-2	0.263	0.387	0.158							0.00	0.00	0.00			0.00	0.00	(0.38)	(0.54)	0.42	0.05	0.00	0.63
-1	0.623	1.010	0.542							0.00	0.00	0.00			0.00	0.00	(0.98)	(1.51)	1.11	0.14	0.00	1.74
1	0.841	1.851	1.507							0.00	0.00	0.00			0.00	0.00	(1.38)	(2.89)	1.69	0.31	0.00	3.43
2	0.896	2.747	2.887							0.00	0.00	0.00			0.00	0.00	(1.58)	(4.45)	2.10	0.54	0.00	5.53
3	0.969	3.716	4.445							0.00	0.00	0.00			0.00	0.00	(1.79)	(6.23)	2.61	0.82	0.00	8.13
4	0.794	4.510	7.784							0.00	0.00	0.00			0.00	0.00	(1.55)	(7.78)	2.69	1.14	0.00	10.82
5	0.492	5.002	8.803				3.90	8.56		0.38	0.00	0.38		0.07	0.00	0.02	(0.19)	(8.80)	2.47	1.45	0.00	13.29
6	0.239	5.241	9.328	4.42			4.05	9.42		1.01	0.00	1.01	0.03	0.08	0.00	0.04	(0.19)	(8.99)	1.89	1.71	0.00	15.18
7	0.235	5.476	9.875	10.77			4.20	10.36		1.37	0.00	1.37	0.03	0.08	0.00	0.05	0.36	(8.63)	1.90	1.91	0.00	16.73
8	0.067	5.543	10.040	13.24			4.20	10.98	7.85	1.50	0.01	1.51	0.03	0.09	0.00	0.13	1.07	(7.55)	1.55	2.07	0.00	17.73
9			10.040	13.70	1.10		4.20	11.64	16.63	1.59	0.02	1.61	0.03	0.09	0.00	0.15	1.29	(6.26)	9.89	2.18	0.00	18.61
10			10.040	13.70	1.10		4.20	12.34	26.45	1.67	0.03	1.72	0.03	0.10	0.00	0.16	1.58	(4.88)	0.91	2.29	0.00	19.52
11			10.040	13.70	1.10		4.20	13.08	37.38	1.77	0.04	1.83	0.03	0.11	0.00	0.17	1.47	(3.41)	0.93	2.40	0.00	20.45
12			10.040	13.70	1.10		4.20	13.86	49.53	1.90	0.05	1.95	0.03	0.12	0.00	0.18	1.57	(1.84)	0.94	2.51	0.00	21.39
13			10.040	13.70	1.10		4.20	14.70	63.00	2.01	0.07	2.08	0.03	0.13	0.00	0.19	1.68	(0.16)	0.95	2.62	6.46	22.34
14			10.040	13.70	1.10		4.20	15.58	77.71	2.13	0.09	2.22	0.03	0.14	0.00	0.20	1.79	1.63	0.95	2.74	6.90	23.28
15			10.040	13.70	1.10		4.20	16.51	94.38	2.26	0.10	2.37	0.03	0.15	0.00	0.21	1.91	3.54	0.94	2.85	0.00	24.23
16			10.040	13.70	1.10		4.20	17.50	112.55	2.40	0.12	2.52	0.03	0.17	0.00	0.22	2.04	5.57	0.93	2.96	0.00	25.15
17			10.040	13.70	1.10		4.20	18.55	132.56	2.54	0.15	2.69	0.03	0.18	0.00	0.24	2.17	7.74	0.90	3.07	0.00	26.06
18			10.040	13.70	1.10		4.20	19.67	145.20	2.67	0.16	2.85	0.03	0.19	0.00	0.25	2.32	10.06	0.86	3.18	0.00	26.92
19			10.040	13.70	1.10		4.20	20.87	158.88	2.86	0.17	3.03	0.03	0.22	0.00	0.27	2.46	12.52	0.82	3.28	0.00	27.74
20			10.040	13.70	1.10		4.21	22.15	173.67	2.84	0.19	3.03	0.03	0.23	0.00	0.28	2.62	15.14	0.78	3.37	0.00	28.49
21			10.040	13.70	1.10		4.21	23.51	189.67	3.22	0.21	3.43	0.03	0.24	0.00	0.30	2.78	17.92	0.68	3.46	0.00	29.17
22			10.040	13.70	1.10		4.22	24.95	206.97	3.43	0.23	3.43	0.03	0.26	0.00	0.32	2.96	20.88	0.57	3.53	0.00	29.75
23			10.040	13.70	1.10		4.22	26.48	225.65	3.63	0.25	3.88	0.03	0.29	0.00	0.33	3.15	24.03	0.45	3.60	0.00	30.19
24			10.040	13.70	1.10		4.23	28.10	245.83	4.09	0.27	4.12	0.03	0.31	0.00	0.35	3.35	27.37	0.29	3.64	0.00	30.49
25			10.040	13.70	1.10		4.23	29.82	267.63	4.09	0.29	4.38	0.03	0.33	0.00	0.38	3.56	30.94	0.10	3.66	0.12	30.59
26			10.040	13.70	1.10		4.23	31.64	291.15	4.34	0.32	4.66	0.03	0.35	0.00	0.40	3.79	34.72	0.00	3.66	0.40	30.47
27			10.040	13.70	1.10		4.24	33.58	316.53	4.88	0.35	4.95	0.03	0.37	0.00	0.42	4.03	38.75	0.00	3.63	0.40	30.07
28			10.040	13.70	1.10		4.25	35.64	335.52	5.21	0.37	5.25	0.03	0.39	0.00	0.45	4.28	43.03	0.00	3.57	0.72	29.35
29			10.040	13.70	1.10		4.25	38.02	355.46	5.21	0.39	5.25	0.03	0.41	0.00	0.47	4.55	47.58	0.00	3.46	1.09	28.26
30			10.040	13.70	1.10		4.33	40.55	377.00	5.56	0.41	5.97	0.03	0.44	0.00	0.50	4.85	52.43	0.00	3.30	1.55	26.71
31			10.040	13.70	1.10		4.33	43.25	399.61	5.93	0.44	6.37	0.03	0.47	0.00	0.53	5.18	57.61	0.00	3.08	2.10	24.61
32			10.040	13.70	1.10		4.33	49.20	423.59	6.32	0.47	6.79	0.03	0.49	0.00	0.57	5.52	63.13	0.00	2.79	2.74	21.88
33			10.040	13.70	1.10		4.38	49.20	449.01	6.74	0.49	7.23	0.03	0.52	0.00	0.60	5.90	65.03	0.00	2.42	3.48	18.40
34			10.040	13.70	1.10		4.41	52.47	475.95	7.23	0.49	7.71	0.03	0.55	0.00	0.64	6.29	35.32	0.00	1.95	4.34	14.06
35			10.040	13.70	1.10		4.41	55.96	504.56	7.67	0.55	8.22	0.03	0.57	0.00	0.67	6.71	82.03	0.00	1.37	5.75	8.71
36			10.040	13.70	1.10		4.44	59.68	534.77	8.16	0.59	8.76	0.03	0.59	0.00	0.71	7.16	89.19	0.00	6.85	6.51	2.20
37			10.040	13.70	1.10		4.52	63.64	600.87	8.72	0.62	9.34	0.03	0.60	0.00	0.76	7.44	96.83	0.00	0.13	2.20	0.04
38			10.040	13.70	1.10		4.52	67.86	636.93	9.30	0.68	9.98		0.37	0.00	0.80	8.15	104.98	0.00	0.00	0.00	0.00
39			10.040	13.70	1.10		4.52	71.93	675.14	10.45	0.74	11.19		0.39	0.00	0.85	8.69	113.68	0.00	0.00	0.00	0.00
40			10.040	13.70	1.10		4.52	76.25	715.65	11.07	0.79	11.86		0.41	0.00	0.90	9.65	123.33	0.00	0.00	0.00	0.00
41			10.040	13.70	1.10		4.52	80.83	758.59	11.74	0.83	12.57			0.06	0.96	10.23	133.57	0.00	0.00	0.00	0.00
42			10.040	13.70	1.10		4.52	85.68	804.10	12.44	0.88	13.33			0.06	1.01	10.85	144.41	0.00	0.00	0.00	0.00
43			10.040	13.70	1.10		4.52								0.06	1.07	11.50	155.91	0.00	0.00	0.00	0.00
44			10.040	13.70	1.10		4.52								0.06	1.14	12.19	168.10	0.00	0.00	0.00	0.00

Table 9 (continued)

Year	Capital Costs (unesc) (£bn)	Capital Costs (esc) cum. (£bn)	Energy Generated per year (TWh)	Firm Power (GW)	Tariff (unesc) (p/kWh)	Tariff (esc) (p/kWh)	Firm Power Rate (esc) (£/kW)	Tariff Revenue (esc) (£bn)	Firm Power Revenue (£bn)	Total Revenue (£bn)	Operating Costs (unesc) (£bn)	Operating Costs (esc) (£bn)	Net Cash Flow (pre finance) (esc) (£bn)	Net Cash Flow (pre finance) cum (£bn)	Drawdown (£bn)	Interest Rate [%] (£bn)	Interest (£bn)	Repayment (£bn)	Balance Outstanding (£bn)
47	5.543	10.040	13.70	1.10	4.52	96.26	852.35	13.19	0.94	14.13	0.06	1.21	12.92	181.02	0.00	0.00	0.00	0.00	0.00
48	5.543	10.040	13.70	1.10	4.52	102.04	903.49	13.98	0.99	14.97	0.06	1.28	13.69	194.71	0.00	0.00	0.00	0.00	0.00
49	5.543	10.040	13.70	1.10	4.52	108.16	957.70	14.82	1.05	15.87	0.06	1.36	14.52	209.23	0.00	0.00	0.00	0.00	0.00
50	5.543	10.040	13.70	1.10	4.52	114.65	1015.16	15.71	1.12	16.82	0.06	1.44	15.39	224.62	0.00	0.00	0.00	0.00	0.00
51	5.543	10.040	13.70	1.10	4.52	121.53	1076.07	16.65	1.18	17.83	0.06	1.52	16.31	240.93	0.00	0.00	0.00	0.00	0.00
52	5.543	10.040	13.70	1.10	4.52	128.82	1140.64	17.65	1.25	18.90	0.06	1.61	17.29	258.22	0.00	0.00	0.00	0.00	0.00
53	5.543	10.040	13.70	1.10	4.52	136.55	1209.07	18.71	1.33	20.04	0.06	1.71	18.33	276.55	0.00	0.00	0.00	0.00	0.00
54	5.543	10.040	13.70	1.10	4.52	144.75	1281.62	19.83	1.41	21.24	0.06	1.81	19.43	295.97	0.00	0.00	0.00	0.00	0.00
55	5.543	10.040	13.70	1.10	4.52	153.43	1358.52	21.02	1.49	22.51	0.06	1.92	20.59	316.57	0.00	0.00	0.00	0.00	0.00
56	5.543	10.040	13.70	1.10	4.52	162.64	1440.03	22.28	1.58	23.87	0.06	2.04	21.83	338.39	0.00	0.00	0.00	0.00	0.00
57	5.543	10.040	13.70	1.10	4.52	172.40	1526.43	23.62	1.68	25.30	0.06	2.16	23.14	361.53	0.00	0.00	0.00	0.00	0.00
58	5.543	10.040	13.70	1.10	4.52	182.74	1618.01	25.04	1.78	26.82	0.06	2.29	24.53	386.06	0.00	0.00	0.00	0.00	0.00
59	5.543	10.040	13.70	1.10	4.52	193.70	1715.16	26.54	1.89	28.42	0.06	2.43	26.00	412.05	0.00	0.00	0.00	0.00	0.00
60	5.543	10.040	13.70	1.10	4.52	205.33	1818.00	28.13	2.00	30.13	0.06	2.57	27.56	439.61	0.00	0.00	0.00	0.00	0.00
61	5.543	10.040	13.70	1.10	4.52	217.65	1927.08	29.82	2.12	31.94	0.06	2.73	29.21	468.82	0.00	0.00	0.00	0.00	0.00
62	5.543	10.040	13.70	1.10	4.52	230.70	2042.71	31.61	2.25	33.85	0.06	2.89	30.96	499.78	0.00	0.00	0.00	0.00	0.00
63	5.543	10.040	13.70	1.10	4.52	244.55	2165.27	33.50	2.38	35.88	0.06	3.06	32.82	532.60	0.00	0.00	0.00	0.00	0.00
64	5.543	10.040	13.70	1.10	4.52	259.22	2295.18	35.51	2.52	38.04	0.06	3.25	34.79	567.39	0.00	0.00	0.00	0.00	0.00
65	5.543	10.040	13.70	1.10	4.52	274.77	2432.90	37.64	2.68	40.32	0.06	3.44	36.88	604.27	0.00	0.00	0.00	0.00	0.00
66	5.543	10.040	13.70	1.10	4.52	291.26	2578.87	39.90	2.84	42.74	0.06	3.65	39.09	643.36	0.00	0.00	0.00	0.00	0.00
67	5.543	10.040	13.70	1.10	4.52	308.73	2733.60	42.30	3.01	45.30	0.06	3.87	41.44	684.80	0.00	0.00	0.00	0.00	0.00
68	5.543	10.040	13.70	1.10	4.52	327.26	2897.62	44.83	3.19	48.02	0.06	4.10	43.92	728.72	0.00	0.00	0.00	0.00	0.00
69	5.543	10.040	13.70	1.10	4.52	346.89	3071.47	47.52	3.38	50.90	0.06	4.35	46.56	775.28	0.00	0.00	0.00	0.00	0.00
70	5.543	10.040	13.70	1.10	4.52	367.71	3255.76	50.38	3.58	53.96	0.06	4.61	49.35	824.63	0.00	0.00	0.00	0.00	0.00
71	5.543	10.040	13.70	1.10	4.52	389.77	3451.11	53.40	3.80	57.19	0.06	4.88	52.31	876.94	0.00	0.00	0.00	0.00	0.00
72	5.543	10.040	13.70	1.10	4.52	413.16	3658.18	56.60	4.02	60.63	0.06	5.18	55.45	932.39	0.00	0.00	0.00	0.00	0.00
73	5.543	10.040	13.70	1.10	4.52	437.95	3877.67	60.00	4.27	64.26	0.06	5.49	58.78	991.17	0.00	0.00	0.00	0.00	0.00
74	5.543	10.040	13.70	1.10	4.52	464.22	4110.33	63.60	4.52	68.12	0.06	5.82	62.30	1,053.47	0.00	0.00	0.00	0.00	0.00
75	5.543	10.040	13.70	1.10	4.52	492.08	4356.95	67.41	4.79	72.21	0.06	6.17	66.04	1,119.51	0.00	0.00	0.00	0.00	0.00
76	5.543	10.040	13.70	1.10	4.52	521.60	4618.36	71.46	5.08	76.54	0.06	6.53	70.00	1,189.52	0.00	0.00	0.00	0.00	0.00
77	5.543	10.040	13.70	1.10	4.52	552.90	4895.46	75.75	5.39	81.13	0.06	6.93	74.20	1,263.72	0.00	0.00	0.00	0.00	0.00
78	5.543	10.040	13.70	1.10	4.52	586.07	5189.19	80.29	5.71	86.00	0.06	7.34	78.66	1,342.38	0.00	0.00	0.00	0.00	0.00
79	5.543	10.040	13.70	1.10	4.52	621.23	5500.54	85.11	6.05	91.16	0.06	7.78	83.38	1,425.75	0.00	0.00	0.00	0.00	0.00
80	5.543	10.040	13.70	1.10	4.52	658.51	5830.58	90.22	6.41	96.63	0.06	8.25	88.38	1,514.13	0.00	0.00	0.00	0.00	0.00
81	5.543	10.040	13.70	1.10	4.52	698.02	6180.41	95.63	6.80	102.43	0.06	8.75	93.68	1,607.81	0.00	0.00	0.00	0.00	0.00
82	5.543	10.040	13.70	1.10	4.52	739.90	6551.24	101.37	7.21	108.57	0.06	9.27	99.30	1,707.12	0.00	0.00	0.00	0.00	0.00
83	5.543	10.040	13.70	1.10	4.52	784.29	6944.31	107.45	7.64	115.09	0.06	9.83	105.26	1,812.38	0.00	0.00	0.00	0.00	0.00
84	5.543	10.040	13.70	1.10	4.52	831.35	7360.97	113.96	8.10	121.99	0.06	10.42	111.58	1,923.95	0.00	0.00	0.00	0.00	0.00
85	5.543	10.040	13.70	1.10	4.52	881.23	7802.63	120.73	8.58	129.31	0.06	11.04	118.27	2,042.23	0.00	0.00	0.00	0.00	0.00
86	5.543	10.040	13.70	1.10	4.52	934.11	8270.78	127.97	9.10	137.07	0.06	11.70	125.37	2,167.59	0.00	0.00	0.00	0.00	0.00
87	5.543	10.040	13.70	1.10	4.52	990.15	8767.03	135.65	9.64	145.29	0.06	12.41	132.89	2,300.48	0.00	0.00	0.00	0.00	0.00
88	5.543	10.040	13.70	1.10	4.52	1049.56	9293.05	143.79	10.22	154.01	0.06	13.15	140.86	2,441.34	0.00	0.00	0.00	0.00	0.00
89	5.543	10.040	13.70	1.10	4.52	1112.54	9850.64	152.42	10.84	163.25	0.06	13.94	149.31	2,590.66	0.00	0.00	0.00	0.00	0.00
90	5.543	10.040	13.70	1.10	4.52	1179.29	10441.67	161.56	11.49	173.05	0.06	14.77	158.27	2,748.93	0.00	0.00	0.00	0.00	0.00
91	5.543	10.040	13.70	1.10	4.52	1250.05	11068.17	171.26	12.17	183.43	0.06	15.66	167.77	2,916.70	0.00	0.00	0.00	0.00	0.00
92	5.543	10.040	13.70	1.10	4.52	1325.05	11732.26	181.53	12.91	194.44	0.06	16.60	177.84	3,094.54	0.00	0.00	0.00	0.00	0.00
93	5.543	10.040	13.70	1.10	4.52	1404.55	12436.20	192.42	13.68	206.10	0.06	17.60	188.51	3,283.04	0.00	0.00	0.00	0.00	0.00
94	5.543	10.040	13.70	1.10	4.51	1488.82	13182.37	203.97	14.50	218.47	0.06	18.65	199.82	3,482.86	0.00	0.00	0.00	0.00	0.00
95	5.543	10.040	13.70	1.10	4.52	1578.15	13973.32	216.21	15.37	231.58	0.06	19.77	211.81	3,694.67	0.00	0.00	0.00	0.00	0.00

Table 10. English Stones scheme: escalated (at 6% p.a.) cash flow

Year	Capital Costs (unesc)	Capital Costs (esc)	Capital Costs cum.	Energy Gen/Fixed per year	Firm Power	Tariff (unesc)	Tariff (esc)	Firm Power (units)	Firm Power (esc)	Firm Power Revenue	Tariff Revenue	Firm Power Revenue	Total Revenue	CEGB Trans. Payment (esc)	CEGB Trans. Payment (unesc)	Operating Costs (unesc)	Operating Costs (esc)	Net Cash Flow (pre Finance)	Net Cash Flow (pre Finance) cum.	Drawdown	Interest	Interest Repayment	Balance Outstanding
	(£m)	(£m)	(£m)	(TWh)	(GW)	(p/kWh)	(p/kWh)	(£/kW)	(£/kW)	(£m)	(£m)	(£m)	(£m)	(£m)	(unesc)	(£m)	(£m)	(£m)	(£m)	(£m)	(£m)	(£m)	(£m)

Table 10 (continued)

Year	Capital Costs (unsec) cum. (£bn)	Capital Costs (sec) cum. (£bn)	Energy Generated per year (TWh)	Firm Power (GW)	Tariff (unsec) (p/kWh)	Tariff (sec) (p/kWh)	Firm Power Rate (sec) (£/kWh)	Firm Power Price (p/kWh)	Tariff Revenue (£bn)	Firm Power Revenue (£bn)	Total Revenue (£bn)	CEGB Trans. Payment (sec) (£bn)	CEGB Trans. Payment (unsec) (£bn)	Operating Costs (unsec) (£bn)	Operating Costs (sec) (£bn)	Net Cash Flow (Finance) (£bn)	Net Cash Flow (pre Finance) (£bn)	Net Cash Flow (pre Finance) (£bn)	Interest Rate 12% (£bn)	Drawdown (£bn)	Interest Repayment (£bn)	Balance Outstanding (£bn)
45	1.150	2.149	2.70	0.25	4.50	95.87	40.00	852.35	2.59	0.21	2.80			0.01	0.24	2.56	36.79	0.00	0.00	0.00	0.00	0.00
46	1.150	2.149	2.70	0.25	4.50	101.62	40.00	903.49	2.74	0.23	2.97			0.01	0.26	2.71	39.50	0.00	0.00	0.00	0.00	0.00
47	1.150	2.149	2.70	0.25	4.50	107.72	40.00	957.76	2.91	0.24	3.15			0.01	0.27	2.87	42.38	0.00	0.00	0.00	0.00	0.00
48	1.150	2.149	2.70	0.25	4.50	114.19	40.00	1015.16	3.08	0.25	3.34			0.01	0.29	3.05	45.42	0.00	0.00	0.00	0.00	0.00
49	1.150	2.149	2.70	0.25	4.50	121.04	40.00	1076.07	3.27	0.27	3.54			0.01	0.31	3.23	48.65	0.00	0.00	0.00	0.00	0.00
50	1.150	2.149	2.70	0.25	4.50	128.30	40.00	1140.64	3.46	0.29	3.75			0.01	0.33	3.42	52.07	0.00	0.00	0.00	0.00	0.00
51	1.150	2.149	2.70	0.25	4.50	136.00	40.00	1209.07	3.67	0.30	3.97			0.01	0.35	3.63	55.70	0.00	0.00	0.00	0.00	0.00
52	1.150	2.149	2.70	0.25	4.50	144.16	40.00	1281.62	3.89	0.32	4.21			0.01	0.37	3.84	59.55	0.00	0.00	0.00	0.00	0.00
53	1.150	2.149	2.70	0.25	4.50	152.81	40.00	1358.52	4.13	0.34	4.47			0.01	0.39	4.08	63.62	0.00	0.00	0.00	0.00	0.00
54	1.150	2.149	2.70	0.25	4.50	161.97	40.00	1440.03	4.37	0.36	4.73			0.01	0.41	4.32	67.94	0.00	0.00	0.00	0.00	0.00
55	1.150	2.149	2.70	0.25	4.50	171.69	40.00	1526.43	4.64	0.38	5.02			0.01	0.44	4.58	72.52	0.00	0.00	0.00	0.00	0.00
56	1.150	2.149	2.70	0.25	4.50	181.99	40.00	1618.01	4.91	0.40	5.32			0.01	0.46	4.85	77.37	0.00	0.00	0.00	0.00	0.00
57	1.150	2.149	2.70	0.25	4.50	192.91	40.00	1715.10	5.21	0.43	5.64			0.01	0.49	5.15	82.52	0.00	0.00	0.00	0.00	0.00
58	1.150	2.149	2.70	0.25	4.50	204.49	40.00	1818.00	5.52	0.45	5.98			0.01	0.52	5.45	87.97	0.00	0.00	0.00	0.00	0.00
59	1.150	2.149	2.70	0.25	4.50	216.76	40.00	1927.08	5.85	0.48	6.33			0.01	0.55	5.78	93.75	0.00	0.00	0.00	0.00	0.00
60	1.150	2.149	2.70	0.25	4.50	229.76	40.00	2042.71	6.20	0.51	6.71			0.01	0.59	6.13	99.88	0.00	0.00	0.00	0.00	0.00
61	1.150	2.149	2.70	0.25	4.50	243.55	40.00	2165.27	6.58	0.54	7.12			0.01	0.62	6.50	106.38	0.00	0.00	0.00	0.00	0.00
62	1.150	2.149	2.70	0.25	4.50	258.16	40.00	2295.18	6.97	0.57	7.54			0.01	0.66	6.89	113.26	0.00	0.00	0.00	0.00	0.00
63	1.150	2.149	2.70	0.25	4.50	273.65	40.00	2432.90	7.39	0.61	8.00			0.01	0.70	7.30	120.56	0.00	0.00	0.00	0.00	0.00
64	1.150	2.149	2.70	0.25	4.50	290.07	40.00	2578.87	7.83	0.64	8.48			0.01	0.74	7.74	128.30	0.00	0.00	0.00	0.00	0.00
65	1.150	2.149	2.70	0.25	4.50	307.47	40.00	2733.60	8.30	0.68	8.99			0.01	0.78	8.20	136.50	0.00	0.00	0.00	0.00	0.00
66	1.150	2.149	2.70	0.25	4.50	325.92	40.00	2897.62	8.80	0.72	9.52			0.01	0.83	8.69	145.19	0.00	0.00	0.00	0.00	0.00
67	1.150	2.149	2.70	0.25	4.50	345.48	40.00	3071.47	9.33	0.77	10.10			0.01	0.88	9.21	154.41	0.00	0.00	0.00	0.00	0.00
68	1.150	2.149	2.70	0.25	4.50	366.21	40.00	3255.76	9.89	0.81	10.70			0.01	0.93	9.77	164.17	0.00	0.00	0.00	0.00	0.00
69	1.150	2.149	2.70	0.25	4.50	388.18	40.00	3451.11	10.48	0.86	11.34			0.01	0.99	10.35	174.53	0.00	0.00	0.00	0.00	0.00
70	1.150	2.149	2.70	0.25	4.50	411.47	40.00	3658.18	11.11	0.91	12.02			0.01	1.05	10.97	185.50	0.00	0.00	0.00	0.00	0.00
71	1.150	2.149	2.70	0.25	4.50	436.16	40.00	3877.67	11.78	0.97	12.75			0.01	1.11	11.63	197.13	0.00	0.00	0.00	0.00	0.00
72	1.150	2.149	2.70	0.25	4.50	462.33	40.00	4110.33	12.48	1.03	13.51			0.01	1.18	12.33	209.46	0.00	0.00	0.00	0.00	0.00
73	1.150	2.149	2.70	0.25	4.50	490.07	40.00	4356.95	13.23	1.09	14.32			0.01	1.25	13.07	222.53	0.00	0.00	0.00	0.00	0.00
74	1.150	2.149	2.70	0.25	4.50	519.47	40.00	4618.36	14.03	1.15	15.18			0.01	1.33	13.85	236.39	0.00	0.00	0.00	0.00	0.00
75	1.150	2.149	2.70	0.25	4.50	550.64	40.00	4895.46	14.87	1.22	16.09			0.01	1.40	14.69	251.08	0.00	0.00	0.00	0.00	0.00
76	1.150	2.149	2.70	0.25	4.50	583.68	40.00	5189.19	15.76	1.30	17.06			0.01	1.49	15.57	266.64	0.00	0.00	0.00	0.00	0.00
77	1.150	2.149	2.70	0.25	4.50	618.70	40.00	5500.54	16.70	1.38	18.08			0.01	1.58	16.50	283.14	0.00	0.00	0.00	0.00	0.00
78	1.150	2.149	2.70	0.25	4.50	655.82	40.00	5830.58	17.71	1.46	19.16			0.01	1.67	17.49	300.64	0.00	0.00	0.00	0.00	0.00
79	1.150	2.149	2.70	0.25	4.50	695.17	40.00	6180.41	18.77	1.55	20.31			0.01	1.77	18.54	319.18	0.00	0.00	0.00	0.00	0.00
80	1.150	2.149	2.70	0.25	4.50	736.88	40.00	6551.24	19.90	1.64	21.53			0.01	1.88	19.65	338.83	0.00	0.00	0.00	0.00	0.00
81	1.150	2.149	2.70	0.25	4.50	781.09	40.00	6944.31	21.09	1.74	22.83			0.01	1.99	20.83	359.66	0.00	0.00	0.00	0.00	0.00
82	1.150	2.149	2.70	0.25	4.50	827.96	40.00	7360.97	22.35	1.84	24.20			0.01	2.11	22.08	381.74	0.00	0.00	0.00	0.00	0.00
83	1.150	2.149	2.70	0.25	4.50	877.64	40.00	7802.63	23.76	1.95	25.65			0.01	2.24	23.41	405.15	0.00	0.00	0.00	0.00	0.00
84	1.150	2.149	2.70	0.25	4.50	930.29	40.00	8270.78	25.12	2.07	27.19			0.01	2.37	24.81	429.96	0.00	0.00	0.00	0.00	0.00
85	1.150	2.149	2.70	0.25	4.50	986.11	40.00	8767.03	26.63	2.19	28.82			0.01	2.52	26.30	456.26	0.00	0.00	0.00	0.00	0.00
86	1.150	2.149	2.70	0.25	4.50	1045.28	40.00	9293.05	28.22	2.32	30.55			0.01	2.67	27.88	484.14	0.00	0.00	0.00	0.00	0.00
87	1.150	2.149	2.70	0.25	4.50	1108.00	40.00	9850.64	29.92	2.46	32.38			0.01	2.83	29.55	513.70	0.00	0.00	0.00	0.00	0.00
88	1.150	2.149	2.70	0.25	4.50	1174.47	40.00	10441.67	31.71	2.61	34.32			0.01	3.00	31.32	545.02	0.00	0.00	0.00	0.00	0.00
89	1.150	2.149	2.70	0.25	4.50	1244.94	40.00	11068.17	33.61	2.77	36.38			0.01	3.18	33.20	578.22	0.00	0.00	0.00	0.00	0.00
90	1.150	2.149	2.70	0.25	4.50	1319.64	40.00	11732.26	35.63	2.93	38.56			0.01	3.37	35.20	613.42	0.00	0.00	0.00	0.00	0.00
91	1.150	2.149	2.70	0.25	4.50	1398.82	40.00	12434.20	37.77	3.11	40.88			0.01	3.57	37.31	650.73	0.00	0.00	0.00	0.00	0.00
92	1.150	2.149	2.70	0.25	4.50	1482.75	40.00	13182.37	40.03	3.30	43.33			0.01	3.78	39.55	690.27	0.00	0.00	0.00	0.00	0.00
93	1.150	2.149	2.70	0.25	4.50	1571.71	40.00	13973.32	42.44	3.49	45.93			0.01	4.01	41.92	732.19	0.00	0.00	0.00	0.00	0.00

Scenario C Case, 100% debt finance for the CW Barrage
would not· be repaid until year thirty nine. Indeed, no
repayments of principal at all would be made until year
twenty-eight. In the BST Case, debt would never be
repaid. (It is assumed that the interest rate is 12%
throughout, that inflation is 6%, and that revenues are
100% dedicated first to paying interest (and where further
debt is drawn down to pay interest if cash flow is
insufficient), and then to repaying principal. A total of
£30.5 billion of debt is needed, of which £17 billion
covers the construction phase). The corresponding figures
for the smaller Barrage are total debt of £4.7 billion
being finally repaid in year thirty-two, with first
principal repayment in year twenty-one. The debt for the
construction period is £3.25 billion (see Table 10).

50. These parameters are well outside what would be
acceptable to potential lenders. In particular, the fact
that for the larger Barrage some £13 billion of debt is
needed to cover shortfalls of cash flow as against
interest payment during the first seventeen years of the
operational phase - a similar profile arising with the
smaller Barrage - would be unacceptable.

Equity Finance

51. In consideration of what return might be acceptable
to private sector investors, the returns available on
North Sea projects have, in particular, been examined.
There are two primary reasons for this approach. Firstly,
North Sea projects represent those major energy projects
that in this country are carried out in the private
sector. Secondly, two of the major risks associated with
North Sea projects - the completion risk, and the risk of
the movement in energy prices - are also the two major
risks associated with a tidal Barrage project. The
experiences of the contenders for the Channel fixed link
have also been followed closely; in particular, the terms
on which the Channel Tunnel Group have been able to obtain
indications of interest for both debt and equity finance
have been examined.

52. The examination of these projects suggest that, while
there is no obvious threshold, a real rate of return of
less than 10% would certainly be insufficient to attract
private sector investment into the Severn Barrage project,
if the investors are to be exposed to all the attendant
risks. Yet to achieve a return of even 10% would require
a tariff level nearly three times the current level as
determined by the BST. To reach such a level by the time
that the barrage starts generating would require energy
prices to increase in real terms by over 7% per annum
between now and the year 2000, which may be ruled out

53. One has to conclude that for any financing to be
arranged, the risks will have to be significantly reduced,
if not eliminated. Of these, one of the most crucial is

the market risk.

Market Risk

54. A number of discussions have been held with the CEGB as to the elements that might form the basis of an appropriate Sale and Purchase Contract between the CEGB and a company owning either Barrage for the electricity generated by it. The starting point for these discussions was the Energy Act 1983, which provides a statutory basis for the sale and purchase of power generated in the private sector. In particular, Section 7(3) provides that prices proposed by the Electricity Board:

"(a) Will not increase the prices payable by the customers of the Board for the electricity supplied to them by the Board; and

(b) Will reflect the cost that would have been incurred by the Board but for the purchase"

55. With this as the background, the CEGB indicated that they would expect the price basis for such a Sale and Purchase Contract to consist of two elements:

(i) a payment for the electricity generated; and

(ii) a payment for capacity.

56. In the initial stages of the discussions, the CEGB indicated that they would expect the payment for the electricity generated to be on the basis of the Bulk Supply Tariff (BST) in force at the time. Thus, with a Sale and Purchase Contract on this principle, the owning Company and hence whoever provides finance to it would be exposed to the risk of how the tariffs might move. (For the present purpose, future movements in the BST may be regarded as directly reflecting shifts in the real price of energy; in fact, it should be noted that no real increase in the BST would imply small real increases in underlying energy prices, which would be compensated for by efficiency increases in the CEGB system). It is judged that potential investors in the project would not be prepared to rely on any possible future real increases in the level of the BST in order to generate a satisfactory return. The economic analysis shows that if the tariffs are based on the current BST, and remain constant in real terms, the two Barrages each generate rates of return of approximately 4%. This basis is therefore clearly inadequate to secure private sector support.

57. Notwithstanding the CEGB expectation that the BST formula would apply, it is noteworthy that the CEGB have in their case for the Sizewell B Power Station used a number of Scenarios (A to E) of the future, which they investigated for high, medium and low nuclear backgrounds. They selected Scenario C as the one against which to perform most of the analysis, as it represents the middle of the Scenarios. It is appropriate in evaluating the merits of the proposed Severn Barrage to make judgments against the same economic Scenarios as the

CEGB itself has used in evaluating the plant that they propose to build for its own system. Thus, the economics of the barrages were analysed against these various Scenarios. In summary, the middle economic Scenario (C), against a low nuclear background, implies a real rate of return of 7.38% for the CW Barrage and 7.75% for the ES Barrage over the life of the project.

58. Against this background, a number of discussions were held with the CEGB to seek to establish the possibility of formulating a contract for the sale and purchase of the electricity generated by the Barrage whereby the risk to the investor, which is the risk of lower than expected tariffs, largely as a result of lower than expected energy prices, is removed from the investor and taken by the CEGB, in return for the risk to the CEGB, which is the risk that energy prices will be higher than expected, and is the investors' upside, is also removed.

59. In particular, the concept of a sale and purchase contract that would effectively guarantee tariffs based on Scenario C was discussed. The CEGB said that it would be prepared to contemplate some degree of mutual risk reduction, and has indicated to us that it would be prepared to consider entering into a sale and purchase contract, with prices linked to coal prices, subject to a minimum real price reduction of 1% per annum, and a maximum real price increase of 2% per annum. Thirty years ahead is as far as they could contemplate such a price commitment. Beyond then, while the CEGB would expect to take the electricity produced from the Barrage, it could not consider any commitment now on the price at which they would take it. The economic analysis shows that tariffs that track the floor and ceiling levels will produce rates of return of 2.1% and 7.2% respectively for the CW Barrage, and 2.3% and 7.5% for the ES Barrage, _if they were continued over the whole life of the Project_.

60. Thus the conclusion is that the bases of power pricing that the CEGB are at this time prepared to contemplate are not sufficient to enable the Project to be financed in the private sector.

Comparison with Public Sector Investment

61. As noted above, the investment criteria for a project of this nature would be fundamentally different in the public sector. The cost of generation from the barrage compares favourably with other forms of generation. It is only slightly more than the cost of generation projected by the CEGB for the proposed Sizewell B PWR, and significantly less than a coal-fired power station.

This, taken together with the fact that the electricity produced by the barrage is essentially valued against the other forms of generation within the CEGB, and hence against public sector investment criteria, suggests that

any power station project, whatever its technology, would
be difficult to finance on a stand alone basis in the
private sector, particularly if tariffs could only be set
on an annual basis.

Other Benefits

62.   There are a number of benefits from a Severn Barrage
which would not accrue directly to a private sector owning
company.   They have not, therefore, been taken into
account in the economic analysis.   They are nevertheless
important, and relevant to any consideration of whether
the project should be undertaken within the public sector.
The principal benefits are as follows:-

(i)   Strategic Diversity of Energy Sources.   The
       barrage would provide the CEGB with an
       alternative to coal and nuclear in the event of
       difficulties with either, and add substantial
       flexibility to the system.   Such strategic
       diversity has been a declared policy of the CEGB
       over the years;

(ii)   Alternative Investment.   The alternative
       investment decisions are not equally available
       e.g. there are delays and objectors for a nuclear
       project and, even assuming Sizewell B goes ahead
       on the planned programme, the expenditure of
       £6000M on a sequence of nuclear stations makes
       assumptions about continuing goodwill from
       Government and public opinion;

(iii) Insurance.   The cost of generation will be
       largely independent of energy prices and thus the
       barrage will provide insurance against
       unpredictable increases in primary energy costs;

(iv)   Long Life.   The technique of discounted cash flow
       does not give any significant credit to the very
       long operating life of the Barrage;

(v)   Technological Improvements.   It is possible that
       technological advance during the life of the
       Barrage may reduce the problems caused by the
       phased nature of its electrical output;

(vi)   Environment.   The generation of electricity by
       tidal power does not pollute the environment;

(vii) Employment Creation.   As discussed in Dr. Shaw's
       paper (number 14) it is estimated that 280,000
       man years of direct, and 160,000 man years of
       indirect employment would be created by the
       construction of the Cardiff-Weston Barrage,
       followed by 3,000 permanent jobs on barrage and
       associated works and an estimated 25,000 to
       30,000 permanent jobs in spin-off activities;

(viii) Regional Benefits.   Benefits to the region such

as increased rate income and higher industrial activity, as well as residential, sports and amenities and tourism benefits.

Minimum Conditions for Finance

63. The analysis above has shown that neither Barrage is financeable wholly in the private sector. Considerable attention has been devoted to analysing what conditions by way of improved power value, or risk reduction through public sector support, would be the minimum necessary for any private sector finance to be possible. Since the substantial increases that would be required in the power sales value is ruled out, it is judged that the two necessary conditions would be:-

(i)   A long term take-if-tendered contract with the CEGB for the sale and purchase of the electricity with a predictable pricing formula (such as is found in co-generation and other private schemes in the USA and elsewhere); and

(ii)   A completion guarantee of last resort.

64. As indicated by the analysis in Paragraph 59 above, the long term take-if-tendered contract would provide that the CEGB would buy all the electricity generated by the barrage at a price that reflected the system marginal costs of generation implied by economic Scenario C, after allowances had been made for off load heat losses, transmission losses, and costs of transmission. The tariff would be indexed to a measure of inflation.

65. As far as a completion guarantee is concerned, analysis suggests that the likelihood of a cost overrun exceeding £850M in amount for the CW Barrage and 6 months in time was comparatively small, with comparable figures for the ES Barrage. Even so, the risk that there will be an overrun beyond this could not be covered within the private sector. It would require the Government to act as a completion guarantor of last resort, undertaking to fund the costs incurred if the project overran by more time or money than stipulated.

66. While these two elements are the minimum necessary for any private sector finance, they may not be sufficient. One of the problems of considering the viability of any scheme based on these requirements is that the construction phase cannot start until 1991 at the earliest, although a degree of commitment in principle would be needed earlier. Moreover, any scheme built upon these two conditions would depend upon assumptions about public sector support which cannot at this stage be validated. It was thus not possible to obtain formal reaction from financial institutions. However, informal soundings were taken. These suggested that the minimum level of public sector support specified might make the

barrage produce a balance between the return available to
an investor and the risk on his return that was not
unacceptable. However, there would still be other
problems. These mainly concerned the unprecedently long
time scales involved, both in terms of the period before
redemption, (or alternatively, the return in an acceptable
period for redemption would be too low), and the period to
any return. A further concern was that the form of the
Government completion guarantee described above related to
the costs of the project in real terms. In principle,
therefore, while the return may be all but guaranteed in
real terms, during the construction phase there would be,
in some shape or form, a commitment to find cash that
would be open ended. Finally, there was also some concern
expressed at the risk of the project being abandoned once
construction had started.

67. As far as the balance between period to redemption
and the return is concerned, it has not been possible to
determine whether there is any point at which financing
would be possible. There may well not be any such point.
The nature of the liabilities of institutions is such that
they are unlikely to want significant assets with a time
scale to redemption longer than about thirty years.
Indeed, with sums of the magnitude involved here, they
would certainly want a spread of periods, with probably
the bulk having a maturity of no more than twenty-five
years. Thus, it may well be necessary to fund for shorter
periods, and the rollover funding would need to be
guaranteed by the Government.

68. Such rollover funding could also cater for the
concern about the period to any return. Many
institutional investors prefer to invest in instruments
that provide them with a running yield. Banks, however,
are more use to providing debt for projects, where, during
the construction phase, the interest is capitalised. It
would be possible to envisage a scheme whereby the
construction costs were met by debt, with an intention to
refinance the debt post-completion by means of some form
of longer term revenue bonds. While it would probably be
possible to structure debt whose drawdown and repayment
schedules were drawn up on the expectation that this is
what would happen, lenders would not at the time of
provision of funds be prepared to rely on such refinancing.

69. Concern at the possibility of an open ended cash
commitment would need to be met by some part of the
Government completion guarantee of last resort including
some limit on the actual cash amount of funds that
investors would have to provide, as well as on the cost in
real terms, with the Government providing the necessary
cash over and above this point.

70. As far as the abandonment risk is concerned, it is
unlikely that the project would be abandoned for reasons

that were associated with the factors internal to the project. Clearly, there could be no financial commitment to the construction phase if at that point any serious risk remained of the project being abandoned for reasons external to the project itself.

Pre-Construction Phase

71. Whether or not the financing of the construction phase is feasible, there is a need to find equity funding for the pre-construction phase, while there is a real possibility that the project could be abandoned, and investment to date lost. Such risk money could only be raised against an expectation of very significant rewards if the project is pursued to a successful conclusion. Clearly, the rewards from this project are inadequate in themselves for this purpose, and need to be boosted for organisations taking an equity shareholding in the owning company at the pre- construction phase. In particular, consideration was given to the possibility of other parties taking such an equity shareholding, where the return is boosted by the tax allowances that become available.

72. An investment only in the pre-construction phase purely in return for the tax allowances has a rate of return of about 40%. However there are a number of risks attached to this investment. In particular there is a significant risk that changes in the fiscal regime would seriously reduce the returns of such an investment. Over this time scale, the fiscal risk could well be an insuperable barrier, unless underwritten by Government.

Conclusions

73. The analysis in this paper has shown that:

(i)   Neither barrage is financeable wholly in the private sector;

(ii)   The primary reason for this conclusion is that the commodity produced by the barrage is being priced according to public sector investment criteria;

(iii)   The economics of the barrage compare favourably with other forms of generation;

(iv)   Significant indirect benefits would arise, but since these would not flow to a private sector owning company, they have not been taken into account in the economic analysis;

(v)   The minimum conditions necessary for some private sector finance would be a significant reduction in the market and completion risks, although, even if these were achieved, significant problems would remain in devising a financing scheme.

REFERENCES

1 Department of Energy. Energy Paper No.46. Tidal Power from the Severn Estuary.
2 Central Electricity Generating Board. Sizewell B Public Enquiry. CEGB P5.
3 Central Electricity Generating Board. Analysis of Generating Costs (1983/84 update).

10. The Mersey Barrage

E. T. HAWS, MA, FICE, FIPENZ, *Managing Director, Rendel–Parkman,*
G. R. CARR, BSc(Eng), ACGI, FICE, *Assistant Director, Rendel Palmer
& Tritton,* and B. I. JONES, BEng, MICE, MIWES, *Associate Director,
Ward Ashcroft and Parkman*

SYNOPSIS The Paper describes the outcome of desk studies to investigate the possible construction of a tidal power barrage near the mouth of the River Mersey at Liverpool. These studies have indicated that a barrage having an installed capacity of some 500-600 MW or greater appears likely to be feasible, and to have the potential for producing energy at costs which would be competitive with any alternative tidal energy source in the UK. Such a barrage could also provide significant non-energy benefits, although evaluation of these awaits further studies. The environmental and navigation aspects are briefly outlined, and the estimated construction costs, programmes and economics for two alternative sites are summarised.

INTRODUCTION

1.   The Mersey Estuary possesses characteristics which make it an attractive location for a tidal power barrage. The river discharges to Liverpool Bay through a channel which is deep and narrow (Fig.1). It therefore could be closed with a barrage of economical length, which would provide the water depths needed to set suitably sized turbines and sluices at effective submergence levels, and would impound an upstream basin of large area. In addition, the tidal range, whilst less than in the Severn Estuary, is nevertheless favourable, with a mean spring tide range of 8.4m.

2.   The Paper gives the outcome of studies undertaken in 1983 and 1985 (Refs. 1 and 2) and information on the current programme of work.

PROJECT DESCRIPTION
General

3.   Two sites identified in the studies as of significance (Lines 1 and 3) are generally located as in Fig. 1. In both cases a straight alignment across the Estuary is assumed.

4.   The barrage length at Line 1 would be about 1.7km, and at Line 3 about 1.8km. Where the strong tidal currents have scoured the channel at Line 1, the estuary bed is locally some 24 metres below MSL. The deepest part of the bed at Line 3 is about 17 metres below MSL, with bedrock overlain by several metres of loose deposits.

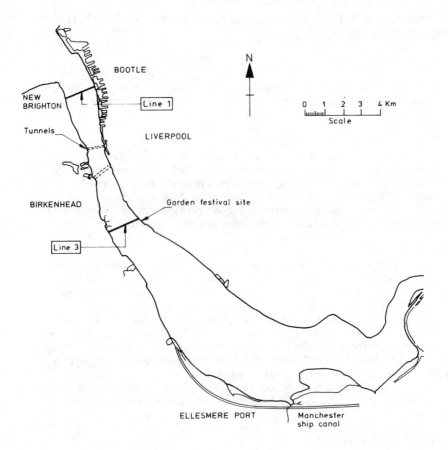

Fig.1   The Mersey Estuary and Location of Barrage Lines

5.   Whatever form of construction might be adopted, the disposition of the various components of the barrage would be generally similar for a particular site. Figs. 2 and 3 indicate the layouts.

6.   With the deepest water towards the middle of the channel the turbines would be located centrally to afford them adequate submergence for the avoidance of cavitation. The sluices would be disposed on either side of the turbines. 27 turbines of 7.6m diameter and 18 sluices 12m square are indicated for Line 1 and 21 such turbines with 15 sluices for Line 3. The installed capacities of the two alternative schemes on this basis would be 621MW and 483MW respectively.

7.   A single lock on the western side is deemed adequate for Line 3 but two locks may be required for Line 1 if all the aspirations of the navigating parties are realised.

Geology
8.   Line 1. On Line 1 the rock profile follows a very flat 'U' shape, which on the Liverpool shore falls from -17m OD at the shore line to -25m OD at the mid-point of the channel. From this point the rock level rises up to ordnance datum at the New Brighton shore.

9.   The bedrock is Bunter sandstone, susceptible near the surface to deterioration on exposure.

10.   The deposits overlying the bedrock are generally composed of sands, gravels and clays. At the Liverpool shore, these deposits are some 10m thick, thinning to rock level at a point some 200 metres from the shore. From there to the mid-point of the deepwater channel, the river bed is rock, alluvial deposits being absent. Between this deepest section and the New Brighton shore, the surface deposits increase in thickness to a maximum of some 12 metres until, at the shore itself, the rock surface again forms the bed of the river.

11.   Line 3. On Line 3 the rock profile again follows a flat 'U' shape, being at a level of about -3m OD at the Liverpool shore line, -22m OD at the midpoint of the channel and rising to a level of approximately -3m OD at the Rock Ferry shore line.

12.   As on Line 1, the bedrock is a red Bunter sandstone, and the surface deposits are again composed of sands, gravels and clays. On the Liverpool half of the channel the deposits vary in thickness between 7 and 12 metres, except at the shore where the rock profile rises to bed level. On the Rock Ferry side, at a point some 400 metres from the shore, the bed profile rises sharply from the edge of the deep water channel, and then gradually slopes to the shore. The deposits in this area are generally 5m deep.

ENVIRONMENTAL CONSIDERATIONS
Regional Development and Sociology
13.   It is considered that the impounded estuary could contribute to the potential for economic regeneration of the whole area. In place of the present expanse of inner estuary which largely dries out on each tide, the extensive lake created would form a unique regional feature. It could attract varied types of development and could bring

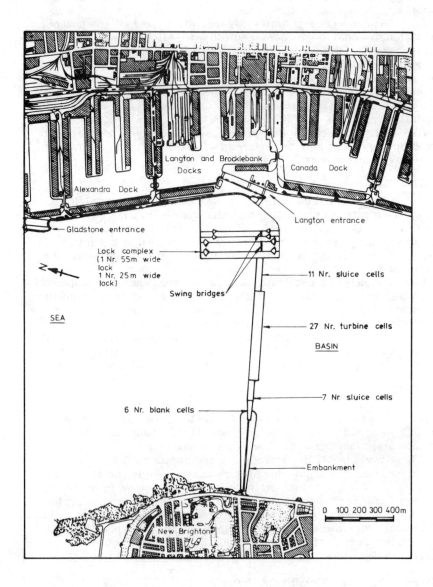

Fig.2 Barrage layout for Line 1

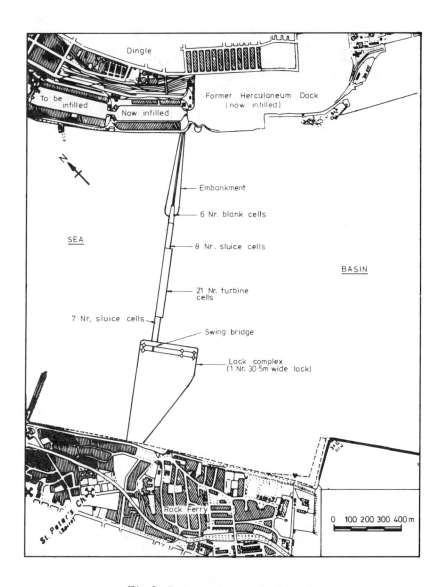

Fig.3 Barrage layout for Line 3

considerable social and financial benefits to the area. Such development would complement steps already being taken to develop the disused South Docks area and to re-develop New Brighton.

14. Water recreation activities are regarded as a potentially large growth area, giving opportunities in the dock areas and elsewhere for the establishment of marinas, aqua-sports centres and the like. In conjunction with these activities, it would be expected that the barrage itself and the enhanced attractiveness of the estuary would encourage the growth of tourism.

15. The use of the barrage as a road link is thought unlikely to be justified in view of the excess capacity of the existing river tunnels and the minimal traffic growth projected, as well as the problems involved in providing suitable approach links.

Sedimentation and Water Movement

16. Studies relating to the circulation of water, and to waves and tides, in Liverpool Bay and the Inner Estuary are continuing at the Institute of Oceanographic Sciences, Bidston, where a computer model of these processes has been developed. Hydraulics Research, Wallingford, and Liverpool University are also engaged in physical modelling, numerical simulations and other investigations in these fields.

Water Quality

17. The catchment area contains five million people and is heavily industrialised. Currently, large pollution loads are carried into the Estuary from the Rivers Mersey and Weaver and the Manchester Ship Canal. Numerous outfalls discharge untreated sewage and industrial pollutants.

18. The North West Water Authority is engaged in a 15-year plan to reduce the pollution load of the Estuary, involving the construction of new sewage works, together with interceptor sewers along both shores. Earlier measures have already improved the oxygen content of the Inner and Upper Estuary and have led to an increase in fish and bird populations.

19. The barrage would limit the flushing action of the tide and cause increased retention times, leading to greater concentration of any pollutants present and higher oxygen demand. Longer retention times could also result in an increase of phytoplankton, which could add to the oxygen demand.

Flora and Fauna

20. The Mersey is the most important estuary in the British Isles for pintail and teal, and attracts large populations of other wildfowl and other species.

21. Species such as pintail, teal and dunlin which feed on invertebrates mainly inhabit the tidal mudflats. The changes in tidal levels in the Inner Estuary would largely flood these present feeding grounds, and it is uncertain whether others would form to take their place.

22.   The small reduction in high tide level in the Inner Estuary is seen as being of considerable significance, as it would affect the various duck species which feed on the salt-marsh plant seeds or graze on the salt water grasses. It is not evident whether a similar environment would become established at the different level.

Fisheries
23.   From the 1930's the Inner Estuary declined as a fishery, presumably as a result of increasing pollution and reducing oxygen levels, and by 1948 the various species previously caught had disappeared.   However, there is evidence that with recent improvement in water quality fish are beginning to return.
24.   Any adverse change in water quality in the Inner Estuary arising from the barrage would obviously diminish the prospects of a fishery re-establishing itself there.

Land Drainage and Groundwater Supplies
25.   The North West Water Authority confirmed that the increase in the upstream low water level would not prevent adequate functioning of most of the outlets entering the River. Exceptions would include the streams passing under the Manchester Ship Canal, which would require to be pumped.
26.   High spring tides and storm surges can at present overtop the Ship Canal and back up watercourses. With the operation of the barrage the lower maximum tide level would alleviate these situations.
27.   The Water Authority's only freshwater supply boreholes are located near Warrington and are too remote to be affected by any saline intrusion of the groundwater caused by the saltwater basin impoundment.

Sea Defences
28.   The barrage would improve sea defences in the areas upstream of it as a result of lower high water levels and exclusion of storm surges.   It is considered that there would be no significant effect on the coastal defences to seaward.

ALTERNATIVE CONSTRUCTION METHODS
Schemes Examined
29.   The alternative schemes costed (for both Lines 1 and 3) were:

a)   turbine-generator and sluice housings of pre-fabricated concrete caissons floated in and sunk on prepared foundations.
b)   as for (a), but caissons of steel construction.
c)   turbine-generator and sluice housings formed by in-situ shells of diaphragm wall construction, and subsequent structural fitting out with precast or in-situ concrete.

30. For each of these alternatives (as applied to a particular location) the same lock complex would be required. This was assumed to be constructed by the diaphragm wall method, as was the perimeter wall of the permanent lock island.

31. The permanent embankment connecting to the shore would also be a feature common to all the alternative schemes at any one site and hence represents a similar cost.

32. Discounting over a 120-year life (with replanting every 40 years) the capital costs and the operating and maintenance costs for the three alternatives at the two sites (assuming those numbers of turbine and sluice units which had earlier been indicated as optimum), it was found that the steel caisson alternative was the most expensive and the diaphragm wall alternative the cheapest.

33. The subsequent economic analyses were therefore restricted to schemes incorporating the diaphragm wall method of construction.

The Diaphragm Wall Method and associated Sand Islands

34. The technique of constructing concrete walls below ground level by the diaphragm wall method has been widely used since the 1950's. It involves excavating a trench which is stabilised by being kept full of bentonite clay slurry, and subsequently displacing the slurry with concrete fed in through "tremie" pipes.

35. Using this method, the turbine and sluice cells would be constructed on temporary sand islands contained between the sunken hulls of very large crude carrier ships (VLCC's), of which there are numbers laid up throughout the world and available at scrap prices. A pair of these vessels positioned in parallel on the prepared riverbed, and with the open ends of the enclosure closed off with bulkheads, would enable a sand island to be formed. When the construction of the series of diaphragm wall cell structures so contained had reached a sufficient stage, the surrounding sand would be pumped out and the hulls floated for repositioning for the construction of the next series of cells. The remaining structural work in the part-completed cells and their subsequent fitting out with equipment would follow.It is envisaged that internal floors, walls and roof would as far as practicable utilise precast concrete units to reduce the extent and duration of on-site work.

36. Figs. 4 and 5 give cross-sectional outlines of the turbine and sluice structures.

NAVIGATION REQUIREMENTS

37. Excluding dredgers and small craft, some 11,000 ship movements per year would need to be handled by the Line 1 locks (based on 1984 figures). At Line 3, the equivalent figure is 8,000.

38. It was assumed for the purpose of estimating construction costs that two locks would be needed at Line 1, one 360m x 55m, the other 270m x 25m. At Line 3, the assumption was for a single lock 270m x 30.5m.

39. The large lock at Line 1 would be able to take a part-laden 250,000 tonne tanker should the Tranmere terminal continue to be

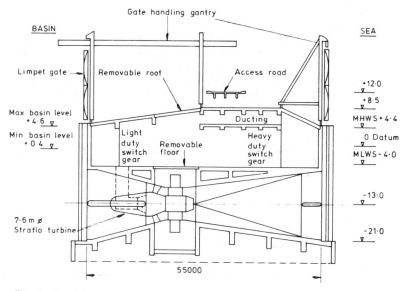

Fig.4　Turbine cell cross-section

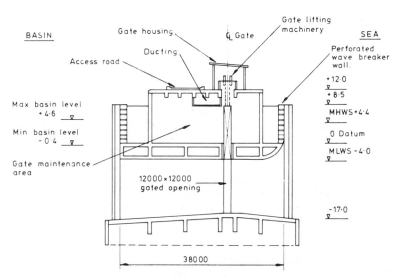

Fig.5　Sluice cell cross-section

served by vessels of this size. A semi-submersible drilling platform similar to that already built at Birkenhead by Cammell Laird could pass through, either as "dry cargo" or, using the two locks together, in floating mode.

40. The walls of the locks, including the recesses for the sector gates, would be constructed using diaphragm wall methods, as would the perimeter walls of the lock island.

41. For the Line 1 scheme, more sheltered conditions and less interference with dock entrances might make it preferable to site the lock complex on the New Brighton side.

ENERGY GENERATION AND OPTIMISATION OF INSTALLATION

42. The turbines (of bulb or Straflo type) were assumed to be 7.6m dia., this size being the maximum likely to be practical from considerations of submergence requirements in relation to the river bed level. The sluice units were taken to be of venturi-type, nominally with 12m x 12m openings, controlled by vertical lift gates.

43. Using turbine characteristics selected by optimisation studies as the most appropriate for the tidal pattern of the estuary, computer simulation of the barrage operation over a near-average annual pattern of tides was undertaken for a wide-ranging series of combinations of numbers of turbine and sluice units for each of the site locations, to evaluate the amounts of energy producible for each case. The 1985 study updated the analyses of energy output to take into account such refinements as reductions in the tidal range as predicted by the Institute of Oceanographic Sciences computer model for river closure at Lines 1 or 3; revised turbine performance characteristics; draft tube losses; improved sluice coefficients; service availability of generating sets; freshwater inflow; and local power losses. Increased output available from a pumping option has not yet been included.

44. The computations established relationships between output, number of installed turbine-generators (23 MW rating assumed) and number of sluices. Relating these in turn to the estimated unit energy cost for a range of combinations of turbines and sluices confirmed the optimum installation.

45. The average annual outputs so determined for the optimum installations at Lines 1 and 3 are 1.20 TWh and 0.965 TWh, and the related energy costs are 2.89 and 2.71 p/kWh (based on discounting to December 1984 at 5 per cent).

46. If the objective were to be the fullest exploitation of the energy resource, minimum unit energy cost would not be the decisive criterion. The chosen scheme might then be one which accommodated a greater number of turbines and sluices, to the extent that site limitations would allow. Extending the length of the barrage by adapting a skewed or "dog-legged" alignment might however prove unacceptable for hydraulic reasons.

CONSTRUCTION PROGRAMME

47. It is estimated that construction of the barrage at Line 1

would take some six years, and at Line 3 a little less. A critical feature in the programme would be completion of the locks before progressive closure of the estuary reached a stage at which navigation conditions through the gap became unacceptable.

48. The programme for the barrage located at Line 1 is given in Fig.6.

49. In association with the study, a limited series of tests relating to the later stages of the construction sequence for each Line was carried out by Hydraulics Research Limited using their existing hydraulic model of the Mersey Estuary (Ref. 3). The results were used in conjunction with a simple computer analysis to obtain a preliminary assessment of hydraulic conditions during closure operations.

ECONOMIC ANALYSIS

50. The estimated construction and operating costs are summarised for each of the sites in Table 1. The estimated basic capital cost is £444 million for Line 1 and £336 million for Line 3 (at a base date of December 1984).For the purpose of the optimisation of installation size and the derivation of the notional unit energy costs referred to earlier, the "standard" post-approval contingency of $17\frac{1}{2}$% was added to these capital costs, but it had no place in and was omitted from the sensitivity and risk analyses subsequently undertaken.

51. The values for the energy produced were assessed in the context of the country's power system to analyse for a range of "scenarios" the cost benefits to the system over the period 1996 to 2030 arising from inclusion of the barrage. The results were adjusted for an assessed firm power credit and also for a 6% value reduction to allow for additional cycling costs involved in tidal plant use, as advised by CEGB.

52. The value of energy adopted as the "Reference Case" for the subsequent economic analysis was £46 million/TWh (or 4.6 p/kWh),this being based on the scenario of "middle nuclear growth, upper middle demand growth and middle fuel costs". The results of assuming other scenarios indicated a range of variation of +15% to -25% from the Reference Case value.

53. The economics were evaluated using a computer programme designed to model the interaction of time, resources, costs and revenue. In addition to the basic "single estimate" (deterministic) analysis, sensitivity and probabilistic risk analyses were carried out for each Line.

54. The sensitivity analyses investigated the effects on the economics of variations in the construction costs and durations, maintenance costs, energy output and energy value. In the risk analyses, different values of all these factors (other than energy value) were combined in a Monte Carlo simulation to assess the probabilities attaching to particular levels of economic out-turn.

55. The main conclusions from the economic analyses were as follows:

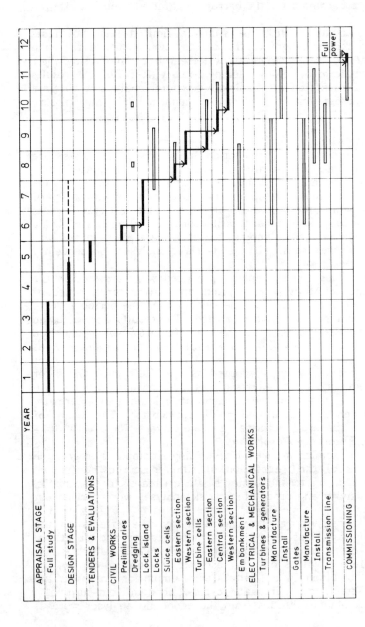

Fig.6 Construction programme, Line 1

TABLE 1

ESTIMATED COSTS

(FOR SCHEMES INCORPORATING DIAPHRAGM WALL METHOD)

		Line 1	Line 3
No. of turbines		27	21
No. of sluices		18	15
No. of locks		2	1

CAPITAL COSTS — £million

Civil Works

		Line 1	Line 3
Preliminaries		16.0	11.0
Dredging		0.8	1.3
Lock complex		75.8	51.5
Sluice cells		35.4	30.4
Turbine cells		78.3	62.2
Blank cells		2.0	2.0
Embankments		7.6	9.1
	Sub-total	215.9	167.5
	10% Civil contingency	21.6	16.7
	TOTAL (1)	237.5	184.2

Electrical & Mechanical Works

		Line 1	Line 3
Turbo generators		116.9	91.0
Electrical and mechanical equipment		18.7	14.5
Gates and cranes		23.9	17.0
Transmission		18.0	7.5
	TOTAL (2)	177.5	130.0
	Total civil and E & M(1) + (2)	415.0	314.2
	7% Engineering	29.0	22.0
	Basic capital cost	444.0	336.2
	$17\frac{1}{2}$% post-approval costs	77.7	58.8
	TOTAL CAPITAL COST	521.7	395.0

		Line 1	Line 3
Operation &	Civil	1.8	1.4
Maintenance	E & M	1.8	1.3
Plant replacement	After 40 years	116.9	91.0
	After 80 years	116.9	91.0

Note: The costs relate to a base date of December 1984

i) Both Line 1 and Line 3 produced an internal rate of return in excess of 9% for the deterministic case based on the Reference Case Energy Value of 4.6 p/kWh.

ii) The minimum returns indicated by the risk analyses exceeded the Government target rate of 5% (6.4% for Line 1 and 6.9% for Line 3).

iii) From the sensitivity analyses based on the Reference Case Energy Value, the most sensitive variables are energy output, the duration and cost of installing and commissioning the turbines and generators, the cost of diaphragm walling, and the cost and duration of installation of electrical/mechanical equipment.

iv) Both Line 1 and Line 3 are "economically robust" across the range of energy values.

56. Fig. 7 illustrates the sensitivity of the rate of return to the energy value.

COMPARISON OF BARRAGE SITES

57. Comparing the two favoured sites, Line 1 would produce the greatest amount of energy and would be only slightly inferior to Line 3 in terms of unit energy cost and internal rate of return on investment. It would also offer the greatest potential for industrial, recreational and amenity developments. However, Line 1 would be the most expensive scheme, would present the greater construction risk, and would involve more interactions with shipping and industry.

58. The following table summarises the estimated main parameters for the sites:

	Line 1	Line 3
Barrage length (km)	1.74	1.80
Installed capacity (MW)	621	483
No. of turbine-generators (rated at 23 MW)	27	21
No. of sluices (12m x 12m)	18	15
Annual energy (TWh)	1.20	0.97
Capital cost (£m)	522	395
Unit energy cost (p/kWh)	2.89	2.71
Internal rate of return (IRR) (%)	8.0	8.3
Probability of 6% IRR (%)	100	100

Notes :

1. Capital costs, unit energy costs and IRR assessments include $17\frac{1}{2}$% post-approval contingency and are at December 1984 base. Costs are based on the diaphragm wall method of construction.

2. Internal rates of return are for energy benefit alone.

3. Energy benefit based on the "Reference Case", i.e. medium increase in fossil fuel prices; medium growth of nuclear; upper-middle load growth.

4. Probability based on all risks, including output risk.

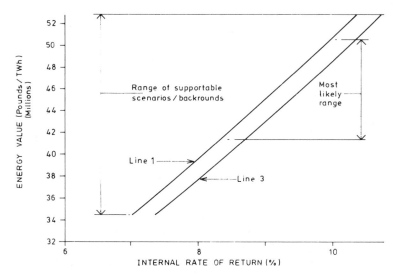

Fig.7 Sensitivity of internal rate of return to energy value

CURRENT STUDIES
General
59. A project on the scale of the barrage would exert a substantial influence on the tidal regime and environment of the Mersey Estuary and would impinge heavily upon the operations of the Port of Liverpool and other river users. As such matters had not featured prominently in the earlier work, it was felt that the initial stages of the current feasibility studies should address these issues which, while being peripheral to the main power generating concept of the project, could potentially be of such significance that they could constitute an over-riding impediment to the implementation of the barrage scheme. In addition, it was considered desirable that one or other of the general barrage locations, Lines 1 and 3, be positively identified as the preferred project site.

Stage 1
60. In accordance with these general objectives, a programme for Stage 1 of the required feasibility studies was compiled, with the main work elements comprising the following:-

(i)	Hydraulic and sedimentation studies
(ii)	Environmental/ecological studies
(iii)	Navigation/locks studies
(iv)	Social/industrial studies
(v)	Geotechnical investigations

61. In recognition of the possibility that these studies could reveal the Mersey Barrage scheme to be unacceptable in a general

context, it was considered essential to contain expenditure at this early stage of the work to the minimum level compatible with the need for confident predictions and recommendations concerning the various subject issues. The work scopes of the proposed study elements were not therefore necessarily exhaustive and, in the event that the overall acceptability of the scheme were to be confirmed, further more detailed studies would subsequently be required.

62. It is anticipated that the duration of the Stage 1 feasibility studies will be some 14 months, with a firm indication of the preferred line being available approximately half through the period.

Stage 2 Studies

63. In the event that the Stage 1 feasibility studies confirm that no over-riding impediment exists to the construction of the Mersey Barrage, it is anticipated that work would proceed immediately to the Stage 2 Studies, which will comprise essentially a re-appraisal of the engineering considerations, together with re-evaluation of the overall economic viability of the project. These re-assessments would, of course, draw on the results of the Stage 1 studies, allowing a comprehensive review of the costs and benefits offered by the project and their combined economic implications.

CONCLUSIONS

64. The preliminary economic evaluations of the Mersey Barrage indicate that the scheme is likely to prove commercially viable, and in any event justifies the more detailed stage of investigation now being undertaken. The desk studies undertaken to date have not revealed any obvious physical or other factors which might rule out the feasibility of constructing and operating the Barrage to generate a significant quantity of energy as well as to offer scope for bringing important recreational, commercial and social benefits to the area.

65. The Line 3 site appears to be slightly more economical, but the Line 1 site could produce a significantly greater amount of energy, probably at only slightly greater unit cost. The Line 1 site would also offer greater benefit to local amenities.

66. The Mersey Barrage would meet about $\frac{1}{2}\%$ of the energy requirements of England and Wales. It is approximately one tenth the scale of the Severn Barrage and would be an ideal prototype before launching on the major investment for the Severn.

ACKNOWLEDGEMENTS

The completed studies described in the Paper were carried out on behalf of the Merseyside County Council, which prior to its dissolution transferred its rights in the intellectual material of the studies to the Mersey Barrage Company Limited. The Authors wish to thank all these parties for approval to publish the Paper. The new work is being carried out on behalf of the Mersey Barrage Company Limited with the likely continued support of the Department of Energy and the EEC, this support now being enhanced by the participation of the CEGB.

The Authors are grateful for the assistance received during the preparation and reviewing of the Paper from colleagues in Rendel Palmer & Tritton and Ward Ashcroft & Parkman. The contributions made by Marinetech North West to the past studies and by Tarmac Construction Ltd in reviewing various aspects of the 1985 study are also acknowledged.

REFERENCES
1. MERSEYSIDE COUNTY COUNCIL: "Mersey Barrage Pre-Feasibility Study" Report in 3 vols. by Marinetech North West, Nov. 1983.
2. MERSEYSIDE COUNTY COUNCIL: "Mersey Barrage - A Re-Examination of the Economics". Joint Report by Marinetech North West and Rendel-Parkman, Nov. 1985.
3. HYDRAULICS RESEARCH LIMITED: "Mersey Barrage, Exploratory Model Tests". Report EX 1339, Aug. 1985.

Discussion on Papers 9 and 10

MR J. TAYLOR, Merz and McLellan
It is important not to confuse economic with commercial
viability in assessing the conclusions of Paper 9. The Author
has concentrated on commercial viability and it is hardly
surprising that the result is not encouraging. It would be
asking much of any scheme to produce a return that is
appropriate to a risky investment in a market where the price
is determined by the costs of a monopoly supplier earning a low
risk return.
     This poses the question why the Central Electricity
Generating Board (CEGB) itself should not fund the project, if
indeed the unit costs compare reasonably favourably with other
forms of generation. From the point of view of economic
viability, it is not appropriate that transfer payments, such
as rates, be added in the assessment of generating costs
(unless for comparisons with the Jizewell B evidence if it
includes such payments), and furthermore the payments for
transmission reinforcement would have to be reduced to remove
double discounting. Paragraph 38 shows the effect of removing
the transfer payment element of operating costs, and the
resulting unit cost, at 2.79p/kWh for the Cardiff-Weston
barrage, is lower than any of the values for coal-fired or
nuclear generation produced by the CEGB. Is it to be concluded
that the CEGB is dubious about the relative costs revealed in
this Paper - fearing, perhaps, that capital costs on a barrage
will overrun substantially - or is it that the costs of
alternative plant are expected to come down? How do the
opposing influences of the doubts over nuclear development
(post Chernobyl) and the expectations of lower fuel prices
alter the economics of tidal power?

MR J. G. CARR
The Paper does concentrate on commercial viability. The reason
for such concentration is that a major objective of the study
undertaken by the Severn Tidal Power Group for the Government
was to determine whether the Severn barrage could be financed

as a private sector project. It would indeed be asking much of
a scheme to produce a return that is appropriate to a risky
investment in a market where the price is determined by the
cost of a monopoly supplier who is earning a low risk return.

The question why the Central Electricity Generating Board
(CEGB) itself should not fund the project should be directed to
the CEGB. One of the areas of work that will be undertaken in
the next study phase on the Severn barrage is to examine
jointly with the CEGB in detail the economics of the Severn
barrage compared with other forms of electricity generation.
As well as considering the unit costs, such an examination will
consider the economics of the barrage on the system as a
whole.

MR R. J. MOULE, Central Electricity Generating Board
The Central Electricity Generating Board (CEGB) is not
currently in a position to be able to verify the studies and
conclusions reported recently by the Severn Tidal Power Group
(STPG) and thus not able to make such a decision. The reason
for this is that, although closely associated with the Bondi
studies, the CEGB assisted in the subsequent STPG work only by
request in a restricted number of areas and would wish to
examine more closely with the STPG all the areas covered before
being able to form a considered opinion.

Additionally, the STPG report recommended a substantial
programme of further work before a decision to build can be
made and the CEGB will be participating fully in the initial
part of this programme. (The point has been made previously
that the CEGB, the Department of Energy and the STPG were to be
equal partners in the next study phase.) The question of
whether or not the CEGB is prepared to build the barrage is
hence premature since the work is not yet sufficiently
advanced.

MR W. J. CARLYLE, Binnie & Partners
As the value of energy is the fundamental variable, has the
possibility of giving up a large element of the energy in
return for having the energy available at a more appropriate
time by employing a multibasin scheme been critically
examined?

DR J. G. CARR
The possibility was examined, for a single-basin scheme, of
changing the timing of generation to maximize not energy output
but the value of that energy output, in effect whether, by
producing the energy at a 'more appropriate time' the increased
value of that energy would more than compensate for producing
less energy. As is discussed in the report of the Severn Tidal
Power Group (STPG), the short answer to the question is no.

The benefit gained from value optimization as opposed to energy optimization is very small.

The question is wider than this with regard to the issue of value optimization by employing a multibasin scheme. The report of the Severn Barrage Committee (SBC) under Sir Herman Bondi came down clearly in favour of a single-barrage scheme. The work of the STPG built on the work of the SBC.

MR H. J. MOORHEAD, EPD Consultants Ltd
Engineers involved in the appraisal of power generation developments will be well aware of the effect of varying discount rates on the economic performance of certain projects. At low discount rates the high capital cost/low operating cost hydropower project will be the best option while at high discount rates the low capital cost/high operating cost thermal power-station will be the best option.

In 1978 the Watt Committee on Energy published a report dealing with the rational use of energy and in the introduction to this report there was a section dealing with the choice of discount rate. Two main approaches were recognized which economists have called 'opportunity cost' and 'social time preference'.

The opportunity cost would indicate a discount rate in excess of 10%, a rate of return acceptable to the private sector.

Social time preference cost would indicate a rate thought to be about 3-5%. In this case the criterion may be considered to be how far the community as a whole is prepared to give up present benefits for the good of future generations.

It is appreciated that this is a philosophical and political question: would the Author of Paper 9 like to comment?

MR M. J. GRUBB, Imperial College of Science and Technology
Dr Carr's Paper and the subsequent discussion have emphasized the many problems concerning financial discounting for projects such as the Severn barrage, and I would like to add a few observations stemming from the now extensive economic literature on discounting theory.

In terms of national economic benefits, cash flows over time should ideally be assessed at a rate which reflects the mean social rate of time preference (SRTP) across the population. Estimates of this vary, but are generally around 2-5%. In the idealized free-market economy of Adam Smith and Marshall, the average rate used in the private sector should be equal to this, with a slight increment for entrepreneurial risk, after allowing for any corporation tax. In fact, the private market rate is very much greater than this ideal.

In theory, the Government, taking decisions on behalf of the public interest, should assess projects at the SRTP (it is immune from bankruptcy and so there should be no allowance for risk - general uncertainty should be taken into account in assessing the mean expected benefits). However, the use of a different rate could lead to a misallocation of resources, as higher returns could be gained by investing in the private sector. Most senior economists in the field now agree that the only way to make logically consistent investment decisions is by performing all public sector cash flow analyses at the SRTP, but using a 'shadow cost of capital' for all cash flows. The shadow cost is determined by the ratio of SRTr to private sector rate of return and is generally greater than the real cost - it reflects the opportunity cost of investing at the lower rate.

The net effect is that projects which give a rapid return will generally be more attractive on private sector criteria, but that longer-term projects of overall national benefit, which would not be touched by private capital, may still be seen as economic public investments (ref. 1).

A clear and unambiguous distortion is introduced by attempting to assess a major project on the financial criteria of the private sector, when the benefits flow and are competing against those assessed in the public sector. If it is desired that the project be run in the private sector, the only way to avoid irrational decision criteria is for the Government to introduce capital subsidies to a degree at which the private sector decisions will reflect the public sector discount rate.

The economic refinements discussed earlier do not necessarily solve the problems of very long-term projects, for which major benefits or disbenefits may be discounted to almost zero. A logically consistent treatment is possible, however. This is based on the recognition that discounting within a given generation is entirely different from discounting across different generations. The former reflects judgements on the preferred 'consumption plans' of the current generation. The latter is essentially an ethical judgement of responsibility to future generations. There is no reason why the two should be the same.

Therefore, the SRTP contains two independent factors. ₊rojecting into the future, the effective rate will change as the current generation dies out and is replaced by future generations: the discounting of benefits should be weighted accordingly. This treatment does not introduce logical distortions and has the great advantage that judgements on intergenerational weighting can be separated from all the complications of current time preference and the opportunity cost of capital in a consistent manner (refs 2 and 3).

Because of its very long lifetime, the Severn barrage is exceptionally sensitive to the treatment of cash flows over time. It may be concluded, rather unfortunately, that if an analysis of national economic benefits of the project is carried out in a logically rigorous manner an entirely different result might be obtained. It should be emphasized that this issue may dwarf all the uncertainties surrounding the magnitude of capital costs and benefits.

DR J. G. CARR
I do not fundamentally disagree with any of these comments.
I would only re-emphasize the comment in response to Mr
Taylor that the project was assessed on the financial
criteria of the private sector because this was what was
requested by the Government. In the full report to the
Government, the point was made that it was inappropriate to
assess the Severn barrage on the financial criteria of the
private sector when it was in essence being judged against
public sector investment criteria.

PAPER 10

MRS M. P. KENDRICK, Hydraulics Research
The questions that present themselves at this stage in the
examination of a project like the Mersey barrage are legion,
but it is to be hoped that the proposed feasibility studies
will be addressing the most important problems and that, if the
combined findings do not rule against a barrage, then the
remainder will be tackled in the next study stage. Therefore I
shall expand slightly on those aspects of the proposed studies
that Hydraulics Research now know it will be involved in.
Firstly, however, there is one minor point which arises in
paragraph 16 of the Paper. The computational tidal models of
water circulation in Liverpool Bay and the inner estuary of the
Mersey were developed at Hydraulics Research, Wallingford, and
not at the Institute of Oceanographic Sciences, Bidston, and it
is these models that will be used in conjunction with the
existing physical model of the inner estuary to examine the
effect of a barrage on the distribution of flow and sediment in
Liverpool Bay and the Mersey Estuary.
    The two-dimensional in plan, two-layer mathematical models of
Liverpool Bay comprise a two-layer hydrodynamic model which
simulates tidal levels and discharges between each layer and a
complementary two-layer sediment transport model (driven by the
stored results of the hydrodynamic model) which simulates
suspended sediment concentration, deposition and erosion. The
model boundaries at the seaward limit pass through the Isle of
Man and the model uses a patched grid system which increases in
two stages from 300 m in the Mersey Estuary to 2.7 km in the
outer bay (Fig. 1).
    The physical model of the Mersey Estuary (Fig. 2) reproduces
50 km of tidal river extending from Seaforth, at the entrance
to the Narrows, to Howley Weir, the tidal limit at Warrington,
and 34 km of the Manchester Ship Canal from its entrance at
Eastham Locks to its tidal limit at Latchford Locks.
Constructed to scales of 1/500 (horizontal) and 1/80
(vertical), the model is equipped to reproduce the 27 tides of
the spring-neap cycle and the natural salinity distribution
prevailing in the estuary for a variety of freshwater
discharges down the River Mersey. A microcomputer programmes
the model control equipment and a minicomputer is used for data

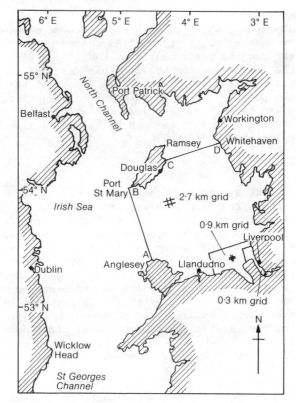

Fig. 1. Limits of model grid zones

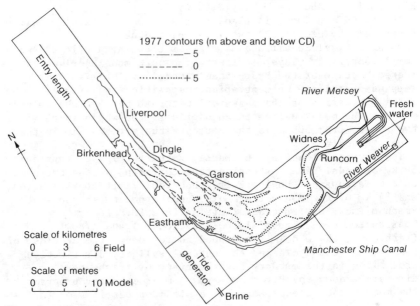

Fig. 2. Physical model layout

Fig. 3. Mersey model

retrieval and the monitoring of model instrumentation.

Measurements of tidal level, current velocity and flow distribution made with the model reproducing existing conditions will be followed by similar measurements undertaken first with a barrage on line 1 and then with a barrage on line 3.

The combined results of the mathematical and physical model studies will establish the relative changes in flow and sediment distribution of the two barrage locations. The mathematical model will give information on likely conditions in the sea channels and banks of Liverpool Bay and the entrance to the Narrows. The physical model will provide similar data for the remainder of the Narrows, the Eastham and Garston channels and the intertidal banks of the inner estuary, in addition to indicating the changes in salinity distribution to be expected with the preferred barrage. A photograph of the model viewed from the entrance to the Narrows is shown in Fig.3.

MR D. KERR, Sir Robert McAlpine and Sons Ltd
The use of supertankers is an excellent piece of lateral thinking. Have

- (a) sea bed preparation to create an even bed and to prevent scouring beneath the tanker
- (b) forces on the tanker during placement in tidal currents

been considered?

Diaphragm walling is a well-proven technique, but to construct within the cells it will be necessary to de-water. How will this be achieved with sandstone beneath?

Sediment movement during construction is a key issue: will this be a part of future studies?

The members of the Severn Tidal Power Group were puzzled by the comparative costs given to the diaphragm walling compared with caisson construction. Could values for total concrete volumes and pounds per cubic metre of concrete be given?

MR E. T. HAWS, MR G. R. CARR and MR B. I. JONES
Bed preparation to receive the very large crude carriers (VLCCs) will consist of dredging, replacement with gravel and grading to profile. Scour will be a potential problem at the bows and stern of the VLCCs and protection by skirts or other means is contemplated.

Forces will be minimized by positioning at slack tide. As many moorings as necessary will be used to hold position during sinking which will be effected by rapid flooding on a falling tide.

The permeability of the sandstone is low, but water inflow through fissures will be reduced to acceptable limits by means of a grouting programme. Drainage wells and tremie plugs will be used in any particularly difficult cases.

Sediment movement during and after construction is of crucial importance and will be the subject of a major study.

The total amount of concrete in the barrage is about 570 000 m^3. The cost of concrete in the diaphragm has been taken as £160/m^3 and of reinforced concrete as £260/m^3 (net of contingencies).

MR K. S. GUINEY, BSC General Steels
In Paper 15 a cost comparison between steel and concrete caisson construction shows significant cost savings in favour of steel caisson construction with 'most of the saving achieved through the reduction in fabrication costs'.

Can the Authors of Paper 10 explain the special features of the Mersey barrage which led them to conclude that steel caisson construction would be more expensive than concrete caisson construction?

What would be the effect on costs of diaphragm walling construction if by the time that construction begins the world shipping market has turned round and very large crude carriers (VLCCs) are not available at scrap prices? What alternatives to VLCCs would be used for diaphragm walling?

MR E. T. HAWS, MR G. R. CARR and MR B. I. JONES
The Severn Tidal Power Group's report also believes it unlikely that steel caissons would offer construction cost savings and

points to the higher maintenance costs for steel caissons.

It is unlikely in the forseeable future that the glut of very large crude carriers will disappear. Even if there were an upturn, it would be feasible to purchase the necessary ships from among those nearing the end of their useful lives.

MR M. HORDYK, Steel Construction Institute
The harbour wall at Westray, Orkney Islands, comprises earth-filled cellular coffer-dams formed in sheet steel piling. A comparison of its cost with that for the embankment for line 1 indicated a potential halving of costs and construction time-scale.

Has this form of construction been considered?

MR E. T HAWS, MR G. R. CARR and MR B. I. JONES
The information about the costs of the harbour wall at Westray is interesting, but there has been a case where clearly a cellular coffer-dam solution was comparatively expensive.

REFERENCES
1. LIND R. C. et al. (eds). Discounting for time and risk in energy policy. Resources for the Future Inc., 1982.
2. COLLARD D. Faustian projects and the social rate of discount. University of Bath papers in political economy no. 179, 1979.
3. GRUBB M. A note on discounting for major energy projects. Cambridge Energy Research Group, Cavendish Laboratory, Cambridge, CERG internal report, June 1985.

11. Changes to the fine sediment regime in the Severn Estuary arising from two current barrage schemes

R. KIRBY, BSc, PhD, MIGeol, *Director, Ravensrodd Consultants Ltd*

SYNOPSIS. Although both barrage schemes envisage ebb generation, and consequently similar overall hydrodynamic changes in the regions affected by the structures, the differing geographical locations would result in marked contrasts in the fine sediment response. For the Cardiff-Weston site concerns over a range of adverse engineering and environmental consequences are largely due to the insensitivity of the 1981 design. An alternative scheme going some way to overcoming these deficiences is presented. In contrast, for the English Stones scheme, fine sediment inputs from both seaward and the rivers seem likely to result in a range of less tractable fine sediment problems.

INTRODUCTION

1. A traditional site for constructing barrages for water supply (Fuljames in 1849) and, more recently, tidal energy, (De Coeur in 1910, Severn Barrage Committee in 1933, Ministry of Fuel and Power in 1945 and Wimpey-Atkins in 1983), is close to the English Stones (Fig. 1). In the last twenty years many other sites have been considered (Severn Barrage Committee in 1981). These "down-estuary" schemes are now almost entirely concentrated in the general area of Cardiff-Weston (Fig. 1). In 1983/5 the Severn Tidal Power Group and Department of Energy funded a broad range of studies to progress an assessment of the feasibility of building tidal power barrages on the Cardiff-Weston and English Stones lines. One area of study was to assess the sedimentological consequences of construction and operation at the two sites.

2. The Severn Estuary has an extremely high fine sediment load, and it was recognised that construction at either site would inevitably lead to major changes in the distribution, stability, and depth of bed sediments, as well as changes to the suspended sediment load and its distribution. These changes were recognised as likely to have very wide implications, embracing aspects of engineering ranging from barrage construction techniques to closure sequences and reservoir lifespan, to navigation, and to coast and subtidal erosion and deposition. The fine sediment is also one of the dominant environmental controls under the present regime. Consequently environmental issues also had to be addressed.

3. Previous investigations at English Stones (ref. 1)
concluded that construction at this site would result in a
basin with an "improved water regime", an implied effectively
infinite lifespan (ref. 1 p 161), and no worse sedimentological
problems outwith the basin than those currently experienced.
In respect of Cardiff-Weston HR Ltd (ref. 2) have modelled the
line 5 and 5S schemes proposed by the Severn Barrage Committee
(ref. 3) during construction and on completion. These studies
showed the local velocity enhancements and dead water areas
which arose during construction, and also the fall in regional
velocity following closure. The method and sequence of construction envisaged in that report gave rise to concern over a
number of sedimentological issues, not least for the large and
readily erodible mud area in Bridgwater Bay immediately outside the barrage.

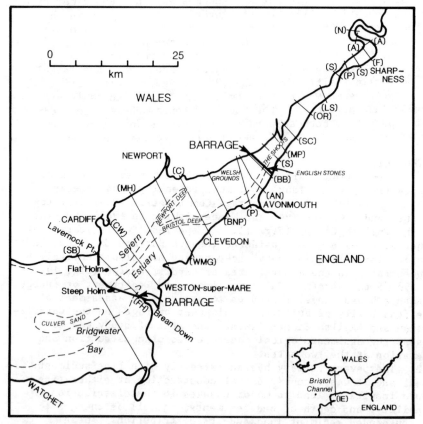

Fig. 1. Proposed sites for tidal power barrages at Cardiff-
Weston and English Stones. Also shown are sections used in
one dimensional modelling for English Stones scheme and
localities mentioned in the text

4. The present study is based on existing literature, unpublished data from the Institute of Oceanographic Sciences,

and mathematical modelling results of Binnie and Partners. The study has been interactive with those responsible for STPG's revised designs.

SEDIMENT DISTRIBUTION AND TRANSPORT IN THE SEVERN ESTUARY

5. Fine sediment may be contributed to the Severn from seaward following subtidal erosion and transport (refs. 4-5), from coast erosion and from the rivers. The relative importance of these sources has not been established. The Bridgwater Bay settled mud area provides both a source and a sink for fine sediment (ref. 6 and Fig. 2). Sediment eroded from this source is transported eastwards into the Severn, chiefly being confined to the English side of the estuary (ref. 7). The suspended solids load of the estuary section between Watchet and The Shoots is close to 12.5 million tonnes (M.t.) on an average spring tide, and 4.3 M.t. on an average neap tide. Owing to the large tidal range the total quantities involved may fluctuate from around 2.5 M.t. on a small neap tide in the summer, to possibly in excess of 30 M.t. on an extreme spring tide in the winter. Mobile fine sediment is eventually exchanged with seabed sinks in Bridgwater Bay and elsewhere. The Bridgwater Bay area is estimated to contain in excess of 600 M.t. of deposited fine sediment, with further quantities in marginal subtidal and intertidal sites in the main estuary, in the tributary estuaries, and in Newport Deep.

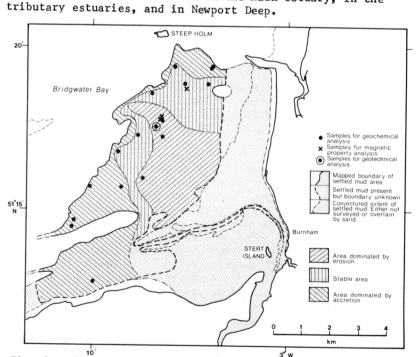

Fig. 2. The Bridgwater Bay settled mud area showing areas of erosion, stability and accretion. Interpretation based on analysis of acoustic records, and on geochemical, geotechnical and geophysical determinations on core samples

CARDIFF-WESTON BARRAGE

6. This proposed barrage would be sited in the region of a line from Brean Down to Lavernock Point. The site truncates the main fine sediment circulation system extending between Bridgwater Bay and the Severn Estuary, and lies close to the erosion-sensitive Bridgwater Bay mud area. As a consequence the orientation, construction sequence and construction method to be employed require very careful design. The site does, however, have the advantage that the size of reservoir created is several orders of magnitude greater than the volume of fine sediment potentially available, and also that it is sited virtually at the downstream limit of the "turbidity maximum" in the region.

7. The sedimentological criteria to be satisfied therefore are:

(a) to ensure that the final design causes the least disturbance to the vulnerable Bridgwater Bay mud area and the adjacent erosional coastline. Minimising erosion in Bridgwater Bay also limits the quantities of fine sediment moved into the basin

(b) to ensure that as far as is possible no slack water areas are created in which rapid sediment deposition can occur.

8. <u>Sedimentological perspectives on closure sequence and construction methods.</u> Preliminary designs assessed for the Severn Barrage Committee (ref. 3), herein called SBC (1981) designs, took only engineering and cost factors into account (Fig. 3). The designs envisaged starting on the Welsh side with the ship locks, progressing to the turbine housings and closing the estuary finally between Brean Down and Steep Holm. Much of this design involved end tipping to create solid embankments. Both the closure sequence and construction method are highly undesirable sedimentologically. During years 4 to 7 of construction, when Bridgwater Bay remained open and velocities in the region would be increased, regional erosion of the subtidal very soft muds would progress at an enhanced rate (ref. 2). In addition, local erosion close to and at the tips of advancing embankments would contribute additional sediment. Juxtaposition of extremely powerful flows and solid embankments would result in the creation of large and sedimentologically undesirable eddies. At the same time the presence of solid embankments would result in pronounced lateral velocity gradients both at the barrage and within the basin, leading to enhanced deposition in low velocity marginal areas. The implication of increased carrying capacity of the currents and creation at the same time of large sediment traps, would be a rapid transfer of sediment from the Bridgwater Bay source to sinks in the Severn Estuary.

9. To minimise the entrainment of Bridgwater Bay muds the section between the English coast and Steep Holm needs to be constructed as rapidly and as early as other engineering considerations will allow. Secondly, to minimise the sedimentological impact the barrage orientation should avoid crossing

or approaching unconsolidated sediment areas. Thirdly, there
are many benefits in building, where practicable, using prefabricated and initially open sluice caissons.

10. The benefit of using open sluice caissons is that a
physical crossing of the estuary is created with minimal interference to tidal flows. After the crossing is complete, but
prior to closure of the openings, the barrage would, along
part of its length, merely constrict the flow locally for a few
hundred metres in way of these openings. After passing the
barrage the flow would expand vertically until it reached the
water surface and estuary bed, and would be largely continuous
across the estuary.

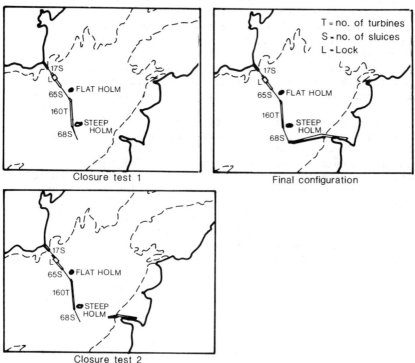

Fig 3. Proposed layout for Cardiff-Weston barrage considered
by Severn Barrage Committee (1981). Figure shows two closure
tests and final configuration of alignment 5

11. Away from the zone immediately influenced by the barrage,
wave effects would be diminished and regional velocities may be
slightly reduced, leading to limited deposition, as opposed to
erosion, in areas such as Bridgwater Bay. Construction largely
using sluice caissons would therefore avoid regional velocity
enhancements, minimise the tip velocities, reduce large scale
eddy formation and prevent the creation of large dead water
areas. One major benefit of using extensive sluice caisson
construction is therefore the minimisation of both high and
low energy extremes.

12. A revised engineering scheme involving an optimised barrage orientation and employing largely prefabricated sluice caissons, devised, in part, to take more account of the sedimentological criteria, is shown in Fig. 4. The design involves 3.5 km of turbines, 3.8 km of prefabricated opening sluices, and 2.7 km of prefabricated, plain, initially-open, caissons. The remaining length now comprises 2.0 km of solid embankment traversing the shallow intertidal area of the English coast, and only 3.8 km of main embankment, part of which is taken up with the ship locks on the Welsh side.

13. The orientation adopted and the construction method proposed provide a great improvement over the SBC (1981) proposal in sedimentological terms. The least satisfactory aspect of the STPG design is the construction sequence suggested, which leaves Bridgwater Bay, and especially the section close to the English coast, open until a very late stage. The first 3.5-4.0 km out from the English coast traverses the readily erodible Bridgwater Bay settled mud and it would be preferable if this section were crossed at an earlier stage.

14. Another important attribute of using a sluice caisson construction is that when a complete crossing of the estuary has been achieved 15% of the cross-sectional area of the estuary would remain open. Fig. 5 shows that the tidal range in the basin, hence regional velocities, would be virtually unchanged at this stage. If the orientation, construction method, and closure sequence are optimised, then one implication is that only a very small quantity of the present mobile fine sediment load would be re-distributed during the construction period.

15. Following completion of the crossing the temporary sluices will be sealed and the controllable sluices and turbines brought into operation. The timescale and sequence of this closure can be modelled with a view to optimising the resulting fine sediment redistribution. Instead of local, sequential deposition throughout the construction period, inevitable with the SBC (1981) design, deposition would be a widespread, major and "once off" event. Such large scale, short term deposition would be almost exactly analogous to an existing neap tide situation. The present variation of tidal energy between springs and neaps is sufficiently large (maximum predicted spring range 14.6 m - minimum predicted neap range 4.3 m at Avonmouth) that on neap tides most of the fine sediment population suspended on spring tides settles to the bed. The quantities involved are so great that extensive, dense, under-consolidated pools of stationary suspended sediment are created in the estuary channels and Bridgwater Bay.

16. The new tidal regime within the basin following closure would have a "spring" range equivalent to an existing neap tide, with the result that the stationary layers deposited at closure will not be re-entrained, but instead would continue to settle and create a normally consolidated settled mud bed.

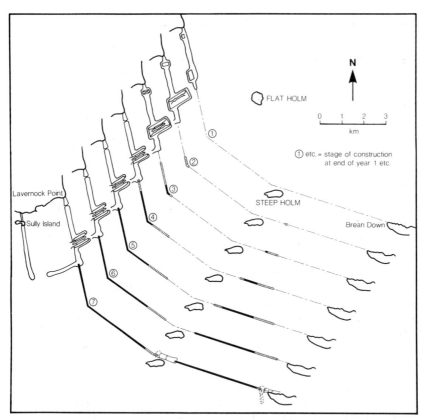

Fig. 4. Proposed layout for Cardiff-Weston barrage considered
by Severn Tidal Power Group (TWC/7424/2). Design envisaged
extensive use of prefabricated caissons. Seven yearly stages
in a construction sequence are portrayed. Alternative landfalls, timing and sequence of construction in Bridgwater Bay
are discussed in the text

17. Subtidal and coastal stability in Bridgwater Bay. In
the longer term, following closure, it was predicted by SBC
that the tidal range outside the barrage would be slightly
reduced and the tidal current regime greatly reduced (ref. 2).
This would reduce the erosion potential and increase the
potential for deposition in Bridgwater Bay. Continued supply
of fine material from cliff retreat and subtidal erosion to
seaward may, depending on the revised water circulation and
rate of fine sediment supply, result in slow but progressive
deposition in Bridgwater Bay. Existing erosional, subtidal and
intertidal areas may be stabilised or even become accretionary.

18. Navigational considerations. At the present time the
major channels of the estuary tend to be self-scouring with
respect to fine sediment. The net up-estuary residual circulation leads to rapid fine sediment deposition in the tri-

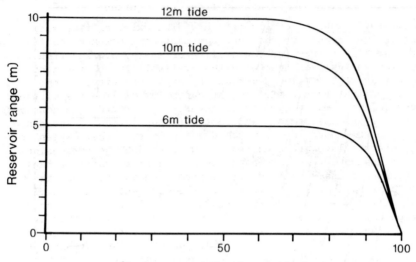

Fig. 5. Decrease in tidal range inside Cardiff-Weston barrage during construction. Using combination of temporary and openable sluice caisson construction the crossing is completed when the estuary is 80-85% blocked. 100% is achieved with all sluices and turbines closed. Data supplied by T L Shaw

butary estuaries and sediment traps formed by the lock entrances. Avonmouth, Newport and Cardiff all have large annual fine sediment maintenance dredging requirements. Large scale, short term deposition of fine sediment at neap tides in the Severn Estuary Channels hint at the sites most favourable for increased deposition if the tidal regime were moderated by a barrage.

19. The SBC (1981) scheme involved potentially serious enhancements in fine sediment deposition locally in docks, locks and approach channels. It was difficult to predict the increase in maintenance dredging need, but this could have been high. Similarly, ascertaining appropriate spoil disposal sites presented problems.

20. Further refinement of the STPG scheme should result in little change to the total fine sediment load, or its distribution, during construction. Consequently during this phase there is no reason to suppose that the dredging need would differ significantly from that at present. Similarly, existing spoil disposal practices could continue. At the short period of closure between 8 and 30 M.t. of fine sediment are likely to be permanently deposited. A certain proportion of the deposition would occur in lock entrances and the major channels of the Severn and its tributaries, as on an existing neap tide. The total deposition following consolidation would be much less than the rise in mean water level inside the barrage. Any temporary additional dredging need could be coped with by contract dredgers if required. Subsequent to closure the annual dredging need at the major ports, Cardiff, Newport and Avonmouth, is

likely to decline by perhaps an order of magnitude.

21. <u>Environmental considerations</u>. The existing fine sediment regime is one, if not the, dominant environmental control. Both the high energy and the cycling of energy are important. The high energy, combined with the availability of fine sediment, leads to prolonged periods of high turbidity, which limit sunlight penetration to the upper few millimetres of the water column. Water quality is adversely affected. In contrast, on neap tides the organic and bacterial population of the dense near-bed suspended layers which develop is such that oxygen in the interstitial waters is rapidly used up and the suspensions become anaerobic, and hence barren. Expansive tracts of the underlying estuary bed are also barren beneath and along the transport path of such layers. On the next entrainment cycle, especially in the tributary estuaries, the layers create an oxygen sag, which may persist for up to 24 hours, as they are remixed into the water column. Intertidal substrates are, in places, hostile to benthic organisms owing to the exposure of overconsolidated substrates, the presence of dense anaerobic suspended layers on neap tides, or both.

22. Consequent upon this all-pervading environmental stress factor, intertidal and subtidal substrates have a fauna generally impoverished, in both species diversity and abundance, whilst planktonic micro- and macro-organisms also tend to be depleted.

23. During the construction phase, provided that the scheme in Fig. 4 can be refined further, the environmental consequences could be minimal. At the time of closure, depending on the final closure sequence adopted, dense stationary layers may form. It is likely that in some localities, such as Bristol Deep and outside the barrage, these would overlie sand substrates. It is unlikely that this deposition will lead to significant mortality of the indigenous fauna.

24. After closure the water column would experience suspended solids concentrations of merely tens, to at maximum a few hundred milligrams per litre ($mg\ l^{-1}$); and would become the site for a wide range of planktonic micro- and macroscopic forms, which, in this area, presently die whilst still in their larval form. The substrate would become stabilised and normally consolidated, the proportion of the bed being occupied by mud increasing at the expense of sand. A more typical profuse and varied estuarine flora and fauna would develop. Although the overall area of tidal flats will decline, those remaining, and any new areas which develop, would be more amenable to colonisation by burrowing invertebrates. The implication is that the dependent bird population would not inevitably decline. There is the prospect of salt marsh encroachment on the higher levels of the intertidal areas, but on balance the environmental benefits would considerably outweigh the disbenefits (ref. 8).

ENGLISH STONES BARRAGE

25. The level of understanding of sediment movements close to this site, and the details of the design of the scheme itself, are less well advanced than at Cardiff-Weston. The

Shoots Channel, which truncates the English Stones, is the site
of the highest tidal range and strongest tidal currents in the
Severn Estuary. As a result of efficient mixing the concentration of fine sediment in suspension, particularly in the
surface layers, is the highest in the estuary.

26. Generalised forecasts of the changed hydrodynamic regime
can be made on the basis of a one dimensional mathematical
model study carried out by Binnie and Partners (ref. 9) (Fig 1)
(Table 1). The general pattern of modification to tidal ranges
and velocities is typical of any ebb generating barrage. The
forecast slight decrease in tidal range to seaward, combined
with the reduced exchange of water across the barrage would,
to varying degrees, reduce current velocities seaward of the
barrage. West of Clevedon the modifications to the regime
would be small, and beyond Cardiff-Weston insignificant, in
sedimentological terms.

27. In the region between Clevedon and the barrage the
decrease in flood tide velocities would generally exceed ebb
tide decreases. Nevertheless the dominance of the flood tide
over the ebb is generally maintained. This is especially
noticeable at the barrage itself, where the maximum cross-
sectionally averaged velocity falls to 0.96 m s^{-1} on spring
floods, but to 0.36 m s^{-1} on spring ebbs. Although there are
local exceptions, it is likely that the existing flood residual
transport of fine sediment will continue.

28. Inside the basin mean water level would rise and the
duration of slack water would increase. The maximum spring
and neap ebb velocities would experience a major decline,
especially close to the barrage, whilst flood velocities would
be minimally altered. Flood tide maxima are presently stronger
than ebb throughout this section of the estuary, and closure
would exaggerate this inequality. Immediately above the barrage
site the present flood/ebb inequality is 27% on springs, and
would rise to 166% following closure.

29. On completion the hydrodynamic conditions in Bridgwater
Bay would be indistinguishable from the existing regime, both
in respect of tidal currents and waves. Thus the suspended
solids regime in the estuary seaward of Clevedon would experience minimal changes. Similarly the wave climate operating
on the broad intertidal mud areas would be unchanged. As a
result large quantities of suspended fine sediment would continue to be supplied to the region in front of the barrage.
Sand and also fine sediment is likely to be deposited in
sheltered areas between Clevedon and the barrage. Such areas
would include wide zones on The Welsh Grounds, the corners of
the barrage and the Avonmouth lock entrances. In contrast,
little fine sediment would be supplied to Bridgwater Bay and
other mud areas from up-estuary.

30. Much fine sediment would, in part, pass the barrage via
the turbines and sluices on the flood tide. A pronounced flood
dominance within the basin would transport fine sediment landwards (Table 1). This factor would be further accentuated by
another, not so far taken into account in the modelling, namely

TABLE 1   "ENGLISH STONES" BARRAGE DATA
from Binnie & Partners' Model

	Spring Tides						Neap Tides					
	Max ebb[+] velocities (m s^{-1})		Diff	Max flood[+] velocities (m s^{-1})		Diff	Max ebb velocities (m s^{-1})		Diff	Max flood velocities (m s^{-1})		Diff
	B	A	A-B	B	A	A-B	B	A	A-B	B	A	A-B
Ilfracombe East (IE)	1.33	1.30	-.03	1.30	1.29	-.01	0.67	0.67	0	0.71	0.70	-.01
Sully Bay (SB)	1.05	0.95	-.10	1.14	1.02	-.12	0.59	0.51	-.08	0.66	0.59	-.07
Flat Holm (FH)	1.45	1.28	-.17	1.62	1.47	-.15	0.82	0.69	-.13	0.91	0.84	-.07
Cardiff-Weston (CW)	1.39	1.20	-.19	1.56	1.47	-.09	0.79	0.65	-.14	0.88	0.79	-.09
Middle Hope (MH)	1.16	0.98	-.18	1.32	1.18	-.14	0.67	0.52	-.15	0.75	0.67	-.08
West Middle Ground (WMG)	1.18	0.96	-.22	1.35	1.10	-.25	0.69	0.60	-.09	0.77	0.66	-.11
Clevedon (C)	1.22	0.95	-.27	1.37	0.93	-.44	0.74	0.64	-.10	0.81	0.63	-.18
Black Nore Point (BNP)	1.68	1.14	-.53	2.03	1.31	-.72	1.04	0.84	-.20	1.12	0.91	-.21
Portbury (P)	1.43	1.29	-.14	1.60	0.85	-.75	0.82	0.77	-.05	0.90	0.68	-.22
Avonmouth N (AN)	1.38	1.00	-.38	1.77	1.13	-.64	0.89	0.73	-.16	0.98	0.80	-.18
Black Bedwins (BB)	1.15	0.36	-.79	1.46	0.96	-.50	0.76	0.50	-.26	0.82	0.66	-.17
BARRAGE												
Sudbrook (S)	1.68	0.52	-1.16	2.12	1.40	-.72	1.08	0.73	-.35	1.20	0.94	-.26
Mathern Pill (MP)	1.40	0.41	-.99	1.61	1.20	-.41	0.79	0.49	-.30	0.84	0.75	-.09
Sedbury Cliffs (SC)	1.15	0.35	-.80	1.34	1.00	-.34	0.68	0.44	-.24	0.71	0.63	-.08
Oldbury Reservoir (OR)	1.17	0.41	-.76	1.47	1.03	-.44	0.78	0.52	-.26	0.79	0.71	-.08
Lydney Sands (LS)	1.31	0.67	-.64	2.25	1.30	-.95	1.18	0.76	-.42	1.30	1.01	-.29
Sharpness (S)	1.82	0.79	-1.03	2.49	1.80	-.69	0.86	0.81	-.05	1.03	0.92	-.11
Purton (P)	1.70	0.79	-.91	2.70	1.81	-.89	1.12	0.96	-.16	1.18	1.07	-.11
Slimbridge (S)	0.93	0.46	-.47	1.45	1.20	-.25	0.59	0.60	+.01	0.66	0.60	-.60
Frampton (F)	1.56	0.85	-.71	2.50	1.75	-.75	1.27	1.53	+.26	1.03	1.02	-.01
Awre (A)	1.51	0.90	-.61	2.85	1.80	-1.05	0.94	0.99	+.05	1.22	1.32	+.10
Arlingham (A)	1.29	0.80	-.49	1.50	1.45	-.05	0.85	1.01	+.16	0.97	1.36	+.39
Newnham (N)	2.30	1.52	-.78	3.10	2.60	-.50	3.23	2.81	-.42	0.80	0.52	-.28

*B = Before Barrage   *A = After Barrage   + = Averaged over estuary cross-section

the increased salinity stratification, leading to increased
residual up-basin sediment transport at the bed. The slack
water periods would also be prolonged, providing the opportunity
for much of the suspended material to settle. As a consequence
of increased consolidation deposited material would be less
readily entrained on the succeeding ebb tide. Sediment retention
is likely to be especially great on the spring down to neap
energy cascade, when each succeeding tide is of slightly lower
amplitude and energy than the preceeding one. The various
factors influencing sediment introduced through the barrage
would largely apply to riverborne sediment.

31. The inputs to the basin and the retention factor can be
predicted only in general terms. The input from the rivers
Severn and Wye lies in the region 0.5-0.75 M.t. yr^{-1}. The quan-

tities entering the basin from the seaward direction are more difficult to quantify. Using rudimentary data the best that can be achieved at present is to scale the problem as shown below:

Average tidal range in basin = 4.0 m
Volume of water passing barrage = 320 $M.m^3 tide^{-1}$
Average suspended load in lower part of water column = 1.0 g l^{-1}
Average suspended load in upper part of water column = 0.05 g l^{-1}
(Comparison of present and future tide regime suggests, from existing sediment data, post-construction solids content not less than these)
Percentage of flood flows passing turbines = 20%
Percentage of flood flows passing sluices = 80%
Sediment input through turbines based on 1.0 g l^{-1} & 20% of flow
= 128,000 t d^{-1}
Sediment input through sluices based on 0.05 g l^{-1} & 80% of flow
= 25,600 t d^{-1}
Total sediment input = 153,600 t d^{-1} or = 56,064,000 t yr^{-1}

32. Estimating the retention rate and the long term supply rate for the basin also presents severe problems. In terms of the retention rate, similar types of barrage elsewhere in the world have a dissimilar sedimentary regime, whilst reservoirs with a high fine sediment input are hydrodynamically and operationally dissimilar and hence not comparable with tidal power barrages. The various factors favouring retention are itemised above and indicate a likely high retention rate. For this type of barrage, with a high exchange rate for water, a high rate may be 20% or more of the fine sediment passing the barrage at the seaward end of the basin. For the purposes of this exercise a factor of 20% has been used for retention of sediment passing the barrage whilst 100% retention of riverine supply has been assumed as follows:

	tonnes yr^{-1}
via turbines	9,344,000
via sluices	1,868,800
river	750,000
	11,962,800
volume at 1.5 t m^3	8,000,000 $m^3 yr^{-1}$

This prediction does not seem unreasonably high when compared to the present spring tide suspended load and to the mud maintenance dredging need at Cardiff, Newport and Avonmouth, which is close to 8 M.t. yr^{-1} (5.3 $M.m^3 yr^{-1}$).

33. The rate of filling of the basin also depends on the sediment supply rate through time. In the early life of the basin the supply of sediment would be unconstrained by availability, because the losses through retention would be less than the readily available existing fine sediment load. The question to be addressed is whether, and at what rate, the availability of fine sediment would decline as the most readily transported material was deposited inside and in front of the basin.

34. There is no doubt that resupply of fine sediment from the Bristol Channel to the Severn Estuary, and wave erosion of tidal flat sediment would continue at its existing, though

unquantified, rate. Density and shear strength profiles of
Bridgwater Bay sediment (IOS unpublished) indicate a normally
consolidated material in which the erodibility probably increases only slowly with depth. The implication may therefore·
be of a relatively slow decline in availability of fine sediment at the barrage through time.

35. The basin capacity is 330 $M.m^3$ at OD and 650 $M.m^3$ at high
water springs. Assuming constant availability the lifespan of
the basin is as follows:-

Filled to HW assuming supply rate estimated above   81 yr

Filled to OD assuming supply rate estimated above   41 yr

Thus, although no firm predictions on the lifespan of the
barrage are currently possible, if the interpretation proposed
above is broadly correct the sedimentation rate in the basin
is an important factor to be taken into account in assessing
the economics of the scheme.

36. In addition to the lifespan of the scheme a number of
other fundamental issues arise. These are the navigability
of the basin up to Sharpness, problems of dredging and dredge
spoil disposal both in front of and within the basin, land
drainage,and rises in bed level in respect of flood water discharges down the river Severn. These issues are not dealt with
here.

CONCLUSIONS

37. Both schemes would have a major effect on bed sediment
distributions, sediment circulations and suspended sediment
concentrations within the estuary. The mobile, and potentially
mobile, populations of fine sediment are sufficiently high that
the fine sediment impinges upon almost every engineering and
environmental aspect of the scheme. The Cardiff-Weston barrage
would be located close to the seaward limit of the turbidity
maximum, construction would result in a major decrease in
suspended solids levels and the barrage would have an infinite
lifespan compared to the quantities of erodible fine sediment
in the potential sources.

38. A construction sequence involving crossing Bridgwater
Bay first would protect the large, vulnerable, fine sediment
area and the adjacent coastline. Use of temporary sluice
caissons in construction would also decrease fine sediment
erosion and redistribution. Such techniques would result in a
negligible increase in total erosion and the retention in suspension of the greater proportion of the mobile fine sediment
population until the last days or weeks of construction. Deposition would then resemble an existing neap tide and closure
techniques should seek to spread the sediment widely in down-
estuary locations, which are neither navigationally nor environmentally significant. Following closure the Bridgwater Bay area
would be dominated by deposition. Such a fine sediment response,
if further work confirms it can be achieved, would counter many
earlier and justifiable concerns about the scheme. Thus, the
engineering-related problems currently look more tractable than

hitherto, whilst the environmental consequences would predominantly lead to a more amenable water and seabed regime.

39. The English Stones scheme is located towards the head of the turbidity maximum and the volume of the basin created would be relatively small compared to the mobile and potentially mobile fine sediment load. Seaward of Clevedon the changes to the hydrodynamic regime would be minimal and fine sediment eroded from intertidal localities and Bridgwater Bay would continue to reach the barrage and the basin beyond following closure. The effect of the placing of the barrage will be to provide a nodal point for fine sediment deposition to either side of it. Further work is necessary to improve the accuracy of predictions of the rate of infilling, but present evidence suggests that a range of difficult problems may arise within a relatively short timescale. The conclusion contrasts with previous predictions for a barrage at this site made by Gibson (ref. 1).

ACKNOWLEDGEMENTS

40. The author wishes to thank the Department of Energy, the Severn Tidal Power Group, Institute of Oceanographic Sciences, and Binnie and Partners for permission to quote from unpublished work and internal reports prepared in connection with Severn Barrage schemes.

REFERENCES

1. GIBSON A.H. Construction and operation of a tidal model of the Severn Estuary. Severn Barrage Committee Report. Economic Advisory Council. HMSO London 1933 Vol I 1-173 Vol II 173-304.
2. Hydraulics Research Station. Severn Tidal Power. Two-dimensional water movement model study. Report No. EX985, 1981, STP41, 1-31pp.
3. Severn Barrage Committee. Tidal power from the Severn Estuary. 2 volumes. 1981, London HMSO.
4. MURRAY J.W. and HAWKINS A.B. Sediment transport in the Severn Estuary during the past 8,000-9,000 years. Journal of the Geological Society, Vol 132, 1976, 385-398.
5. MURRAY L.A., NORTON M.G., NUNNY R.S. and ROLFE M.S. The field assessment of effects of dumping waste at sea : 7. Sewage sludge and industrial waste disposal in the Bristol Channel. MAFF. Fisheries Research Technical Report No. 59, 1980. 40pp.
6. KIRBY R. and PARKER W.R. Settled mud deposits in Bridgwater Bay, Bristol Channel. Institute of Oceanographic Sciences Report No. 107, 1980, 1-65.
7. KIRBY R. and PARKER W.R. Distribution and behaviour of fine sediment in the Severn Estuary and Inner Bristol Channel, UK. Canadian Journal of Fisheries and Aquatic Sciences, Vol. 40, No. 1, 1983, 83-95.
8. SHAW T.L. Environmental aspects of tidal power barrages in the Severn Estuary, Paper 12 ICE/IEE symposium on Tidal Power 1986. Thomas Telford, London, 1987.
9. BINNIE AND PARTNERS. Report on confidential studies of energy potential of a barrage at English Stones, 1983.

12. Environmental aspects of tidal power barrages in the Severn Estuary

T. L. SHAW, BSc(Eng), PhD, *Sir Robert McAlpine & Sons Ltd*

SUMMARY. A general review is given of the information used by the Severn Tidal Power Group to assess the possible environmental effects of tidal power barrages at the proposed Cardiff-Weston and English Stones sites in the Severn Estuary.

Working from the assumption that physical aspects of an estuarine regime are the major determinants of its character, the effect of each barrage on the tide regime and hence on the currents, turbidity and salinity is reviewed. The differences between the two schemes, especially with respect to turbidity and hence to patterns of sedimentation are explained.

Aspects of water quality are reviewed and found to be both important and capable of proper recognition within the supplementary arrangements made to implement either scheme.

The ecological changes are, by their complexity, less amenable to solution. The changed inter- and sub-tidal substrates, and their associated turbidity and salinity regimes would significantly affect the productivity of the whole estuary, from the primary fauna and flora through to birds and fish. An indication of the state of knowledge of some of the more important issues controlling the relevant food chains is given. The likely change in the productivity of the éstuary is related to relevant aspects of the design and operation of either scheme.

Reference is made to the relevance of the conclusions to other proposed barrage schemes.

INTRODUCTION

1. Tidal power projects have much in common with hydro-electric projects though there are also some substantial differences between them. For example, whereas schemes of both types are more likely to be amongst the most economic of their type when operating under the highest heads, this means many hundreds of metres for hydro but not more than about ten for tides. Also, whereas hydro schemes use fresh water and uni- directional flows, tidal schemes depend on saline water and periodically reversing flows.

2. This Paper concerns the environmental impact of tidal

power barrages, with particular reference to schemes in the
Severn Estuary. In just the same way as two hydro schemes
are unlikely to have identical environmental effects and
hence must be assessed on an individual basis, so it is
necessary to consider each tidal barrage on its own merits.
In the latter case, however, the fact that similar physical
changes in tide regime are caused by each barrage scheme
means that the generally accepted method of operation can be
expected to have different consequences in each case.

3. The environment of each estuary is unique. It is
determined by a variety of factors, principal among which are
the tide range, currents, sediments and salinity, though this
is not to imply that all chemical and biological issues are
determined solely by physical parameters and hence are of
only secondary influence. The Rance Barrage in Brittany
shows how a new regime may develop in the basin formed by
such a scheme, and the smaller and more recent project at
Annapolis on the Bay of Fundy is also being monitored from
this point of view (Section 5). Whereas the results from
both are instructive as to the changes which can occur, there
is no case for anticipating that either is likely to be more
than broadly indicative of the consequences of a barrage in
the Severn or any other estuary.

4. In fact, the Rance Barrage could give a particularly
misleading impression of the environmental effects of this
type of project. For any scheme these effects may be grouped
into three phases, namely those which occur during
construction, those which characterise the basin in the
longer term when the regime has essentially stabilised
(according to the way in which the scheme is operated), and
those which constitute the transitional phase between the end
of construction and the time when the new regime is
established.

5. It might be thought that the ultimate regime would not
depend on those which precede it, the latter only determining
the time before the 'stable' state is reached. Whereas this
applies to tides, currents and salinity, it does not
necessarily include sediments. The construction sequence
would have a greater or lesser effect on mobile material
located close to the line of the proposed works depending on
the chosen design and method of project implementation. In
this respect, the use of floated-in caissons for construction
is now generally regarded as preferable to forming a barrage
in-situ (Refs 1,2,3). This significantly affects the likely
impact of the project, not only in the short term but also
through the subsequent transitional phase and perhaps
thereafter depending on the nature of the changes which
occurred at the outset.

6. The total effect of any scheme must therefore be
assessed as a whole and on a largely individual basis.
Whereas this does not negate the benefit of looking both to
other similar schemes and other estuaries which naturally
have a similar tide range to that which a barrage would

create in its basin, the information gained cannot be expected to do more than generally guide an assessment of the consequences of each proposed scheme. However, the steps which have to be followed are not scheme-dependent. They are set out in this Paper in the context of how they were followed by the Severn Tidal Power Group during their studies of the two barrage schemes shown in Fig. 1 (Ref 4).

PHYSICAL PARAMETERS

7. Fig. 2 shows the present normal variation in tide range in the Rance basin on spring and neap tides. These are established by generating power on each ebb tide phase and pumping around high tide on neaps to raise the basin level higher than would be possible by gravity alone. The merits in doing this are explained in Ref 5. If pumping is not done and generation is confined to the ebb tide phase, the levels in the basin would remain as shown in Fig. 2 except that the neap high water level would not be above that of the tide. In either case, both the minimum and mean water levels would be increased significantly whereas high water levels may be marginally reduced.

8. The actual extremes in level reached in the basin vary along its length according to surge and shallow water effects. Fig. 3 shows the loci of high and low water levels on spring and neap tides through the Severn Estuary for the two schemes shown in Fig. 1, operated without pumping. The regimes seaward as well as landward of each barrage are included. (Ref 4 explains how these results were produced.) The reversal in the positions of low waters on springs and neaps in the basin is significant. It indicates that the neap and spring ranges in the basin are more nearly the same than naturally occur, a similarity which is furthered if pumping is done on neaps (Fig. 2).

9. Fig. 3 shows that the tidal changes are not limited to the basin but also occur seaward of each scheme, though not in the same way for each. The difference between them is due to the shallow channel bed level seaward of the smaller, English Stones proposal. This constraint would principally be felt at low water. It is not indicated in Fig. 2 for the larger, Cardiff-Weston scheme, where the forecast reduction in seaward range is effectively symmetric about the same mid tide position.

10. In the upper parts of the basins where water levels are determined by bed levels, the effects of either barrage on tide range progressively diminish. Fig. 3 includes data for the upper Severn Estuary which illustrates this point. A similar situation applies in each river estuary landward of either barrage, whereas for estuaries seaward of each, the tide conditions would change by relatively small though not necessarily insignificant amounts.

11. The shapes of the tide curves in each basin would be determined principally by the protracted ebb flow period compared with the little-changed flood tide refilling rate.

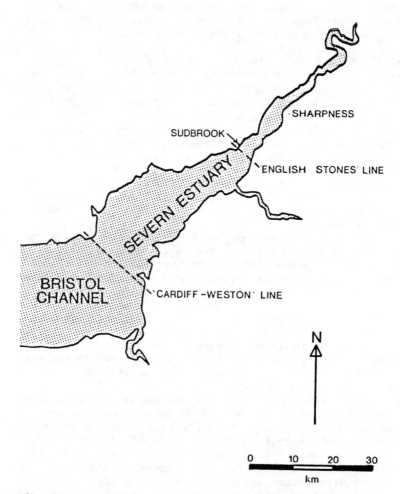

Fig. 1

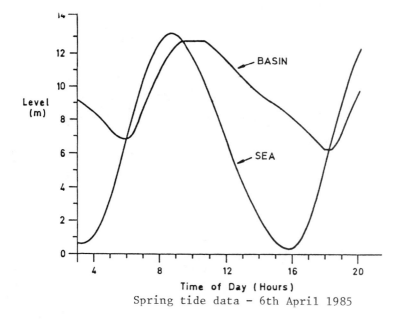

Spring tide data – 6th April 1985

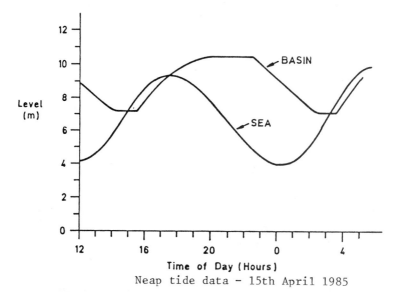

Neap tide data – 15th April 1985

Fig. 2. Examples of operations at the Rance barrier

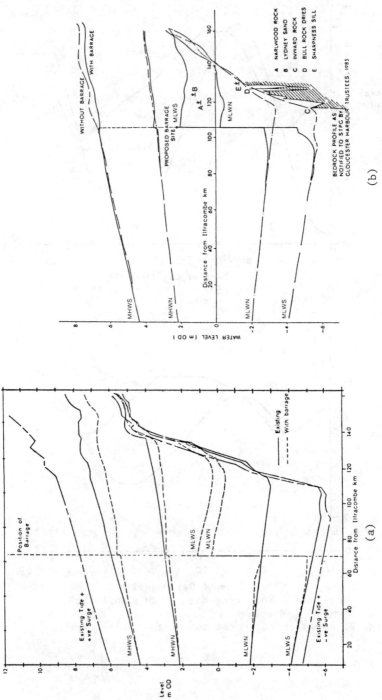

Fig. 3. Extreme water levels for spring and neap tides: (a) Cardiff—Weston barrage; (b) English Stones barrage

Fig. 2 shows this alteration. Rapid refilling is essential once sea level rises above low water level in the basin, in order that maximum volume and head are available when generation starts. This phase therefore takes about 3 hours. It is followed by a lengthened high water period during which sea level falls sufficiently for generation to start under an appropriate difference in head. This period is longer on neaps than on springs and is reduced by pumping (Fig. 2). It is followed by the generating period of up to about 7 hours, this depending principally on tide range. It ends when the head is again insufficient. The optimum flow rate at which the turbines should discharge, i.e., the slope of the ebb tide curve in the basin, is a compromise determined by energy output between releasing less water under a greater average head, and vice versa.

12. The slopes of the flood and ebb tide phases determine the strength of the currents on each phase at any place. Fig. 4 illustrates 4 examples of this aspect of the regime. It shows how barrage operation reduces both peak flood and ebb currents in the basin, the former because depths are then greater and refilling rates slightly reduced, and the latter because depths are greater and emptying rates much less. The net effect is to accentuate strongly the bias for potential up-Estuary sediment drift, i.e., the capacity for the currents to erode and transport material of an appropriate grain size and type. (The subject of sediments is considered later in this Section.)

13. Fig. 4 also shows that, higher in each river estuary where the tide range would be less affected by either barrage, the currents would also be relatively unchanged. However, as with water levels, the fact that currents may be little modified in these reaches does not necessarily mean that the consequences would be correspondingly less. In this case, a small reduction in tide range and current strength at any place could change the salinity by a sufficient amount for this parameter alone to have a major effect on the established ecology.

14. Fig. 5 shows 5 examples of salinity variations on spring and neap tides measured at the locations shown (Ref 6). This information clearly indicates that large variations in salinity per tide cycle occur in the upper reaches of the estuary, where the tide range is a large proportion of high water depth but where much of the flow is tidal rather than riverine. At times of high river flow, however, the mean salinity at these locations decreases greatly and the reach in which conditions vary strongly moves seaward. Correspondingly, during protracted low flow periods, this zones moves landward.

15. Ref 7 analyses mean salinity variations through the estuary as a function of river flow. It clearly shows the strong relationship which exists between them. The same study was extended to post-barrage conditions, for which it predicted reduced mean salinity values at any place, these

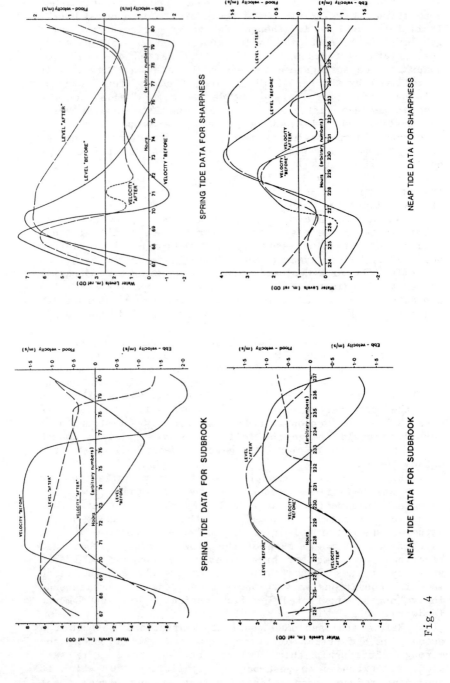

Fig. 4

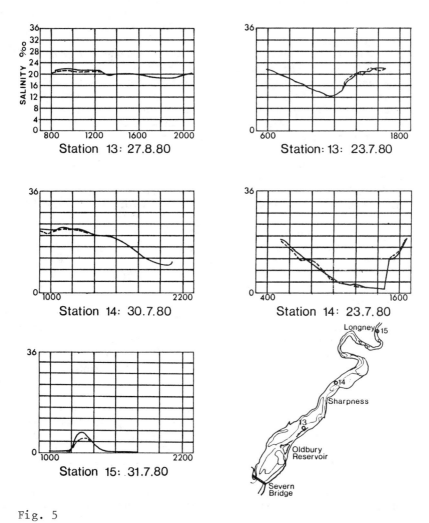

Fig. 5

also depending on river flow. The fact that the variation per tide cycle would be reduced because of the smaller tidal amplitude is also pertinent to determining its ecological implications (Section 4).

16. While it is probably neither possible nor necessary to attempt to assess which change in a physical parameter would have the greater environmental effect, it must nevertheless be true that the sediment regime is a major determinant of the productivity (biological capacity) of a turbid estuary. Ref 4 deals with this important topic in considerable detail. In addition to the changes which could occur in each part of the estuary, including those from further seaward up to the tidal limits, it also necessarily covers the prominent sand and fine sediment (essentially mud) fractions which characterise the Severn Estuary. These have their origins in both the sea and the rivers, with most if not all of the sand moving slowly up-Estuary and much of the mud, though seemingly of unknown proportion relative to the whole input, coming from the rivers.

17. Compared with the sand, the fine sediment fraction is highly mobile. Detailed measurements of its distribution over the Severn Estuary through a full range of tide cycles were made by the Institute of Oceanographic Sciences (Ref 8). From continuously recorded vertical depth profiles of turbidity, estimates of suspended loads have been made. These correlate closely with tide range, i.e., with the strength of the currents responsible for erosion and transport. Between the lowest neap and highest spring tides (a factor of about 1.8 in peak currents), the peak suspended loads differ by a factor of about 10. Fig. 6 shows 4 typical turbidity distributions throughout the estuary.

18. The close correlation found to exist between the strength of the currents (i.e., the tide range) and the amount of sediment carried strongly suggests that each flow phase persists for long enough to erode sufficient material for that current to approach and perhaps reach its carrying capacity. Furthermore, although existing peak flood flows on any tide cycle are generally stronger than those on the ebb of that cycle, the latter have sufficient strength to ensure that material carried landward on the previous flood tide is taken seaward on the subsequent ebb. The main exception to this is probably that which at high water finds its way into sheltered coastal and deep-water locations from which it is not readily remobilised during the ebb, but which may re-enter general circulation as a result of either storm wave action or heavy rainfall. (The latter force is able to cause locally high runoff in creeks and in other sheltered coastal features.)

19. A maximum of over 30 Mt (million tonnes) of fine sediment is suspended throughout the Severn Estuary during a high spring tide. Much of it starts its up-Estuary (flood) phase from deposits in Bridgwater Bay. On the neap-to-spring tide sequence, some material not previously in suspension is

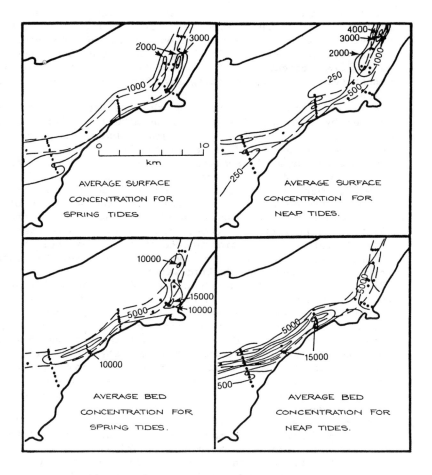

Fig. 6. Distribution of average surface and bed, spring and neap suspended solids concentrations (in mg l^{-1}) in the region between The Shoots and Clevedon (data available for sample stations only)

known to be added from this deposit one each tide to that which returned to this part of the estuary during the previous ebb phase. Significantly, very little material is taken further seaward, probably because ebb currents have by then decreased to such an extent as to prevent this. The field studies reported in Ref 8 therefore show that Bridgwater Bay includes some areas of net erosion and others of net accretion. It is not the only source of fine sediment in the estuary though it is thought to be the most important as far as concentrations carried on each tide cycle are concerned, as well as being by far the largest single deposit.

20.   Ref 9 suggests that even the largest sediment load carried by a spring tide may not be more than about 10% of the total transported at some stage during a period of perhaps a year or so. This period is short compared with the time needed for muddy sediments to de-water and consolidate to the point where they become significantly more difficult to erode when exposed to tidal currents and/or wave forces. The fine sediment picture is therefore one which involves many complex features superimposed on a basic tidal rhythm which is relatively well understood. Fig. 6 shows that the main elements of the motion are confined to the English side of the estuary, and that the highest recorded concentrations occur off Avonmouth and in Bridgwater Bay.

21.   As far as barrages are concerned, the part played by the Bridgwater Bay deposits in the overall sediment circulation process appears to be most important. The proposed Cardiff-Weston scheme comes close to its up-Estuary edge. The reduced tidal flow rate past the barrage confirms that the currents which occur for some distance seaward, this much exceeding the extent of Bridgwater Bay, would be generally reduced. Close to the barrage, however, the strength and distribution of the currents would depend more on the layout of the sluice and turbine openings across the width of the estuary. Clearly there is a case for minimising flood flows across the more sensitive parts of Bridgwater Bay, and this appears to be possible (Ref 4).

22.   By reducing both the significance of this source of fine material and the general strength and duration of flood currents in the basin to less than existing peak values on neaps, it has been estimated that typical turbidity concentrations would fall to about 1% of those which at present occur (Ref 10). Whether or not this is an accurate value, and the effect it would have on the ultimate disposition of that proportion of the sediments which could be moved by that current regime, have not been determined. Ref 4 sets out procedures by which this may be done. Its wider implications are considered in Section 4.

23.   The balance between the post-barrage fine sediment transport processes seaward and landward of an English Stones Barrage is likely to be significantly different from that for a Cardiff-Weston scheme. This is because, whereas the latter

scheme is close to the seaward limit of this zone, the smaller one would be towards its landward end. The basin of the English Stones scheme comprises an area which, at present, retains very little of the muddy sediments brought into it from seaward deposits by flood tides because these are mostly taken sedward again by the comparatively strong ebb currents (relative to those which occur at down-Estuary locations). As for the Cardiff-Weston scheme, an English Stones barrage would weaken both flood and ebb currents on its seaward side though this effect would have largely died out towards the seaward end of the fine sediment zone. The consequences of this for transport processes seaward of English Stones are considered in Ref 4.

24. Although the tide ranges within the basins of the two schemes would be similar for the same seaward range (Fig. 3), the significantly smaller mean depth landward of the English Stones scheme means that currents in that basin would typically be considerably greater than for the Cardiff-Weston scheme. Sediment mobility would increase accordingly, and the bias for flood velocities to exceed ebb suggests that the net sediment transport direction would be up-Estuary. The rates involved and the form in which these may in time stabilise to create a new regime (in which the action of waves, freshwater floods, etc. would also play a part) would determine the way in which the environment of the basin area would change. This aspect of sedimentation is also considered in Ref 4.

25. Superficially, the process of constructing a barrage may appear likely to throttle an estuary, progressively reducing the tide range on its landward side and creating even higher velocities through its openings. The extent to which this need actually occur in practice depends on the proportion of the total flow area kept open as construction proceeds. Fig. 2 shows that the turbine capacity required to harness the available energy is sufficient to ensure that about half the volume naturally discharged per ebb tide (extended) continues to pass that way through the structure even when these are on-load, i.e., not discharging freely. Furthermore, as previously noted, on the flood tide the sluices and idling turbines must have sufficient capacity to refill the basin at approximately the same rate as naturally occurs.

26. It is therefore reasonable to expect that the tide range in the basin would be little changed during the construction process. Fig. 7 shows the results of one-dimensional mathematical model studies of the dynamics of spring tide propagation in the estuary at intervals throughout the construction of both schemes (Ref 4). These confirm the expected small effect of the structure up to the time when the main civil works are complete and control is imposed on flows passing through the various openings prior to the start of generation. However, in order to achieve this little-changed basin tide regime, velocities close to

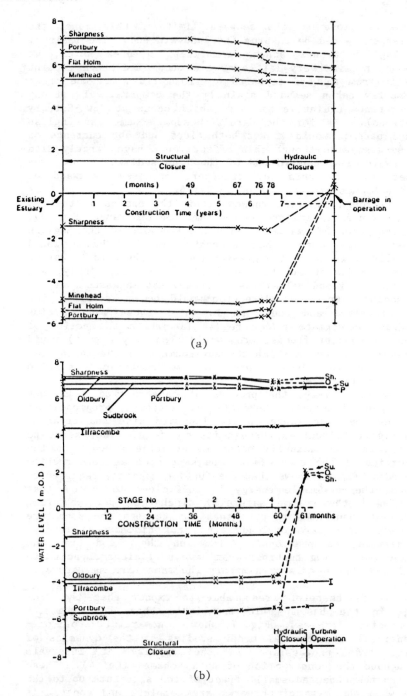

Fig. 7. Water level variations during construction and closure (spring tides): (a) Cardiff-Weston barrage; (b) English Stones barrage

the remaining openings must increase as construction advances. In this respect, the Cardiff-Weston barrage would experience a greater percentage increase in velocity than that at English Stones. This is because, whereas existing velocities are less, the required turbine flow area in each scheme is proportional to each basin area rather than to the width and mean depth of the estuary at each barrage location. One result of this is that, for the Cardiff-Weston scheme, there is more freedom to locate the openings across the estuary (these have to be in deep water) in order to minimise such problems as erosion in Bridgwater Bay.

CHEMISTRY

27. Ref 3 includes forecasts of the effects of a Cardiff-Weston barrage on a wide range of aspects of estuarine water chemistry, and these are reviewed in Ref 11. A number of possibly adverse consequences are identified, including increased concentrations of metals and higher biochemical oxygen demands (B.O.D.). Ref 11 points out that it is the aim of those Water Authorities with responsibility for water quality in the Severn Estuary to ensure that any barrage does not lower the prevailing standards. This suggests that suitable remedial works to improve and/or reduce effluent discharges would be necessary in advance of barrage completion. Ref 3 estimates what would have to be done to maintain the status quo.

28. In this respect it is significant that substantial improvements in outfall conditions have taken place since the Water Authorities came into being in 1974. A continuing investment programme is expected to maintain this trend over a similar period into the future, recognising relevant UK and EEC legislation on coastal water standards. The further remedial works needed to counter the adverse effects of either barrage must therefore be judged in the context of conditions prevailing at that time.

29. The smaller basin volume of the English Stones scheme reduces its potential for effluent dilution, but its larger ratio of tidal volume to total volume compared with the Cardiff-Weston scheme means a higher rate of exchange past the barrage. The net effect of these two factors on the water quality changes caused by the small barrage has not been quantified. However, Ref 3 emphasises that the problems foreseen to be caused by the Cardiff-Weston scheme are more likely to be local rather than total in extent, for example in the river estuaries where high concentrations of suspended solids are created by strong currents (causing high B.O.D.), and close to various outfalls including those discharging at or near to the tidal limits. Although smaller in size, the English Stones scheme may therefore have more than a correspondingly small effect on these aspects of water quality because of the location of the sources responsible for them.

ECOLOGY

30. In order to forecast how either barrage would affect the biological productivity of the estuary, it is necessary to appreciate within an overall framework the significance of each of the various physical and chemical changes which would occur. Whereas each change is important, the fact that the consequences of some are opposed to those of others means that only a total rather than a disaggregated appraisal is meaningful.

31. A clear example of this is the effect of either scheme on the intertidal feeding areas of waders and wildfowl. The reduced tide range in either basin area means that these intertidal zones would be appreciably smaller, especially seaward of the shallower river estuaries where the tide regimes would be less affected (Fig. 3). However, the capacity of these zones to support birds is not simply dependent on their areas. More important is the variety of animals which colonise the surface substrates, the density of each community and the average size to which individuals grow. The limited evidence at present available suggests that, notwithstanding the large intertidal areas created by the high tide range, the total capacity of these is low by general standards in other UK estuaries having a lower tide range, i.e., usually smaller intertidal area. Even so, however, work reported in Ref 12 suggests that existing bird populations do not fully utilise the available food resource in the Severn Estuary.

32. A detailed study is needed to quantify with greater precision the factors which determine bird-carrying capacity. The above evidence suggests that this may not hinge solely on foodstocks, in which case the fact that a barrage would reduce the intertidal area would only significantly affect bird numbers if they depend to a considerable degree for their food on those areas which become sub-tidal. Although some birds certainly feed in this area, it is not thought to be particularly productive. This may be largely the result of the mobility of the surface muds which extend over much of it and which can be insufficiently stable for animals to colonise. The high turbidity of the water column is an additional constraint, limiting photosynthesis and hence affecting the supplies of organic detritus produced within the water body and on which the benthic invertebrates feed. Strong salinity variations per tide cycle, especially on springs, further limit the ability of some animal species to colonise this area. As a result its marine fauna are poorly diversified, though some species which can tolerate the regime thrive for lack of competition.

33. All the physical changes produced by either barrage are expected to greatly increase the variety and density of species which inhabit the reduced intertidal zone. In terms of absolute productivity, the net effect is expected to mean that, from the point of view of foodstocks, the bird carrying capacity of the Severn Estuary would increase many-fold for a

Cardiff—Weston barrage, but probably less so for an English Stones scheme. If, as Ref 12 suggests, the existing regime is not at capacity (a conclusion which Ref 13 disputes), it might be inferred that neither barrage would increase (or decrease) bird numbers. A better understanding of this subject is clearly needed.

34. A similar situation applies in the case of fish. Species which inhabit and/or feed in esturies are in general not found in the Severn Estuary in the numbers expected for an area of this size. The strong currents, high turbidity, variable salinity and, therefore, poor productivity of the benthic fauna combine to dissuade both colonisation and feeding by marine species which otherwise would be expected to come in on the flood tide and move out on the next or later ebb. Either barrage scheme would improve this situation, especially the Cardiff—Weston scheme which would reduce currents, turbidity levels and salinity variations throughout the whole estuary.

35. The principal uncertainty of either barrage for fish is the extent to which they would be able to pass through its openings and what may be done to reduce risks to an acceptable level. The many, large diameter, slowly rotating turbines installed in each scheme are not thought to present either much of a deterrent or a risk to the safe passage of almost all fish species, whether juveniles or adults. Considerable worldwide evidence from hydro-electric schemes involving much higher pressures and smaller and faster turbines gives a generally consistent account of the success with which fish are able to negotiate this type of hazard. The available evidence strongly suggests that tidal power turbines would present no more than a minor obstacle to almost all species, a fact which experience with the slightly smaller and faster turbines at the Rance Barrage appears to confirm.

36. The two most pressing uncertainties are the risk to shad from passing through the turbines (the particular problem being the prospect of their sensitive scales being torn off in strongly sheared and highly turbulent flows), and the possibility that migrating salmon smolt would resist being drawn down through the turbines, instead collecting close to the face of the barrage where they could be preyed on by birds and other fish. Ref 4 proposes a work programme to evaluate the effects of turbine operation on shad, also the inclusion of specially designed surface-level fish 'chutes' at regular intervals along the turbine section of the barrage, these performing the task of now-traditional types of fish ladder, though in the present case suited to conditions in which both water levels change and small head reversal occurs on flood tides.

OTHER ESTUARY BARRAGES

37. The data available from other barrage schemes do not seem to have been well co-ordinated and compared. There are

various reasons for this:

(a) Only one scheme, namely Rance, is sufficiently large and has been operating for long enough to show how the ecology of an estuary adjusts to a new tide regime;

(b) The ecology of the Rance prior to barrage construction was not recorded in detail, and comprehensive studies of it are relatively recent;

(c) Detailed studies have been made of the natural environment of a number of estuaries for which barrages are being considered but, with the exception of the small Annapolis Royal scheme on the Bay of Fundy, Canada, these are still being assessed;

(d) The ecology of every estuary, even those with high tide ranges, displays sufficiently different features to make their direct comparison both difficult and perhaps misleading.

38. Nevertheless, the direction of change is the same for each barrage scheme. A reduced tide range in the basin is a common feature, hence currents, turbidities, salinity variations, inter-tidal areas, etc., are all lessened. The size of each change and hence its consequences must, however, depend on the initial, natural regime. Thus Rance (pre-barrage) and Severn, although of similar tide range, carry very different fine sediment loads, Rance containing little compared with the high loads described in Section 2 for the Severn. This alone is sufficient to produce very different basic ecologies, though the effects of barrages would be to make them more nearly similar.

39. While this may not mean that they can be readily compared, many of the trends observed at Rance are likely to occur at Severn, where they may be more strongly marked because of the relatively suppressed natural state of that environment. However, as previously mentioned, there is no case for regarding the whole tidal area of an estuary as everywhere similar. From the upper tidal reaches of the sub-estuaries where the tide range is most of the high water depth and salinities vary greatly, to the seaward end where conditions border on marine and are less variable, each zone must be assessed separately within a total ecological environment.

40. For any other estuary to yield directly helpful information therefore means that it has much in common with that being assessed. Observations made at Rance following barrage construction, such as larger numbers of fish of a wider range of species, confirm the trend expected at Severn but are not of themselves sufficient to indicate post-barrage populations and species types. Where Rance is of direct help is in such specific matters as the effect of turbine rotation on fish movement. As mentioned in Section 4 and Ref 4, this is one of a number of topics requiring detailed attention, and which will be considerably assisted by experience at

Rance, and probably at Annapolis Royal.

CONCLUSIONS
41. These are not the only studies relating to fish recommended in Ref 4. Together with those on birds and on many other aspects of the food chain, a well consolidated programme of definitive and inter-related ecological work is seen by the Severn Tidal Power Group as essential to the proper planning of either barrage scheme and to the estimation of their likely effects on all aspects of the estuarine environment.

42. The detailed information available to the Severn Barrage Committee (Ref 3) and that collected more recently and presented in Ref 4 does not suggest that either scheme would, on balance, necessarily have an adverse influence on the ecology of the area it would affect, though this depends to a large extent on the regularity with which levels in either basin are permitted to vary. However, until the results of further studies become available, it will not be possible to quantify many of these changes. Even when this becomes possible, it cannot be expected that all the evidence will be equally convincing on all topics and to all people. The balance between environmental gains and losses will therefore remain subjective and hence a matter of individual value judgement, as it must when assessing the consequences of any anticipated change to established habitats.

43. The two proposed barrage schemes would not cause similar environmental changes and the net consequences of each are not expected to relate to their relative sizes. The major difference between them would be the way in which they affect the mobile fine sediment body in the estuary. Whereas the Cardiff-Weston scheme is expected to stabilise much of this material by virtue of its location, the English Stones scheme would have little effect on its general mobility while providing conditions in which that which enters the basin could be largely retained. The environment of each basin area would be strongly influenced by this. Ref 4 assesses these implications in more detail than has been possible in the space available here.

REFERENCES
1. BERNSHTEIN L.B. "Kislogubskaya Tidal Power Plant", FIP Symp. on Concrete Sea Structures, Tbilisi, 1972.
2. DERRINGTON J.A. "The Use of Concrete Caissons for River Barrages", Proc. 30th Symp. Colston Res. Soc. (on Tidal Power and Estuary Management), Scientechnica, 1979.
3. SEVERN BARRAGE COMMITTEE "Tidal Power from the Severn Estuary", Energy Paper No.46, HMSO, 1981.
4. SEVERN TIDAL POWER GROUP "Tidal Power from the Severn", Rpt. to UK Secretary of State for Energy, 1986.
5. HILLAIRET P. "Optimizing Production from the Rance Tidal Power Station", BHRA Symp. on Water for Energy, 1986.

6. HYDRAULICS RESEARCH STATION "The Severn Estuary : Observations on Tidal Currents, Salinities and Suspended Solids Concentrations", Rpt. No.74 to Ref 3 above, also HRS Rpt. No. EX966, 1981.

7. RADFORD P.J. "Predicted Effects of Proposed Tidal Power Schemes upon the Severn Estuary Ecosystem", Rpt. to Ref 3 above, No. STP37, 1980.

8. KIRBY R. "An Assessment of Changes in the Fine Sediment Regime Associated with Construction of a Tidal Power Barrage near the Holms Islands in the Severn Estuary", Supplementary Document to Ref 4 above, No. SD2.5, 1984.

9. KIRBY R. and PARKER W.R. "Settled Mud Deposits in Bridgwater Bay", Institute of Oceanographic Sciences, Rpt. No. 107, pp65, 1980.

10. RADFORD P.J. and YOUNG K.M.E. "Predicted Effects of Proposed Tidal Power Schemes upon the Severn Estuary Ecosystem", Rpt. to Ref 3 above, No. STP30, 1981.

11. OWEN, Morlais "Severn Estuary - an Appraisal of Water Quality", Marine Pollution Bulletin, 15, 2, 41-47, 1984.

12. METTAM C. "Severn Tidal Power : Invertebrate Sediments and Fauna", Rpt. to Ref 3 above, No. STP52, 1980.

13. FERNS P.N. "Research on Wading Birds and Shelduck in the Severn Estuary : Summary and Final Comments", Rpt. to Ref 3 above, No. STP47, 1980.

ACKNOWLEDGEMENTS
On behalf of the Severn Tidal Power Group, the author wishes to record his appreciation of the advice readily given by the many organisations consulted during the course of the studies to which this Paper makes either direct or indirect reference. However, the interpretation of this advice and the conclusions drawn from it are those of the author alone.

Appreciation for permission from the UKAEA to publish this Paper is also recorded. The Authority part-funded the study reported in Ref 4, on which much of the Paper is based.

Discussion on Papers 11 and 12

PAPER 11

MR A. C. BAKER, Binnie & Partners
Dr Kirby states (section 8) that 'the designs prepared for the Severn Barrage Committee (SBC) took only engineering and cost factors into account' and that 'the closure sequence and construction method proposed are highly undesirable sedimentologically'. Fig. 1 shows both the alignments (line 5 and line 5S) that were tested in the Hydraulics Research two-dimensional computer model of water movements (ref. 1) and the alignment now proposed by the Severn Tidal Power Group (STPG). As can be seen, the STPG alignment has a single block of sluices on the English side which is the approximate equivalent of two blocks of sluices in line 5S. Line 5 had no sluices close to Brean Down. In terms of the alignment of the turbine caissons, the STPG scheme is closer to line 5S than to line 5.

Figure 2 presents current vectors for a closure test for line 5 to compare with existing conditions at the same state of the flood tide. It demonstrates that there would be a strengthening of currents in the area around Brean Down with a consequent risk of erosion of the Bridgwater Bay sediments. This diagram is referred to in Dr Kirby's Paper. Because of the clear risk of damage to Bridgwater Bay, a second closure test was carried out (Fig. 3) with a nominal 1.5 km of embankment, constructed early, extending from Brean Down. This arrangement helped to reduce current velocities in Bridgwater Bay to values which would be closer to those existing and thus was incorporated in the recommended construction sequence but is not referred to by Dr Kirby. The construction sequence proposed by the STPG (Kirby, fig. 4) leaves a gap off Brean Down until late in the construction sequence and thus is arguably worse, not better, than the scheme proposed by the SBC.

The effects of installing sluices between Steep Holm and Brean Down on long-term sediment transport can be assessed by comparing Figs 4 and 5. Fig. 4 shows vectors of potential sand transport for line 5. Transport in the area around Brean Down is predicted to be substantially reduced. In contrast, Fig. 5

shows that sluices near Brean Down could greatly increase sediment transport into the basin behind the barrage. Thus the STPG's proposed arrangement of sluices on the English side of the estuary, which is broadly the same as line 5S, could result in more adverse effects on Bridgwater Bay sediments than would be caused by the SBC's line 5.

From this, it follows that Dr Kirby's claim (paragraph 13) of a 'great improvement' in orientation and construction method is premature.

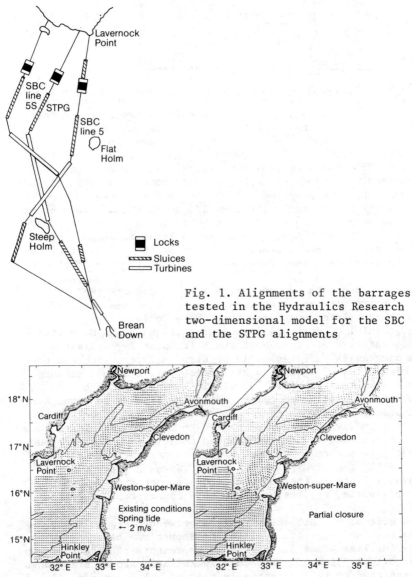

Fig. 1. Alignments of the barrages tested in the Hydraulics Research two-dimensional model for the SBC and the STPG alignments

Fig. 2. Current vectors: flood, alignment 5, closure test 1

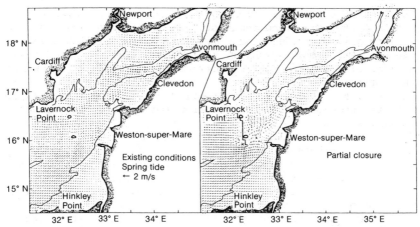

Fig. 3. Current vectors: flood, alignment 5, closure test 2

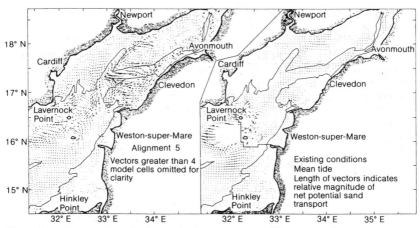

Fig. 4. Net potential sand transport: alignment 5

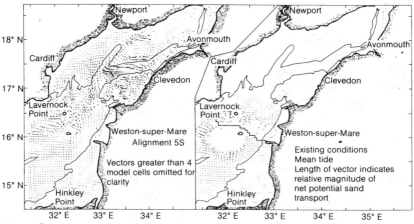

Fig. 5. Net potential sand transport: alignment 5S

DR R. KIRBY

Before Mr Baker's comments I had not appreciated that the
closure test 2 modelling was undertaken at a later date to
closure test 1 modelling, and that it was carried out
specifically as a result of concern which arose with respect to
the potential for increased erosion of Bridgwater Bay settled
mud deposits. Rather I had understood that closure tests 1 and
2 represented stages in a projected construction sequence.
Similarly, I had understood that only engineering and cost
factors had been taken into account in the design.

However, perhaps these are side issues because they do not
cause a change in opinion that many aspects of the line 5 and
line 5S schemes are undesirable sedimentologically. Space
restrictions in the Paper prevented the following itemization
of these problems, identified as sedimentologically undesirable
components in connection with the orientation, construction
method, and construction sequence proposed. Not all comments
apply to both line 5 and line 5S.

With regard to orientation, the smaller the disturbance to
the natural flow regime and the associated sediment circulation
the more localized and limited the sediment redistribution that
will result. As a bonus and corollary, the modelling should be
simpler and the results should have a higher level of
confidence. This arises directly from the fact that sediment
transport investigation is not a pure analytical science.

As can be seen, the orientation of the Severn Tidal Power
Group (STPG) scheme is very closely aligned with the existing
streamlines. In contrast, significant sections of lines 5 and
5S are at 45° or more to existing streamlines. Lines 5 and 5S
both result in cross-flows and eddy formation, which provide
additional problems for modelling and sediment redistribution.
In addition flows from the turbines are, for line 5S, directed
towards the low-lying, unprotected and erosional Bridgwater Bay
coastline. For line 5 the orientation in Bridgwater Bay is
such as to provide the risk of wave focusing or reflection on
to the coast and/or a dead water area due west of Brean
Down.

Both lines 5 and 5S directly overlie or, for long distances,
pass close to existing unconsolidated sediment bodies on both
the English and Welsh sides of the estuary. The erosion
potential of these should be reduced if the barrage line avoids
them as much as possible. It is not by chance that the STPG
line does largely avoid them.

With regard to the construction method, the line 5 and line
5S schemes involve of the order of 9 km of solid embankments,
either formed by end tipping or caisson placement. The use of
extensive solid sections is undesirable sedimentologically
owing to the regional velocity enhancements which progressively
develop, the local tip erosion which arises and the creation of
dead water areas to either side. As a result local, if not
regional, erosion could be expected to be enhanced and the
entrained sediment might not travel far to the newly formed

sinks in the dead water areas created nearby. In this way large amounts of sediment could be redistributed on a short time-scale. A wide range of coastal and navigational problems could arise. It was for this reason that I recommended that as much of the structure as possible should be constructed in open caissons of various types. Nowhere does the Severn Barrage Committee report envisage such a design. As a consequence the STPG scheme now involves 10 km of initially open turbine, sluice and other caissons and only 3.8 km of main embankment, much of which is taken up by the ship locks.

When the crossing is completed about 15% of the cross-section will remain open and the changes to the flow and sediment regime might reasonably be expected to be small until this stage is reached. Closure can then take place under controlled conditions, with many potential sedimentological benefits in both engineering and environmental terms.

With regard to the construction sequence, both line 5 and line 5S are undesirable in that they leave Bridgwater Bay open for almost the entire construction period and close it right at the end. Even if an identical closure sequence were proposed by the STPG the regional velocity enhancement would be smaller with the latter scheme, since a greater proportion of the flow still goes through the barrage line. With the line 5 and 5S schemes both the high velocity and the marginal low velocity areas are undesirable sedimentologically.

In sedimentological terms the TWC/7424/2 scheme is still not optimal, particularly in respect of the Bridgwater Bay section, as is specifically stated in section 13 of the Paper.

The STPG scheme has already been modified to take account of some of the generally held concerns about Bridgwater Bay. More detailed work in the future should give rise to further suggestions for refinement of the design.

I therefore reiterate that the STPG design is sedimentologically more desirable than the original line 5 and 5S designs, of which several of the limitations have been described. The most significant aspect of the existing STPG design is that simply as a result of a small amount of careful thought a whole range of potentially difficult or intractable issues have been reduced in scale, if not designed out of the scheme. This should be regarded as progress and welcomed.

Aside from these sedimentological issues the STPG design is considerably shorter in length, a fact which should assist in keeping down the cost.

MR P. ACKERS, Hydraulics Consultant
The Severn Barrage Committee (SBC) recognized the importance of fine sediments in the Severn Estuary and appreciated the possible problems should the major deposits in Bridgwater Bay be eroded to such an extent that there was permanent loss of a significant part of those intertidal zones, with consequent mobilization of large quantities of sediment which would then

settle elsewhere. Pages 83 onwards of volume 2 of Energy Paper 46 (the Bondi Report) summarize the data and interim conclusions on the basis of the limited amount of work completed by 1981. This includes the data from the Institute of Oceanographic Sciences on the distribution of turbidity and the results of the two-dimensional numerical modelling of water movements carried out by the Hydraulics Research Station (now Hydraulics Research). That was as far as the studies of sediment had progressed at that time, but it was sufficient to draw some interim conclusions about the layout then under consideration, with respect to long-term performance and the shorter-term effects during construction.

The sensitivity of the Bridgwater Bay area was recognized, as careful reading of the 1981 report will show, but there was a limit to the quantitative predictions that were possible. Dr Kirby takes the SBC team to task for not paying sufficient attention to the sediment issues of Bridgwater Bay - but little progress has yet been made in quantitative terms in the last five years.

Figures 2.40 and 2.41 of Energy Paper 46, volume 2, showed how the shear stresses in the estuary would be affected by the Bondi scheme, based on a limited number of two-dimensional model results at the critical stage of construction, with an embankment on the English side as a late element in the works. The report went on to say 'The depth and extent of scour of existing deposits has not been estimated, as to do so would require detailed knowledge of the sediment properties, including their variation with depth'. But it was obvious too that if there was significant scour potential over a small area, for example on the line of the embankment itself, the bed would have been protected, and if there was a major problem over a large area the scheme would require modification. It is a pity, therefore, that the resistance of Bridgwater Bay sediments to any increase in velocities is still not known.

The SBC layout had implied doubling the flow approaching the bay past Brean Down at a late construction phase. The corresponding information for the Severn Tidal Power Group's (STPG's) layout was not available, although it must be clear from the Author's fig. 4 that the blockage in the centre and on the Welsh side must cause some increase in the proportion of flow on the English side. The revised layout and construction sequence now proposed appear to have resulted from a premise that the sediments in Bridgwater Bay are so highly sensitive to any increase in velocity - and over a considerable depth of deposit - that the SBC scheme was not feasible. This premise may seem less justifiable when the necessary field work has been carried out, as this may show that the thickness of highly erodible material is limited. Moreover the actual advantage of the new layout over the SBC scheme remains unquantified in the absence of corresponding two-dimensional model results at critical construction stages.

The 1981 shear stress results were summarized by identifying zones where the maximum stress would be under 1 N/m^2 (potential deposition of fine sediment) and over 3 N/m^2 (potential erosion of soft mud). This showed that with embankment closure on the English side there were areas where at present the maximum stress is below 1 N/m^2 which, during the critical closure phase of the Bondi layout, would have maximum stresses of over 3 N/m^2, and the report went on 'Further study would show the extent of any bed protection or change in closure method needed to limit the volume of sand and mud brought into suspension during the construction phase'. There is little further factual evidence from Bridgwater Bay itself, but I suspect that the bulk of the deposits there are much firmer than a limiting stress of 3 N/m^2 would imply, except for the ephemeral deposits of the spring-neap and seasonal cycles and a surface layer of relatively few centimetres.

Thorn and Parsons (ref. 2) have shown that the resistance of mud to erosion by currents is a function of the density of the exposed surface, and that the density increases with depth in the usual estuarine conditions. Research on several estuary muds, including Grangemouth mud, can be approximated by a shear resistance function as shown in Fig. 6. The density scale here should be noted: a shear resistance of the order of 1-3 N/m^2 is achieved with a surface density of the mud of around 200-300 g/l. These are soft muds rather than clay soils and probably correspond to recent deposits rather than those that have been in position long enough to have de-watered and partly consolidated.

Is it possible to walk over much of the Bridgwater Bay material? I have seen footprints, indicating that perhaps the deposits are not very different from the muddier parts of other estuaries. For example, in muddier parts of the Wash foreshore, one might sink in 10-30 cm. This is consistent with increasing strength with depth, but what does this mean in terms of the strength of the deposits underlying the very soft surface material?

```
weight = 60 kg = 600 N
bearing area = 10 x 30 x 2 = 600 cm2 = 0.06 m2
loading = 10 000 N/m2
implicit shear strength = 5000 N/m2
```

The relationship of shear strength in soil mechanics terms to hydraulic resistance is not direct, nor is it well understood, although Owen (ref. 3) has suggested a ratio of around 7:1. This gives as a very rough indication

hydraulic shear resistance = 500-1000 N/m^2

This crude analysis of the range of possible shear strengths in estuarine deposits - over two orders of magnitude between the very soft surface and the 'human-bearing' elevation - shows how much quantitative information for Bridgwater Bay itself is needed, so it is disappointing that since 1981 this important evidence has not been produced. Incidentally, Peirce and his colleagues (ref. 4) did some field measurements of hydraulic shear resistance in the Brean Down area but did not publish those particular results.

Dr Kirby implies in his Paper that it was the STPG who originated the concept of using caissons with maximum openings to ease construction problems, including any erosion and sedimentation during the misnamed 'closure' stage, and thereby were able to develop a much improved scheme. No other paper at this symposium makes such a claim, because the concept of using caissons opened to the flow after sinking was central to the development of the Bondi scheme. It was carried forward from the previous experience of some members of the SBC team in feasibility studies for the Kimberley tidal power project in Australia (ref. 5). Indeed, in overall concept the STPG scheme remains very close to the SBC scheme, having generally confirmed the installed capacity, the advantages of caissons,

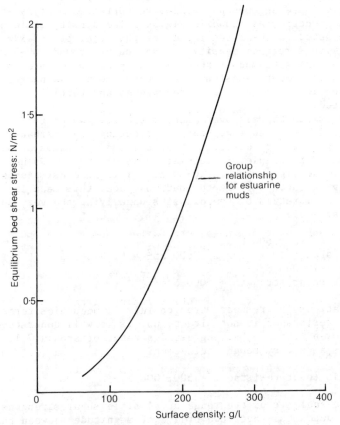

Fig. 6. Shear resistance of a mud bed (after ref. 2)

the broad construction concepts, the location, rather benign sediment effects etc.

It is only to be expected that further work finds areas where there is scope for improvement, and in effect another loop of iteration towards an optimum scheme has now been carried out. Those who have been trying to promote proper studies of tidal power against a great deal of government inertia in the past should take heart from the measure of agreement between the new studies and the work carried out - over a short two-year period - for the SBC.

One final point: the major problem that needed to be resolved when the SBC studies began was the feasibility of handling and sinking large caissons in the severe tidal conditions of the Severn Estuary, already severe but becoming progressively worse as the lock island was built and more caissons were placed. It was because of the team's feelings about minimizing the risks of this operation that caisson placing took precedence over embankment construction, which with the advantage of the large flow passages through the caissons could proceed with minimum risk. It is important before final decisions are made that risks are properly balanced: it would be a mistake to allow the potential erosion in Bridgwater Bay to dominate the construction sequence unless and until the risks and extent of such erosion are quantified.

DR R. KIRBY
I agree with Mr Ackers on the crucial importance of assessing the erodibility of Bridgwater Bay settled mud. I have consistently drawn attention to the importance of this issue (see refs 6 and 7 and proposals currently being considered for the feasibility study).

In consideration of Mr Ackers' disappointment that so little has, in his view, been achieved by the Severn Tidal Power Group's (STPG's) studies of this issue during the last five years, it is important to establish a perspective on the objectives and input of the Severn Barrage Committee (SBC) and STPG studies. The SBC were charged with carrying out, over a two-year time-scale, a pre-feasibility study into the problems to be faced in constructing barrages at various locations in the Severn Estuary and Inner Bristol Channel. During this study many separate investigations relating directly or indirectly to sediments were carried out. A specific element of the studies was that they involved extensive field investigations and hydraulic modelling.

In contrast the STPG were charged with ascertaining the prospects of attracting private finance to fund the construction of a barrage at either Cardiff-Weston or English Stones. The time-scale for the entire study was also two years. At the same time several small studies were initiated by the STPG to assist the decision-making process. There was neither the time nor the money to undertake additional field work or laboratory testing. In view of the acknowledged importance of sediments, but constrained by the financial and time restrictions, two studies to provide decision information on the Cardiff—Weston and English Stones barrages were commissioned. The studies relied largely on unpublished data collected by the Institute of Oceanographic Sciences (IOS) and previously not fully worked up.

The results of these two studies were submitted as commissioned reports to the STPG, having been reviewed at draft stage by Binnie & Partners (refs 7 and 8). It is not for me to judge whether the STPG or the general scientific community at large derived value from these studies. Nevertheless, I look forward to the work being so judged. The evaluation of the English Stones scheme resulted in a completely different estimate of the likely sedimentological regime within and in front of the basin from that previously accepted (ref. 9), and the information was one of the major factors leading to STPG recommendations on this scheme. The Cardiff—Weston report evaluated a range of issues that had not previously been addressed and raised a large number of aspects (some outlined in response to the comments of Mr Baker) in which the original design could be improved. It showed how issues which had previously given rise to concern could be reduced in their effect, or overcome, by the simple expedient of changing aspects of the design. In the search for better guidance on the sedimentological regime of the Severn, the STPG have also supported an Energy Technology Support Unit commission to complete the interpretation and reporting of outstanding IOS data on the turbidity regime of the estuary (ref. 10).

Another closer, and relevant, perspective on the erodibility of Bridgwater Bay settled muds, specifically mentioned by Mr Ackers, can be gained by considering the present state of knowledge in this matter. Thanks largely to chance factors there is already a more detailed knowledge of the nature and rates of changes, which are in progress in both the subtidal and intertidal mud regimes, here than is available at many other estuarine locations. Were it not for these studies, undertaken by a range of geophysical, geotechnical and geochemical techniques not normally used by engineers, and applied to the area over a longer time—scale than that currently available for feasibility studies (ref. 11), present scientific discussions would be on a much less informed basis.

As shown in the Paper (Fig. 2) and in ref. 11, Bridgwater Bay is a complicated area in which the settled mud area has an accretionary, peripheral seaward margin, in which the deposition rate is known with great precision in a few areas. This zone merges landwards with a stable zone in which there is neither erosion nor deposition. Landwards from the stable aureole is another zone, whose presence appears to have escaped Mr Ackers' attention. He maintains that if the barrage construction sequence adopted led to regional acceleration of flows across Bridgwater Bay, after an initial period of, perhaps, rapid erosion, an underlying firm substrate largely resistant to erosion would be exposed. What he does not acknowledge is that a significant area of the Bridgwater Bay settled mud patch is already an overconsolidated mud, which is nevertheless experiencing continued erosion at present. This is the area across which, with the SBC scheme, the maximum bed stress was envisaged to more than triple on a regional scale, as explained by Mr Ackers.

Unlike accretion rates in the subtidal regime, which can be quantified with precision, erosion rates are difficult to quantify. Hydrographic survey techniques do not have the resolution and most other techniques are too difficult to deploy. In this case the total absence of artificial radionuclides even at the sea bed surface testifies that the erosion is still in progress, while the most sophisticated geotechnical determinations devised produced an answer that of the order of 1 m may have been lost from the surface over an unknown time-scale. Even these measurements are now viewed with scepticism. However, there is little doubt that considerably more than the 'few centimetres' which Mr Ackers suspects would be lost is involved.

Mr Ackers' remarks are principally directed, in view of his comments on trafficability, not to the subtidal zone discussed earlier, but to the intertidal mud flats. Knowledge of the sedimentological regime here is less advanced than that of the subtidal zone. The work of Kidson and Heyworth (ref. 12) in Bridgwater Bay and of Mettam (ref. 13) on the Wentlooge Levels indicate that the present intertidal zone of the Severn is characterized by a dynamic regime, but with a long-term erosional trend. On the Wentlooge Levels data provided by engineering works indicate that in excess of 2.0 m of tidal mud flat have been removed in less than 50 years.

All this information, readily available in the literature, poses a range of questions which engineers should now be addressing. Some of these areas are as follows.

    (a) Can existing engineering tests recognize any difference in the hydrodynamic regimes over the subtidal accretionary, stable and erosional areas in Bridgwater Bay? I suggest that the answer is no.

    (b) Can existing laboratory engineering tests measure the erodibility of the substrate surface in the normally or overconsolidated subtidal areas and predict these quantitatively with confidence? There are several reasons for a lack of confidence in this.

If I am correct, how can engineers hope to predict by how much the erodibility might change under various revised schemes for the construction of a barrage?

I am well aware of the need to determine the erodibility of Bridgwater Bay sediment. Since 1981 the STPG has neither had the financial capability nor the remit from the Government to undertake detailed studies of erodibility. Even if money had been available it would not have been well spent in this direction, since an appropriate technology to study erodibility of muddy substrates in the field did not exist at that time.

Remoulding such sediment and placing it into a flume to test its erodibility has intrinsic limitations. In addition, many current measurements of the rheological properties of cohesive sediments are inappropriate to predicting the strengths of these materials in terms which have a direct relationship to natural conditions at the bed of an estuary.

Fortunately suitable equipment to make measurements of the erodibility of clay substrates in the natural environment does now exist. The importance of the problem demands a pilot programme of comparative measurements to assess the response of the same substrates to standard engineering laboratory tests compared with field determinations. At its most basic this will permit an assessment of the value of laboratory test procedures.

In summary I share Mr Ackers' concern, but have pointed out why the STPG did not address this issue previously. There is already some knowledge of the field situation in Bridgwater Bay and of the limitations of current engineering tests when applied to such regimes. These limitations imply that it remains to be proven that current engineering tests can reliably predict whether local or regional erosion can be anticipated. Consequently they cannot provide a quantitative predictive basis on which to make decisions on the need for differing styles of bed protection, or the necessity for altering barrage designs in consequence.

My stance has not therefore been that Bridgwater Bay settled mud is 'highly sensitive' to erosion, but rather that it has been 'highly resistant to making reliable quantitative predictions'. Notwithstanding this issue, if revisions to the barrage design, as a result of which these issues cease to be important or assume a lower priority, can be made, this must be an important step forward (hence my comments on the advantages of the STPG design in response to Mr Baker).

With regard to the issue of the more extensive use of initially open caissons, it is not apparent from published SBC reports, nor from Binnie & Partners' review of my report in 1984, that the issue of the greater use of initially open caissons had been considered by these organizations.

DR G. MILES, Hydraulics Research
The question of what is likely to happen to muddy sediments in the Severn Estuary after construction of a barrage is very important because there are implications on

(a) siltation in ports and shipping channels
(b) water quality (via the biochemical oxygen demand and heavy metal content of the mud)
(c) ecology (via the turbidity of the water)

(d) environment (via the nature of bed deposits).

The only satisfactory way of predicting the new mud regime is by a dynamic computer model including facilities to simulate

(a) erosion of bed material
(b) transport in the water body
(c) settling in slack water
(d) resuspension of new deposits on the next run of the tide.

One-, two- and three-dimensional models are all available, but for the Severn Estuary a one-dimensional model would involve too many approximations for it to be of any value. However, a three-dimensional model study would take much longer and need more resources than are apparently available in the current stage. Under these circumstances it is sensible to use a two-dimensional model. A model of this type includes the

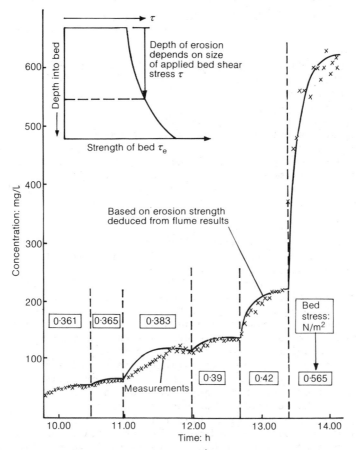

Fig. 7. Numerical simulation of the erosion of marine mud

main mud transport processes referred to earlier. It could
also resolve the gross features of the lateral and longitudinal
distributions mentioned in the papers. Could Dr Kirby comment
on this approach for tackling the problems?

I am concerned about the amount of erodible material that is
available in the estuary to feed the progressive siltation that
Dr Kirby refers to in his Paper. It is normal for the density
(and hence the strength) of muddy deposits to increase with
depth below the surface of the bed. An example of this is
given in Fig. 7 which shows some results from the Hydraulics
Research erosion flume. Tests in this flume show how erosion
is limited by the exposure of denser (stronger) deposits. The
example also shows how the erosion process can be reproduced in
a computer model. This simulation was possible because the
density structure of the bed in the flume was known. A major
problem in practice is to obtain the density structure of the
deposits, and in this context could Dr Kirby say what
information there is on the muddy deposits in the Severn
Estuary? This aspect will need careful consideration at the
next stage of tidal barrage studies.

DR R. KIRBY
Mathematical models are the best way forward; a two-dimensional
model of the type suggested is the most cost-effective tool at
present and moreover Hydraulics Research are best placed to
carry out this modelling.

However, undue emphasis should not be given to mud transport
modelling at present, to the possible exclusion of other
important work on aspects of fine sediment behaviour. Although
current two-dimensional models often work well in relatively
low turbidity regimes, there are serious difficulties with
their application to high turbidity regimes, such as that of
the Severn. Aspects of mud behaviour which will require
special modelling techniques, or modification to existing
techniques, are the vertical segregation and lateral
partitioning of the suspended solids load (ref. 14). In a
general sense also there are recognized deficiencies in the
deposition parameter in current mud models, certainly
numerically if not conceptually. Even more serious there are
acknowledged difficulties in parameterizing the erosion
process. It is generally agreed that at present, and for
current models, the answers provided will be of a qualitative
nature only. Research is therefore necessary to improve the
predictive capability of mud models in such regimes.

The second issue to which Dr Miles refers is the erodibility
of deposited fine sediment in the estuary, which will need
careful consideration at the next stage of barrage studies, and
I agree with this opinion. Two comments are relevant here.
Firstly, there are questions which relate to the total
potentially erodible fine sediment that the estuary has access
to. I have crudely estimated that the settled mud population
of the estuary, i.e. the Holocene unconsolidated fine-grained

bed deposits, outside the river walls in Bridgwater Bay,
Newport Deep, the tributary estuaries and the tidal flats
marginal to the main estuary may amount to 900-1000 million t.
To this must be added the annual river inputs from natural and
artificial sources, any occasional capital dredging schemes,
the input from coastal erosion and the input from subtidal
erosion seawards. Some of these can be quantified but,
significantly perhaps, although it has been known for 150 years
that sources to seaward contribute fine sediment particles to
the estuary, no progress has been made scientifically to be
able to quantify this.

The second aspect is that, even if it can be ascertained
accurately what the amounts of fine sediment in the various
potential sources are, the technical state is too poor to
estimate the erodibility at present or by how much the regime
would need to change to influence significantly the balance of
accretion and erosion which exists at present.

To be specific the settled mud area in Bridgwater Bay
consists of a very weak, frequently gas-rich, deposit.
Considerable areas of it, of the order of 7.5 km^2 in the
subtidal regime (ref. 15) are known to be overconsolidated
deposits which are experiencing erosion under the present
regime. They are not normally consolidated beds of the type in
Fig. 7, although normally consolidated substrates also occur.

At present not enough is known to be able to judge how
sensitive to erosion the settled muds are, and therefore how
serious the problem is. For this reason judgement should be
reserved while emphasis is given to finding out. There is an
urgent need for a pilot field study of the erodibility of
Severn Estuary settled muds, especially those of Bridgwater
Bay, to stand beside any flume or laboratory measurements. It
will be crucially important to see how the results compare.
Any new knowledge will be of prime importance in calibrating
mathematical models, as Hydraulics Research will recognize.

MR A. V. HOOKER, Consulting Engineer
Dr Kirby shows in his Paper how essential it is to take into
account the problems of siltation in designing a Severn
barrage. Changes have already been made in the configuration
and design of the proposed Cardiff-Weston barrage as a result
of sedimentation studies, but it is to the English Stones
scheme that these remarks are directed.

Dr Kirby predicts that annually about 8 million m^3 of fine
sediment might be retained above the barrage and this is many
times the volume estimated about 60 years ago by Professor
Gibson on the basis of a physical model. Apart from the more
accurate methods of measurement that are now available other
reasons which might explain this wide discrepancy should be
looked for. The use of the low level turbine openings as
sluices is shown to account for more than half the siltation.
Closure of the turbine draft tubes during refilling, and the
provision of an equivalent area of additional sluice openings,

would increase the cost of the barrage but if the need, for reasons of siltation, to keep the sluice sills at a relatively high level is accepted, changes in the alignment of the barrage could lead to a more economical solution.

In Gibson's scheme the long sluice dam was sited on the English Stones shelf where the level of the rock is only 2-3 m below ordnance datum (OD) and is exposed at the low water period of spring and neap tides. He proposed open channel sluices with a sill level of 2.7 m below OD, permitting in situ construction between tides.

Upstream of the present alignment there is one point where the deep channel widens sufficiently to accommodate the turbine dam with a reduced volume of rock dredging (Figs 8 and 9). The crossing distance at this point is shortened from 7.1 km to about 5.4 km and it is likely that the savings on the embankment, and on the much shallower in situ sluices, would more than outweigh the cost of the additional sluices. The power output would be slightly reduced but the criterion for

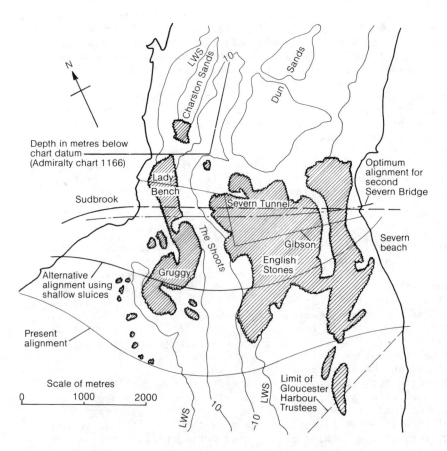

Fig. 8. English Stones barrage: alternative alignments

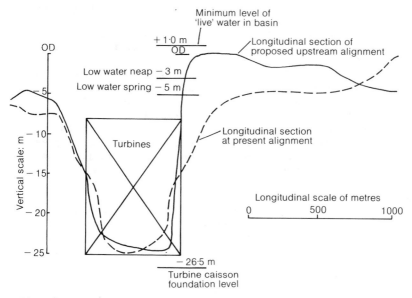

Fig. 9. English Stones barrage: sections

the English Stones scheme has always been cost effectiveness rather than output.

If the much-needed second Severn crossing were carried over the barrage, at least part of the £50 million or more saving should be credited to the owning company and there would also, presumably, be an income from tolls. The combination of two very large civil engineering projects, making each more cost effective, might make all the difference in attracting private capital. It would also reduce their environmental impact.

Dr Kirby's work has already led to a reappraisal of the Cardiff-Weston proposal and is a challenge to engineers to carry out a similar exercise for the English Stones scheme.

DR R. KIRBY
The annual deposition rates now predicted for the scheme are an order of magnitude greater than those estimated by Gibson in 1933 (ref. 9), principally because these early physical model studies envisaged only contributions from riverine sources to the deposition, whereas at present it seems inevitable, with the operation sequence envisaged and the modelling results available, that the major input will arise from seawards. This new perception largely arises from a greatly increased knowledge of sediment transport in coastal areas, in this case the Severn Estuary, combined with the mathematical modelling capability, neither of which were available to Gibson.

Nevertheless it should be pointed out that it has been known

271

for 150 years (refs 16 and 17) that a certain component of the fine sediment fraction in the inner estuary was derived from the sea, and that Gibson failed to appreciate this. This perspective should assist in explaining why the quantities predicted differ by such a large amount.

In my report to the Severn Tidal Power Group (ref. 8) on the English Stones scheme I drew attention to the reduction in silt penetration which could arise if the turbines were not used to assist filling on the flood. I do not know whether further detailed consideration was given to the question of revising the scheme, or aspects of its operation, to assess the degree to which silt penetration from the estuary could be minimized. Simply increasing the number and size of the sluices to compensate for the relatively small loss of the turbine cross-section for filling may not result in sufficiently reduced accretion rates. For example, significantly increasing the length of the sluice gate bank may involve transgressing into shallow water. In these zones major enhancements in the suspended solids load frequently occur due to wave erosion of the marginal tidal and subtidal flats. Although not always present, wave entrainment could result in larger quantities being moved through the barrage than have been calculated for the daily input through the turbines, but at infrequent intervals. Whether the net result would provide a significant improvement is unclear. Is it not likely that the predicted lifespan of the barrage would have to be considerably enhanced beyond that presently envisaged before the scheme could be considered viable?

I am unable to comment on the alternative sites and cost-benefits.

PAPER 12

DR H. R. SHARMA, Central Electricity Authority, Government of India
Have the environmental studies carried out so far indicated any significant changes in the salinity in the tidal basin? If so, what is the magnitude of the changes?

Has the problem of salt intrusions due to a permanent increase in the water level in the tidal basin been studied? If so, what are the indications of these studies?

According to paragraph 36 of Paper 12, the question of providing surface level fish chutes at regular intervals is still under examination. Is it necessary to provide a fish passage or fish ladder when a very large area is available through the sluice and the turbines for fish migration?

DR T. L. SHAW
The Severn Tidal Power Group's (STPG's) studies have included a careful review of the way in which the salinity of the basin area may be changed by the barrage. Forecasts made by the

Severn Barrage Committee (1981) suggested that concentrations would fall significantly, whereas observations at the Rance barrage in Brittany have shown the opposite in that case. It is not clear why the two projects should display different trends. Dr Sharma has apparently shown from studies of a proposed tidal barrage project in India that an increase in salinity is forecast in that case. The STPG believes that much more work on this topic is needed for a Severn barrage and has plans to implement detailed studies during its forthcoming programme of work with the Department of Energy and the Central Electricity Generating Board (CEGB).

The ecological implications of salinity changes are readily assessed if that is the only relevant factor determining the biological regime. In a tidal estuary, and especially one with a high tidal regime, this is not the case. Firstly, the salinity at any place is neither constant during any tide cycle, nor does its average tidal value remain unchanged throughout the year. Evidence from the Severn Estuary suggests that large parts of the coastline experience variations through a year which can be as high as half of the difference between freshwater and open sea conditions, and that much of this occurs during single tidal events.

The ecological stress which this causes is exacerbated by the high turbidity of the water created by the strong currents necessarily associated with those tides. This leads to unstable and hence difficult to colonize substrates. The combined effect is that the productivity of the intertidal and subtidal zones is severely reduced relative to that of coastal areas that are not so exposed to the various environmental consequences of high tides. The STPG is at present engaged with the Department of Energy and the CEGB in defining studies which will hopefully highlight the net ecological effect of reducing the tide range in the basin area. A less variable salinity regime is only one of the factors which will determine the net result. It remains to be shown whether any shift in average annual salinity value will be a significant factor in determining the direction and extent of that change.

The STPG wishes to distinguish between ecological changes caused by the barrage in the main basin area and those which would occur in the tide reaches of rivers draining to that basin. There is good reason to suspect that the two zones will display some different characteristics. A single judgement for the whole tidal regime may not therefore be possible.

The provision of chutes to assist fish movements past the barrage is seen by the STPG as additional to the substantial capacity of the turbines and sluices for this purpose. On present evidence, the movement of some species, particularly when juveniles, would be assisted by chutes. If their provision is favoured on no more than precautionary grounds, it would be an inexpensive detail to include in the design of the caissons and need have no significant effect on the energy potential of the scheme.

On present evidence it is not possible to state with

conviction that fish chutes are necessary. Evidence from the Rance barrage in Brittany, where there is no such provision, clearly shows that fish movement past the barrage has substantially increased since the barrage was commissioned in 1966. This should not be taken to mean that the same would apply for all species in another case. Individual designs to suit the migration requirements of each estuary may be necessary according to the species present. However, the fact that the Rance estuary now supports many more species than before the barrage was constructed makes foresight essential.

DR K. R. DYER, Institute of Oceanographic Sciences
The anticipated life of the barrage on the Cardiff-Weston line is taken to be about 120 years. Present estimates of the increase in mean sea level due to global warming are generally in the range 1-2 m at the end of the next century. Although this may not in itself affect the operation of the barrage, the accompanying coastal erosion could provide a new source of sediment. Has any asessment been made of the effects of mean sea level rise on the operation of the barrage, the sedimentation and the enviromental concerns in the long term?

DR T. L. SHAW
The Severn Tidal Power Group (STPG) has given much attention to the possible effects of a rise in mean sea level on the operation and viability of a Severn barrage. Further work will be done during the forthcoming programme. A specification of the likely rate of change in level is being sought from the Department of the Environment to establish this programme in context.

The barrage operating procedure preferred by the STPG restricts the highest basin level to about that of normal high water springs. This is about 2 m less than can now occur over much of the main area of the Severn Estuary on an exceptional flood surge, for which existing sea defences are designed.

If the mean sea level were to rise, the implications of increasing the maximum sea level by a corresponding amount would have to be assessed. To do this by 1 m should cause few if any problems, and it should be possible to extend this to 2 m without difficulty. The costs involved with such a change would primarily relate to provisions for land drainage and sea defences. The cost of these is likely to be less than the value of the extra energy that the barrage would then produce due to additional tidal volume.

If a constraint on peak basin level had to be imposed, less energy and hence less revenue would be produced. However, this would not become significant until it exceeded about 1 m, because it is set by barrage operation on the higher spring tides and hence would not immediately affect more than a small proportion of all operations, and even for those it would only constrain operations during a small part of the tide cycle, at the start of the generating period.

The extra basin volume is expected to exceed greatly any increase in coastal erosion rates which may be caused by a rise in sea level. The scheme is designed to minimize sediment input from the sea, and the wave climate established in the basin area will be generally mild compared with the existing regime and is unlikely to cause any significant coastal erosion, irrespective of the basin level.

The cost of any remedial works needed landwards of the barrage following a rise in sea level is likely to be much less with the barrage in place than without it. No allowance for this benefit from the scheme has been taken into account in determining its present value.

MR R. BATE, Council for the Protection of Rural England
I should like to draw the symposium's attention to the views of the Council for the Protection of Rural England (CPRE) and of our sister organization the Council for the Protection of Rural Wales (CPRW). The CPRE and the CPRW fully support the Severn Tidal Power Group's efforts to involve interested organizations, such as these, in their review of the environmental effects of a barrage, which has helped all concerned to identify issues and to discuss them as early as possible. This approach has been quite the reverse of, for example, the Government's attitude to that other huge engineering project – the Channel Tunnel. The environmental consequences of that scheme are likely to be profound but have generally been inadequately studied. The CPRE's and CPRW's discussions with the Severn Tidal Power Group have clearly been useful as their views have been reflected in Dr Shaw's two papers (Papers 12 and 14).

The most obvious impact of any Severn barrage on the countryside would be the structure itself, whether seen from England, Wales or the estuary. This impact would need to be kept to a minimum by careful design and careful control of the landfalls. Any road across the barrage should pass across the shipping locks at low level, avoiding the need for a bridge with a 40 m clearance above high tide levels.

The electricity supply from the barrage to the national grid would also need careful planning, to avoid the march of festooned pylons across the country. The barrage would be a completely fresh development, so consideration should be given to laying cables underground.

The greatest impacts on any Severn barrage on rural England and rural Wales would come from the development pressures generated by the barrage rather than the barrage itself alone. Some of these are touched on in Dr Shaw's Paper 14. As well as the important issues of tourism and recreation which would create extra demands for development, principally on the banks of the estuary, there would be further regional development pressures. The scale and detail of these would depend on where the barrage was positioned and the scope for the movements of people and goods across it.

If, for example, a four-lane highway was built across a Cardiff—Weston barrage and linked into the existing trunk road network, then the transport options across a substantial area would be transformed. Pressure could be expected to mount for development in the corridors of the approach roads. Industrial, commercial, tourist and other types of development would no doubt be proposed, followed by further pressure for housing, services and further consequential developments to absorb the increased resident labour force. The CPRE and the CPRW would like to see any such development pressures diverted into the existing urban areas in Glamorgan and Gwent rather than allowed in the countryside. This would bring welcome opportunities to Welsh towns and cities while preserving open countryside everywhere. Similar considerations apply when reviewing the likely impact of the construction phase of a Severn barrage, whether at the English Stones or between Cardiff and Weston.

The Severn Tidal Power Group's recognition that it does not yet have adequate knowledge of the consequential impacts of any barrage and that a further regional study of these is essential if the project is to be pursued further is much welcomed. This holds out the possibility that any scheme that may eventually be promoted would be the least damaging possible. The CPRE and the CPRW would be pleased to co-operate in any discussions.

DR T. L. SHAW
The Severn Tidal Power Group (STPG) has valued the constructive advice given by the Council for the Protection of Rural England (CPRE) and the Council for the Protection of Rural Wales (CPRW) during its earlier studies of the Severn barrage project and looks forward to continuing that constructive dialogue during its forthcoming programme of investigations, much of which will be relevant to the interests of those organizations. A mechanism to ensure closer liaison forms part of the STPG's proposals to the Department of Energy for these investigations.

The forthcoming studies will include more detailed attention to the need for and, if appropriate, the capacity of a public road across the barrage. On present evidence, the STPG considers that such a road could be justified but that it would not be to motorway standard. If it were to that standard, and if it were deemed necessary for it to be uninterruptable (toll booths introduced this feature at existing major estuary crossings), then it would have to include a high level section over any commercial ship locks incorporated into the barrage.

However, if interruption is accepted, the much cheaper solution of incorporating swing or bascule bridges towards each end of those ship locks would provide an efficient arrangement. Such a system at present operates at the entrance to Bristol Docks. It need not cause more than momentary and occasional delay to road traffic and is not seen by the STPG as out of keeping with the case for both a free-flow road artery across the barrage and unconstrained shipping movement. Further, it would appear to meet CPRE/CPRW concern on this matter.

The case for laying power lines underground has been considered by the STPG, both in terms of added cost and aesthetic benefit. The impact of overhead lines cannot be judged until the preferred landfall positions for the barrage have been chosen. It has hitherto been premature in the STPG's view to attempt to do this. Power line routes could be a factor in making this decision. If their installation underground is favoured, the opportunity to incorporate them in association with the new approach roads which the barrage would need should be contemplated.

These and many other issues form part of the regional study of the wider impact of the barrage which is to form an essential part of the STPG's further studies of the project, in association with the Department of Energy and the Central Electricity Generating Board. The CPRE/CPRW rightly stress the need for this to be directed towards minimizing environmental damage. It is for this reason that the STPG has proposed that it should include an 'impact assessment', not only ecological but on all relevant issues pertaining to the environment. The extent to which this encourages development to focus on the existing urban areas in Glamorgan and Gwent must depend on the scale and nature of the investment foreseen by the regional study to be attracted to the region in association with the promotion of the barrage as a power-station.

MR I. G. RICHARDS, Richards Moorehead & Laing Ltd
As an engineer I have an increasing interest and involvement in environmental matters and following the frequent references in New Civil Engineer etc. to 'environmental concern' in engineering cycles I was hoping to find evidence of a fresh approach to the subject in the papers. I was disappointed. I would like to remind the meeting of Mr Howard Pullan's remarks in his recent Chairman's Address to the South Wales Association in which he advised engineers of the imperative need for them to carry 'the public' with them in the promotion of construction projects.

I would suggest that the 'man in the street' in south Wales will welcome the barrage and be excited by the idea of harnessing nature for the general good. The prospect of another crossing of the estuary will also be seen as an advantage. Much more can and should be done by the engineers involved to ensure that the project gains the overwhelming support of the public. For example, dredging of large quantities of fill material from the estuary is justified in Paper 3 on the grounds of cost and convenience. It is necessary to do better than that and this has been a typical lost opportunity for the promoters of the scheme to win over the man in the street. Surely waste materials which are a nuisance or a hazard elsewhere, for example mine spoil or quarry waste, should be used as fill for the barrage.

Proposals in this vein will not only win over public support but confuse or even silence the environmentalist lobby. It is unfortunate that arguments on this and similar matters are not being rehearsed here, as they will be aired more publicly in due course.

A balance between inherent environmental gains and losses will satisfy no one: positive gains must be created further afield and so avoid rearguard actions against the environmentalists. How much better would it be for all concerned to have these peope queueing for engineering favours?

If this path is followed not only will the way of economic development be smoothed: there will be fewer disappointed engineers and a great many more employed in the construction industry.

REFERENCES
1. HYDRAULICS RESEARCH STATION. Severn tidal power: two dimensional water movement model study. Hydraulics Research Station, Wallingford, July 1981, Report EX 985.
2. THORN M. F. C. and PARSONS J. G. Erosion of cohesive sediments in estuaries: an engineering guide. Proceedings of the 3rd International Symposium on Dredging Technology, Bordeaux, March 1980. British Hydromechanics Research Association, Cranfield, Paper F1.
3. OWEN M. W. Discussion on An experimental study of silt scouring. Proceedings of the Institution of Civil Engineers, 1971, Vol. 48, Apr., 666.
4. PEIRCE T. J. et al. An experimental study of silt scouring. Proceedings of the Institution of Civil Engineers, 1970, Vol. 45, Feb., 231-242.
5. ACKERS P. Tidal power projects - Australia. The Colston Symposium on Tidal Power and Estuary Management, University of Bristol, April 1978.
6. PARKER W. R. and KIRBY R. The behaviour of cohesive Sediment, in the Inner Bristol Channel and Severn Estuary in relation to construction of the Severn barrage. Institute of Oceanographic Sciences, 1981, Report 117.
7. KIRBY R. An assessment of changes in the fine sediment regime associated with construction of a tidal power barrage near the Holms Islands in the Severn Estuary. Report to the Severn Tidal Power Group, 1984, unpublished.
8. KIRBY R. Opinion concerning likely sedimentation issues with a tidal power barrage located at English Stones in the Severn Estuary. Report to Severn Tidal Power Group, February 1985, 64, unpublished.
9. GIBSON A. H. Construction and operation of a tidal model of the Severn Estuary. Severn Barrage Committee Report. Economic Advisory Council. Her Majesty's Stationery Office, London, 1983, vol. I, 1-173; vol. II, 173-304.
10. KIRBY R. Suspended fine cohesive sediment in the Severn estuary and Inner Bristol Channel, UK. Energy Technology Support Unit, 1986, Report ETSU-STP-4042.

11. KIRBY R. and PARKER W. R. Settled mud deposits in Bridgwater Bay, Bristol Channel. Institute of Oceanographic Sciences, 1980, Report 107.
12. KIDSON C. and HEYWORTH A. The Quaternary deposits of the Somerset Levels. Quarterly Journal of Engineering Geology, 1976, vol. 9, 217-235.
13. METTAM C. J. An estuarine mud flat re-surveyed after forty-five years. Oceanologica Acta, 1983, SP 137-140.
14. KIRBY R. and PARKER W. R. Distribution and behaviour of fine sediment in the Severn Estuary and Inner Bristol Channel, UK. Canadian Journal of Fisheries and Aquatic Science, 1983, vol. 40, suppl. 1, 83-95.
15. PARKER W. R. and KIRBY R. Sources and transport patterns of sediment in the Inner Bristol Channel and Severn Estuary Severn Barrage, Thomas Telford, London, 1982, 181-194.
16. HAM J. On the mud deposited by the tidal waters of the Severn, Usk and Avon. Report of the British Association for the Advancement of Science, 1837, 76-77.
17. SOLLAS W. J. The estuaries of the Severn and its tributaries. Quarterly Journal of the Geological Society of London, 1883, vol. 49, 611-626.

13. Implementation, organisation and project planning

W. M. BUCKLAND, *Taylor Woodrow Construction Ltd*

SYNOPSIS. This paper summarises the work of the Severn
Tidal Power Group on project and organisation planning and
considers the venture capital risks of implementing the
project. The conclusions, together with those on
construction risk appraisal and financing, provide other
bases on which the project may be developed in the private
sector.

INTRODUCTION
1. A basis for successful financing depends both on the
project rate of return and on the potential investors'
perception of risk. This latter embraces the risks of the
market for electricity sales, risks associated with the
engineering technologies and construction methods, and the
quality of project planning and organisation for control of
risk.

2. This paper discusses conclusions drawn from work on
project and organisation planning and considers the venture
capital risks of implementing the project. In conjunction
with the studies on construction risk appraisal and
financing which are discussed in other papers, it provides
the basis on which Her Majesty's Government (HMG) can
consider the implications of supporting further development
of the project in the private sector. STPG had to rearrange
certain procedural steps to meet the requirements of the
project. Whilst assumptions had to be made, STPG recognise
the need for these to be the subject of further study and
consultation.

PROGRAMME FOR PROJECT DEVELOPMENT
3. The complete project period for the Cardiff Weston
barrage is planned for 15 years, comprising six years of pre-
construction development, seven years of construction up to
the generation of first power and an additional two years of
construction up to generation of full power. In the case of
English Stones, the corresponding project period is planned
as 12 years, comprising six years development as for Cardiff
Weston but with shorter construction periods, i.e. five years

to first power and one additional year to generation of full
power. The programmes for the project development periods
are shown on Fig. 1 Cardiff Weston and Fig. 2 English Stones.

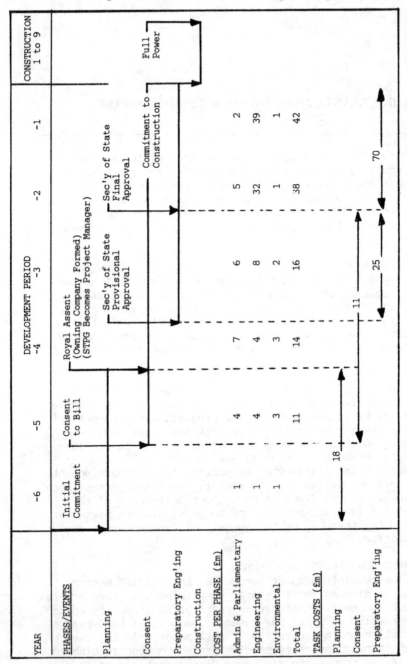

Fig. 1. Cardiff Weston - Programme for Project Management

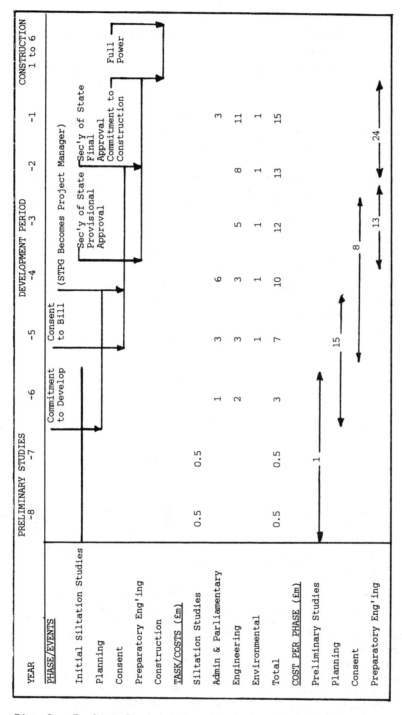

Fig. 2. English Stones - Programme for Project Management

4.   The key events which must be achieved in the six year pre-construction development period for either project are foreseen as follows:

(a) Initial commitment. After consideration of STPG report, agreement by HMG and the STPG on the funding of further development work.

(b) Promotion of a Private Bill. Agreement by HMG to provide parliamentary time for passage of a bill, provided that the development work following initial commitment so justifies.

(c) Secretary of State provisional approval. Approval of the project in principle, subject to further satisfactory development work meeting detailed safeguards incorporated in the Enabling Act.

(d) Secretary of State final approval. Following submission of a Project Report and Works Order provided for parliamentary debate and resolution.

(e) Commitment to construction. Following receipt by the Owning Company of commitments to funding of the construction cost.

5.   The period divides into three separate but overlapping phases:

Planning - from initial commitment until Royal Assent to the Private Bill.

Consent - from promotion of the Private Bill to the Secretary of State's final approval.

Preparatory Engineering - from the Secretary of State's provisional approval until the point of commitment to construction.

6.   At the conclusion of the six year period, given firstly that all legitimate concerns have been satisfied such that the Secretary of State's final approval has been given, and secondly, that firm commitments to funding have been secured, the Owning Company could commit to construciton.

7.   The costs prior to construction are summarised in Table 1 by tasks to be carried out, and in Table 2 in relation to the phases of development. The cost of the pre-construction development period for the Cardiff Weston scheme is £124m with an approximate allocation of £18m, £11m and £95m for the three phases of planning, consent and preparatory engineering. The corresponding costs for the English Stones scheme total £60m consisting of £15m, £8m and £37m for each of the three phases. The preliminary studies prior to initial commitment, which would only by required in connection with the English Stones

scheme, would cost an additional £1m. All costs quoted are related to January 1984 prices.

8. As this expenditure would be risk money (venture capital) with no certainty of reward until approval is secured for construction, the planning must be such that the venture capital would be called for progressively, only as confidence increased in the project being successfully promoted. This balance is fundamental to the rearrangement of the procedures.

Table 1. Costs Prior to Construction (£m)
(Summary)

	-8	-7	-6	-5	-4	-3	-2	-1	Total
				Years					
CARDIFF WESTON SCHEME									
Administration and Parliamentary Procedures			1	4	7	6	5	2	25
Civil Engineering and Related Field Investigations			1	4	4	6	30	35	80
Mechanical and Electrical Engineering						2	2	4	8
Environmental Studies and Related Field Investigations			1	3	3	2	1	1	11
Total for Cardiff Weston			3	11	14	16	38	42	124
ENGLISH STONES SCHEME									
Administration and Parliamentary Procedures			1	3	6	6	4	3	23
Civil Engineering and Related Field Investigations	0.5	0.5	2	3	3	3	6	8	26
Mechanical and Electrical Engineering						2	2	3	7
Environmental Studies and Related Field Measurements				1	1	1	1	1	5
Total for English Stones	0.5	0.5	3	7	10	12	13	15	61

9. Planning: The work involved in the planning phase would be principally (not necessarily in chronological order) as follows:

(a) Further definition of the project scope, with studies to refine, validate or amend the present engineering concepts.

Table 2. Costs Prior to Construction (£m)
         (Summary per Phase)

	-8	-7	-6	-5	-4	-3	-2	-1	Total
					Years				
CARDIFF WESTON SCHEME									
Planning			3	9	6				18
Consent				2	7	2			11
Preparatory Engineering: Up to Sec of State's Final Approval					1	14	10		25
After Sec of State's Final Approval							28	42	70
Total for Cardiff Weston			3	11	14	16	38	42	124
ENGLISH STONES SCHEME									
Preliminary Studies	.5	.5							1
Planning			3	5	7				15
Consent				2	2	4			8
Preparatory Engineering: Up to Sec of State's Final Approval					1	8	4		13
After Sec of State's Final Approval							9	15	24
Total for English Stones	0.5	0.5	3	7	10	12	13	15	61

(b) The planning of environmental and regional studies, including initiation of those of long duration, and the preparation of preliminary employment, regional economic and environmental assessments.

(c) A programme of consultation with interested parties, as an iterative process, for the further definition of the requirements of local interests in order to secure the effective development, construction and operation of the project.

(d) The development of the Owning Company and regional organisation structures necessary for the planning and execution of the project.

(e) Negotiation of critical aspects of the agreement for sale of electricity to the CEGB.

(f) The development of a financing scheme and sourcing of funds to the point of commitment in principle, on the basis of a preliminary prospectus.

(g) Development of the requirements for the Private Bill necessary to secure an Enabling Act.

10.   Until it could be seen that the planning undertaken in this phase provided justification for continuing development, private sector sources of capital, not having a specific interest in the project, might be unwilling to commit the funds necessary to secure consent to construction. Nor could it be expected that HMG would be willing to allocate parliamentary time for the necessary legislation. One solution would be for funding of the planning phase and the first year of the consent phase to be undertaken jointly by HMG and the STPG in agreed proportions, which would recognise, inter alia, that part of the tasks (a) to (g) above were undertaken by the STPG members themselves.

11.   It is assumed that the Owning Company would be function-ing at approximately the time of Royal Assent, and would there-after be responsible for providing the remaining £102m of the venture capital required (£43m for English Stones).

CONSENT

12.   A major consideration in the promotion of large infra-structure projects is that they require close public scrutiny of their advantages and disadvantages, mainly exercised by means of the public inquiry process. The outcome of such inquiries is uncertain until the relevant Secretaries of State give approval or otherwise to the Inspector's report. This high degree of uncertainty, coupled with the heavy expenditure necessary to prepare and present evidence to a major inquiry, effectively rules out the use of that process for private sector promotion of the Severn Barrage scheme. It is therefore proposed that the alternative, well proven, procedure of a Private Bill should be used. This provides the opportunity for parliamentary debate. An Enabling Act, if passed, would have established both the need for the scheme and the requirements for public safeguards which have to be satisfied before the Secretary of State gives final approval.

13.   The concerns of relevant authorities, for example, the Port Authorities, might be met by the negotiation of project agreements between the Owning Company and the authorities concerned. These agreements would determine obligations which the Owning Company would be legally required to observe. It might not be possible to resolve certain key detailed aspects by project agreements. In such a case an appropriate form of local public enquiry would need to be devised to enable these aspects to be resolved by the appropriate Secretary of State.

14. The principal work involved in the consent phase would

therefore be as follows:

(a) The drafting and presentation of, and support for, a Private Bill leading to an Enabling Act.

(b) The negotiation of project agreements between the Owning Company and authorities having specific interests in the region.

(c) The preparation and submission of evidence to local public inquiries.

(d) The further development of financing agreements and the organisation for finance.

(e) The initiation, or continuation for long duration tasks, of environmental studies, including further development of an environmental assessment.

(f) Civil engineering site investigation and design and turbine generator design to provide both definitive bases for tender and the design information necessary to improve confidence in cost estimates and avoid later changes.

15. At the conclusion of the consent phase, the project requirements would have been satisfied to the following extent:

(a) An Enabling Act would exist.

(b) The Owning Company would have been incorporated.

(c) Local Public Inquiries would have provided reports satisfactory to the Secretary of State, possibly with requirements for additional work or other obligations, the cost of which would have to be included in the project or operating budgets.

(d) Project agreements would have been concluded, again with the possibility of additional costs, to be provided for in the project or operating budgets.

(e) An environmental assessment would have been finalised, defining to the Secretary of State the balance of advantage and disadvantage arising for installation of the Barrage.

(f) Engineering design would have been developed in essential respects.

(g) The basis for financing would have been developed, both in terms of the financing organisation and the financing agreements.

16. The project thus defined would, at the conclusion of the consent phase, be submitted to the Secretary of State as a Report and a Works Order would be tabled, seeking approval subject to Parliamentary Resolution.

17. The major part of the estimated expenditure required in the preparatory engineering phase, £70m for the Cardiff Weston scheme and £24m for the English Stones scheme, would be incurred after the Secertary of State's approval, at which time the risk of the project not proceeding would be much less than during the consent phase. This work would be to complete the design work, secure and evaluate tenders for the civil engineering construction and equipment supply, finalise cost and project programme and prepare a formal prospectus on the basis of which the commitments to finance in principle could be converted to firm commitments. The prospectus would update and amend earlier estimates of cost and programme, and provide the detail of the agreement for sale of electricity negotiated with the CEGB during the consent phase.

18. The approximate allocation and timing of expenditure, related to key events, is shown in Fig. 1 and Fig. 2 for the Cardiff Weston and English Stones schemes respectively.

ORGANISATION FOR PROJECT DEVELOPMENT AND REALISATION
19. Reference has been made to the planning phase task of developing the Owning Company and regional organisations necessary for the planning and execution of the project. The organisational structure proposed is illustrated in Fig. 3 and may be summarised as follows:

(a) A promoter to initiate development of the implementing organisations and financing proposals;

(b) An owner company;

(c) A joint committee of directly interested local authorities having all-purpose-as-necessary jurisdiction to be the local authority for the Barrage;

(d) A joint committee of water authorities;

(e) A regional joint committee of interested local authorities with responsibilities for the necessary works and regulation of the use of the water above the Barrage.

20. The policy for the proposals developed has been that they should use, to the maximum, the existing local authority structure and legislation, with the objective that new organisation concepts and working methods do not have to be learnt in the early stages of project development. The same objective has been applied in the development of the Owning Company organisation. Throughout the entire period of project development and realisation, continuity would be provided by the STPG in undertaking the work and in supervising its physical translation into assets during construction.

21. At the stage of initial commitment to further development, when HMG would have agreed in principle the extent of its

● PLANNING

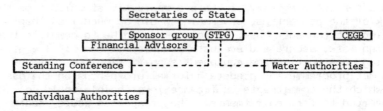

● CONSENT

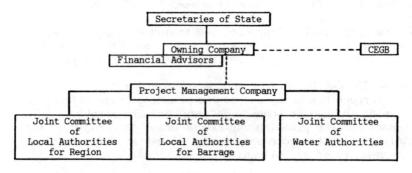

● PREPARATORY ENGINEERING AND CONSTRUCTION

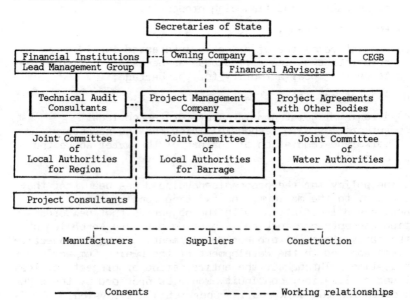

Fig. 3 - Evolution of Organisations

financial and legislative involvement in the project, the STPG
would be acting as Promoter. In this role it could provide
part of the funding required for the planning phase; it could
also seek commitment in principle to equity participation (for
venture capital requirements) in a putative Owning Company.
With this arrangement the appropriate role for the STPG itself
would be to evolve into a limited liability company, responsible for project development, design and project management.
It would be for individual partners in the STPG to decide if
an equity interest in the Owning Company should be taken,
although it is probable that the reward to the STPG for
contributions to the funding of the planning phase would be by
issue of equity in the Owning Company.

22. At the time of Royal Assent, it is expected that the
Owning Company will be functioning with equity commitments
sufficient to fund the remaining work to take the project to
construction commitment, together with the amount necessary to
refund and reward the promotional costs incurred by HMG and the
STPG. It is at this stage that the STPG would be reconstituted
as the company responsible for project design and management,
capitalised by the STPG partners. The STPG would then undertake
the work necessary to secure final consent, on a reimbursable
basis. Following Secretary of State's approval for the
project at the conclusion of the consent phase, the STPG would
undertake management of the project.

Acknowledgement

The author acknowledges the permission of the Department of
Energy to publish this paper and the assistance of colleagues
in STPG in its preparation.

14. Regional infrastructure and employment implications of tidal power barrages in the Severn Estuary

T. L. SHAW, BSc(Eng), PhD, *Sir Robert McAlpine & Sons Ltd*

SUMMARY This paper reviews the work done by the Severn Tidal Power Group in reaching its findings on the regional infrastructure and employment implications of the proposed Cardiff-Weston and English Stones barrage schemes (Ref.1).
By drawing on the engineering work presented in earlier Papers to this Symposium, conclusions are reached regarding the possible scale and extent of the effects of the construction phase in areas away from the Severn Estuary where work is done, as well as in direct association with work on either barrage site. Reference is made to remedial measures made necessary by either project to coastal installations elsewhere within the estuary.
The principal demands made on regional infrastructure following barrage commissioning are assessed in terms of their likely impact on the local community. Growths in tourism and water-based leisure industries are amongst the more likely to emerge rapidly following completion of either project, if not earlier. Industrial, commercial and service sector activities are also expected, depending on associated investment and promotion levels. Improved transportation across the estuary using the barrage to carry a public road could add confidence to the realisation of new opportunities, especially in South Wales.
The Paper concludes with an assessment of employment opportunities created at all phases of each project and the revenues which each would bring to the Government from various forms of taxation and reduced benefit payments.

INTRODUCTION

1. As with any power station, there are many (if not more) reasons why the promotion of a tidal barrage has to recognise its wider implications. These must be brought into focus during both the construction and operation phases, some more so during one than the other. Whereas the issues involved resemble those faced when promoting any power station, important differences must also be recognised. The purpose of this Paper is to explain how these have been assessed in the context of barrage projects in the Severn Estuary, in particular as regards the immediate implications stemming

from the construction of the scheme and, quite separately, from the effects of its presence on wider commercial and public reactions (Ref.1).

2. The price of electricity production by any power station depends on the combined effect of three factors, namely the capital, fuel and operating/maintenance costs of that station. For each type of station, these factors exert a different influence on all-up costs. Tidal power, which is in essence only a variant of hydro-electric power, has the advantage of zero fuel and low o/m costs. Its value is therefore essentially determined by its capital cost - a component particularly worth minimising because it carries the front-end financing penalty. (see Papers 10 and 11)

3. Papers 3-5 described the main elements of a barrage and the forms of construction now preferred. The opportunity to adopt a modular approach, thereby permitting use of quantity production methods, was emphasised. The bias in the total cost estimate towards that for civil works is unusual by thermal power station standards. It is the main reason why the barrage cost, expressed in terms of £/kW installed and recognising plant load factor (Paper 6), is relatively high.

4. Paper 1 reported that the Cardiff-Weston barrage (7200 MW) should be fully operational after about 9 years, whereas the English Stones scheme (1050 MW) would reach this point after 6 years. These periods closely resemble those for similarly sized thermal stations. Capital expenditure patterns are also alike, both display the characteristic S-shape.

5. It follows that the employment created during the construction of either barrage would be of the form shown in Fig. 1. As Papers 3-6 indicate, this would be biased more to civil than to E&M work, though all would be heavily in demand on a repetitive basis. Paragraphs 32-34 below summarise the procedures used in preparing these estimates.

6. It also follows that since the main cost centre for a tidal barrage relates to its civil works content, its total employment implications will largely relate to where the caisson construction, etc. are carried out (Paper 3). However, in terms of total employment created (in person-years) per unit of electricity generated annually, a barrage is more demanding during the construction phase than thermal power stations.

7. Conversely, when operating, a barrage would create relatively little employment (associated with low o/m costs). However, this does not necessarily mean that no other jobs would result, nor that the number of these would be small. In fact quite the opposite may be true. To a certain extent this would depend upon the investment of additional finance in support facilities, which in turn will require that opportunities are seen and promoted.

8. A barrage could therefore affect regional infrastructure in many ways. Some issues, like commercial shipping and agriculture, for which the efficiency of land drainage is

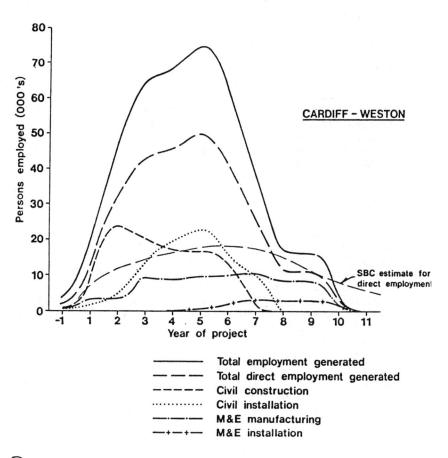

CARDIFF – WESTON

SBC estimate for direct employment

——————— Total employment generated
— — — — — Total direct employment generated
— — — — — Civil construction
················· Civil installation
—·—·—·— M&E manufacturing
—+—+— M&E installation

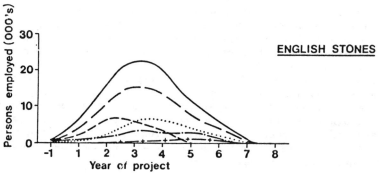

ENGLISH STONES

Fig. 1

determined by water levels in the estuary, would be
influenced directly and unavoidably by the scheme, whether
for better or for worse. Others like land transportation and
tourism would depend on decisions being taken over and above
those of barrage construction for power generation. Between
these extremes there are items like water-based recreation
for which an expansion of facilities would be further
encouraged by the inclusion for suitable lock facilities in
the barrage.

9. The sum total of the wider impact of a barrage is therefore difficult to judge in advance of a comprehensive look at
future pressures for the development of both the area likely
to be influenced by a barrage and the activities which it
either would or could then support. Also, it is necessary to
know how these pressures may be accommodated within individual County Structure Plans which, in their present form,
probably have little or no regard for them because they are
more or less foreign to existing development opportunities.

10. This Paper will not address the important question of
how Structure Plans are best adjusted to deal with the new
situation, nor how revised controls are best applied. The
case for general guidance to be developed at more of a
regional than County level is endorsed by the fact that some
impacts may be on that scale, for example the up-grading of
public highways. On that basis, a less parochial approach to
deciding project impact may be essential for maximum value to
be gained at all levels.

11. In this respect it will take time to realise full
opportunities and subsequently develop new ones, hence to
secure maximum annual returns on 'non-energy' investments.
The significance of these to the Exchequer in terms of
overall returns from the basic investment in the barrage,
however this is carried, could assume considerable importance
in decisions taken about the wider viability of the scheme.
Furthermore, the so-called indirect or 'induced' impact of
each direct effect is likely to be equally important in its
own way in determining the full and changing impact of the
scheme over its lifetime (a term meaningful in the context of
thermal stations but of little significance for hydro, of
which tidal power is a variant).

12. The general observations reported below represent
conclusions reached by the Severn Tidal Power Group (STPG)
from their recent broad assessment of the issues involved.
Although their appraisal was based on discussions with a wide
range of representative organisations, the observations made
remain both preliminary and in no way a commitment on those
who have given advice.

REGIONAL INFRASTRUCTURE - CONSTRUCTION PHASE

13. Paper 3 shows that the site of the barrage bears little
if any relationship to where the caissons and their machinery
are best constructed. That site is simply where all the
components finally come together. Where these are

constructed and (for the plant) most sensibly assembled and tested before transport to site raises totally unrelated issues. STPG sees no reason why the plant has to be tested in the Severn Estuary region, indeed there is a case for this to be done in other locations. The choice of caisson construction sites, while not considered in detail by STPG, appears to be less constrained, i.e., there would be more of an effect regionally if some at least of this work was done more locally. However, similar effects would be felt wherever that work took place, in just the same way as would plant assembly and testing in one or more locations.

14. Until possible caisson construction and plant assembly sites have been explored in detail there is no means of judging where these could sensibly be located. In the Severn Estuary, most activity is likely to focus on the barrage site itself. STPG have estimated the proportions and time-distributions of capital expenditure on civil and E&M works which would be committed by each operation. These are identified on Fig. 1.

15. The impact made by each element of the construction phase on the region in which it is sited would depend on the demands it makes on the supply and movement of suitable labour and materials, including the extent to which each is available locally. Together with the provision of land and other support services including, for the caisson construction sites, the availability of adequately deep water close to shore (Paper 3), these issues will determine the acceptability of specific sites for each operation. The single exception is, of course, the barrage site itself, which may be decided with less reference to these factors than to the overall economics of the project and its environmental implications.

16. In this respect, the use of floated-in caissons for most of the barrage not only permits installation work to start at and proceed in either direction from two or more offshore sites but, in so doing, it ensures that much of that operation is necessarily remote from the coastline and hence largely without direct interaction with the public. It may also help to ensure that those so employed could reach that site from one of several directions, in which case the many established sources of labour around the estuary become accessible and that, where temporary accommodation is needed, a wide range of facilities lies within easy reach.

17. Supplies of construction materials from onshore to either barrage site would then be limited to any stone needed for caisson foundations as well as that required to armour the external faces of the embankments against erosion. To withstand wave attack, layers of suitably heavy stone must protect the sand from which the core of the embankments is most logically formed (Papers 3 and 4). The protective breakwaters at the lock entrances would also most economically be formed of stone, the enclosed area and hence breakwater size depending on the sheltered zone needed, i.e.,

on the ship sizes and numbers expected to pass through the locks. Whereas it may be sufficient to supply all sand from natural marine deposits close to each barrage site, the rock would most logically be delivered by barge from one or more local jetties, having been moved there by rail or perhaps by road. An increasing use of rail and ship-connected quarries may make this the more acceptable option for the barrage by limiting this aspect of the impact of its construction on the community.

18. The likely significance of the 'non-energy' factors which may both affect and be affected by each proposed Severn Barrage scheme is outlined in paragraphs 23-31 below. An estimate is given of their combined economic impact relative to that of the power generated by each scheme (paragraphs 37-38).

19. In addition to this indirect consequence of a barrage, it will be necessary during the construction phase to carry out accommodating works to establish that infrastructure away from the barrage made necessary by the way it would alter the tide regime when commissioned. For example, sewage outfalls, sea defences and land drainage services would all need some attention. Allowances for these are included in the all-up capital cost estimate for each scheme. It is noteworthy that the burden imposed by these items, to which must be added maintenance dredging and other aspects of sedimentation presented in Papers 12 and 13, is relatively small in the context of a Cardiff-Weston barrage compared with that proposed for English Stones. Indeed it may be the case that, when sedimentation is fully evaluated, the cost implications of the latter scheme may outweigh those of the former. Other small barrage schemes built towards the head of highly turbid estuaries may also carry this economic penalty, partly because of their size but also because of their environmental impact in that situation.

REGIONAL INFRASTRUCTURE - OPERATIONS PHASE

20. Towards the end of the main construction phase of either barrage and before full commissioning (i.e., in year 7 for Cardiff-Weston and year 5 for English Stones - Paper 1), the tide ranges landward of either scheme would be changed to post-barrage conditions (Paper 13), a public road across the scheme would be opened (if such was by then incorporated into the structure), new marinas might be brought into use in anticipation of early growth in demand for water-sports, etc. Paragraph 9 noted that it is premature to do more than speculate as to the nature and scale of the non-energy developments which may be attracted to the region if suitable facilities to accommodate them are provided. A review of what these might comprise for each barrage is given in paragraphs 23-31. In addition, barrage operation would create various demands, both directly for the control and maintenance of plant, and less directly in connection with ship locking operations, possible extra effluent treatment

works, pumped drainage outfalls unable to discharge adequately by gravity, etc. Paragraph 32-33 refer to the employment implications of these demands.

21. The modified tide regime created in the basin landward of the barrage is explained in Paper 2. The preference for ebb generation with pumping around high tide, more so on neaps than springs, is now the widely accepted method of barrage operation. The fact that the Rance Barrage is equipped with machines which pump and generate in both directions does not mean that similar flexibility need be incorporated into other tidal schemes. Planned as a demonstration facility, Rance has ably shown the relative merits of all forms of operation. The fact that it is now little used in other than the two modes proposed for newer projects not only confirms the lessons learned in this connection but is said to show that the other two modes are only used because they can be, rather than that the economics of a project designed to omit them would in any way suffer economically.

22. It follows that water levels in the basins of both Severn Barrage schemes would vary on spring tides from about normal high water to about +2 m O.D. (present mid tide level, approx.), and on neap tides from about 1 m above present normal high water to about +1 m O.D., i.e., a smaller variation in spring-neap ranges than occurs naturally. This is a consequence of the pumping operation.

23. Permanently deeper water, much reduced currents and an accurately predictable tide range in the basin (perhaps more so than at present) are expected to improve conditions for commercial navigation to and from the enclosed ports, i.e., Sharpness, Lydney and Chepstow for both barrage schemes, but also Cardiff, Newport, Avonmouth and Royal Portbury in the case of the Cardiff-Weston barrage (Fig. 2). For the latter scheme, the barrage locks could be in sufficiently deep and accessible water for all likely ship movements to and from the barrage on either side to be possible at all states of all tides. Only for ships of over about 100,000 tonnes might it be necessary to impose some constraints for economic reasons, limiting their freedom around low water while still markedly reducing present access problems. Shallow water for some distance seaward of English Stones would make a barrage in that location less accessible in that direction, and continued shallow water around low tide at the ports contained by that scheme would do little to improve local access to them at that time (as also would a Cardiff-Weston scheme).

24. Whether or not this would generally help rejuvenate commercial transport to and from the estuary's ports depends on many factors identified but not yet researched in detail by STPG. The location of the Severn Estuary relative to both Atlantic trade routes and centres of UK demand and production, the motorway and rail links, prospects for a Channel Tunnel and the forecast continued increase in population in the S.West of England (Ref.2) suggests that the

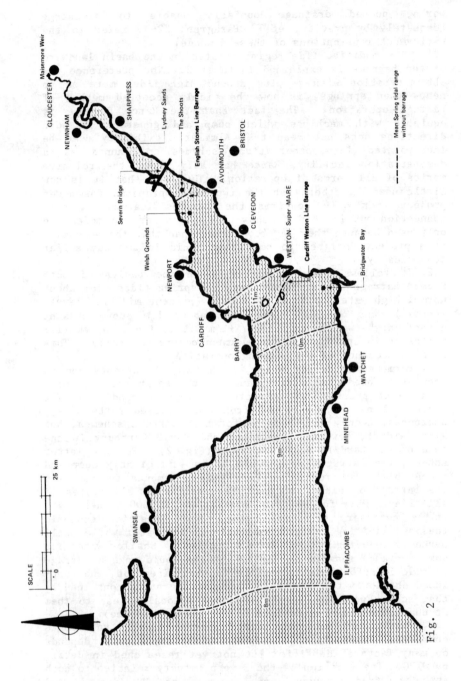

Fig. 2

potential for growth exists, even if trends over the past two decades have mainly brought declining demands. Whereas the barrage itself would not, for its operation, create new trade of this sort, the fact that it would, by its operation, remove the need for vessels to shelter behind locks at individual ports could allow unloading/loading operations to be done at new open jetties off existing docks, unfettered by the timing and range of the tides.

25. The prospects for such changes in both the shorter and longer terms and the case for and means of encouraging them to justify specific levels of capital investment in both the barrage locks and at individual ports remains to be studied in detail. As with other 'non-energy' aspects of a barrage, the full prospects are likely to hinge on the success of marketing operations, which in turn are likely to be more appealing to customers if their commercial interests are seen to be fully acknowledged in the form of management set up to ensure safe and speedy passage for all vessels as and when required between the open sea and individual ports.

26. Among the various problems to be resolved, one may be the segregation of commercial from recreational use of the estuary. The scale and nature of each activity over and seaward of each basin area (about $500 km^2$ for Cardiff-Weston and $100 km^2$ for English Stones) will determine the need for and type of controls required. Although minimal at present because of the high tides, strong currents and wide intertidal areas which inhibit water sports, these could become the largest growth industry stimulated by either barrage. The Cardiff-Weston location has most to offer because it would embrace existing centres like Penarth, Weston and Clevedon at which expansion may be particularly appropriate. The large local population and those from further afield able to use the good communications to Severnside could quickly create major demands in this sector. Comparisons with Southampton Water and The Solent are instructive. As in the Hamble and Beaulieu river mouths, a barrage would transform access to and the pleasure derived from sheltered areas like the estuaries of the rivers Taff, Usk, Wye and Avon, places which at present are made difficult by tide conditions.

27. The infrastructure needed to support leisure activities of this type and the rate at which it may have to be provided should be appreciated in advance of its realisation. Many people believe that this aspect of a barrage will in time come to be publicly regarded as more significant than its energy output, whereas simple economics readily suggests otherwise. With it must be linked all aspects of tourism, whether to the barrage as an attraction in its own right (hydro-power schemes generally hold this appeal and a Severn Barrage could be even more popular because of its ready accessibility to some of the UK's main holiday routes), or because it adds to the existing list of visitor facilities available on both sides of the estuary and could therefore increase their appeal by incorporating them into both formal

and spontaneous tours centred on a barrage.

28. Public access to the barrage and hence generally across and around the estuary would be improved by the addition of a roadway of appropriate standard along the barrage. As with commercial and recreational sailing, the likely demand on such a transport link has to be judged from such time as it could be available, i.e., from about the year 2000 at the earliest (Paper 1). This demand would comprise growth in existing traffic up to and beyond that date plus that added directly and indirectly by the barrage, divided between the various routes then available. In this respect, the location of the barrage relative to existing traffic 'desire lines' and how it might modify these if it provides an additional crossing must be assessed. The M25 is perhaps a relatively extreme example of how desire lines can be largely reorganised by the provision of a new route, though not in that case one which also directly creates new reasons for travelling over and above its intended purpose to reduce journey times and hence encourage mobility.

29. The cost of adding a public road to a barrage may have to be justified by the service it provides. The road standard, number of lanes and provisions made for crossing the ship locks would have a considerable influence on this cost. By being close to sea level and on a robust structure, the road would not be subject to the weather and other delays which occasionally affect traffic on the Severn Bridge. The way in which use of these two routes would interact will determine the traffic carrying capacity of the barrage. The greater additional demands stimulated by a Cardiff-Weston scheme and its complementary location to that of the Severn Bridge in respect of the transport desire lines which may then prevail suggest that a 4-lane, non-motorway link may be justified on that route, similar to that on the Rance Barrage and which now carries about 6M vehicles/year (about 50% of that of Severn Bridge).

30. A general increase in industrial, commercial and service sector activity around the estuary is expected. This would in part be stimulated by local investment directly associated with the construction phase, this continuing and perhaps further expanding thereafter as new opportunities encouraged by improved access, the addition of greater use of the estuary to other local recreational pursuits, and the expected continued growth in population in neighbouring Counties (which the barrage might well amplify) became accepted. The rate of response to these opportunities and hence their implications for the provision of housing and its associated social infrastructure are clearly vital if prospects are to be realised fully and efficiently.

31. Each barrage could affect land use in low-lying areas close to the estuary, especially where this depends on drainage by gravity during the lower part of the existing tide cycle. The higher low tide level created over much of the basin area (smaller increases would occur in the

tributaries) would reduce the efficiency of much of this system and make pumping necessary to maintain the status quo. Agriculturalists may appreciate the advances which this could bring through an ability to control groundwater levels more closely than is possible at present. Conversely, those keen to preserve and encourage the wildlife which inhabits these coastal margins fear that this could lead to 'improved' drainage and a further loss of habitats, though changes now taking place in agricultural management, production and control, coupled with the relative ease and not excessive cost of providing differential drainage standards to suit local land use functions suggests that there is room for various practices to be implemented simultaneously.

EMPLOYMENT IMPLICATIONS

32. Paragraphs 5-6 noted that the economics of a barrage is largely determined by its capital cost. A higher than normal proportion of this by thermal power station standards would be committed to civil works, both at and in support of caisson construction yards as well as at the site of the barrage. Fig. 1 shows the anticipated distributions of expenditure on each part of the work, and Ref.1 shows how this may be converted into employment estimates including jobs created indirectly ('induced') as a consequence of primary expenditure. Totals are given in Table 1.

Table 1 Employment opportunities

	Direct Employment		Induced Employment	
A. CARDIFF-WESTON BARRAGE				
Construction Phase				
(a) Primary	44 600	persons*	22 300	persons*
(b) Secondary	2 100	persons*	1 050	persons*
Operation Phase				
(a) Primary	1 770	persons	570	persons
(b) Secondary	21 500+	persons	6 900+	persons
B. ENGLISH STONES BARRAGE				
Construction Phase				
(a) Primary	12 300	persons*	6 150	persons
(b) Secondary	1 400	persons*	700	persons
Operation Phase				
(a) Primary	350	persons	110	persons
(b) Secondary	3 000+	persons	960+	persons

*These are the average numbers of persons employed during the construction phase. Total employment for either scheme is given approximately by multiplying by 6 years of work for the Cardiff-Weston barrage and 4½ years of work for the English Stones barrage respectively. Some tasks would extend for up to 8 and 6 years, respectively, whereas others (mainly 'secondary') would be of shorter duration.

33. Those other essential works in and around the estuary needed to reorganise services adversely affected by a barrage (paragraph 19) would also create both direct and induced employment. Table 1 gives details.

34. STPG's employment estimates have been questioned on the basis that the cost allocation per person-year of work created is high, also that the translation of primary sector investment into induced employment underestimates the size of workforce thereby sustained. Totals up to 40% higher than those given in Table 1 have been suggested. STPG are prepared to accept that their estimates might be low but, in view of the dependence of the project on heavy and expensive plant, the true employment totals may not be much above those shown in Table 1.

35. Ref.1 also includes detailed consideration of post-construction employment on and in direct association with the barrage, including operation and maintenance of its generating plant, the ship locks and those other directly associated services like effluent treatment and land drainage plant referred to in paragraph 19. An estimated total of 1770 people would be directly associated with these continuing tasks for a Cardiff-Weston barrage, and 350 for the English Stones scheme. In turn, these commitments would induce further employment estimated at 570 and 110 persons for the two schemes.

36. Paragraphs 23-31 covered many of the major infrastructure developments which may take place in and around the estuary because a barrage is in place for power generation. It was emphasised that the rate at which each would grow and the scale duly achieved would, on the one hand, depend on further investment, promotion and good management, and on the other on a planning policy which respects relevant constraints and encourages appropriate opportunities.

37. At this early stage it is difficult to estimate the employment attainable in each secondary area. STPG looked for parallels elsewhere, for example in the water-sporting industry around The Solent. Numbers employed in tourism as a proportion of the residential populations of Severnside were compared (unfavourably) with those so engaged in other similar regions. Increased industrial and commercial activity resulting from improved communications, etc., is less straightforward to judge. Ref.1 sets out some of the relevant factors determining employment levels. Together with the other items referred to in this paragraph, the estimated total annual levels of direct, and hence of induced, employment in this secondary sector during the operations phase are given for each barrage scheme in Table 1. These figures refer to the situation which could prevail 5-10 years after barrage commissioning, following a period of growth and rationalisation in each sector.

38. Investments made during each phase of either project would in part flow to The Exchequer as personal taxation,

Corporation Tax, rates, reduced benefit payments and VAT
payable by the 'induced employment' sector.   (VAT is not
levied on capital works.)   Provisional estimates of these
items suggest that total Government revenues during the
construction phase of either project would amount to 40-50%
of capital cost estimates. When operating, and assuming the
secondary employment estimates given in Table 1, these items
would add up to a similar proportion of the annual revenues
received from the generation of electricity.

CONCLUSIONS

39. It cannot be said that the implications of each barrage
project on regional infrastructure and employment are in all
respects in proportion to their capital costs, though this
factor has a close bearing on the numbers of people involved
during the construction phase and with barrage operation.

40. The location of each scheme relative to the main centres
of population and to the ports may have a significant bearing
on the extent to which the public is influenced by water-
based activities, whether commercial or recreational.   A
public road on either barrage may be justified on the grounds
that it increases cross-estuary transport capacity at a time
when pressures caused by the existence of the barrage are
likely to add appreciably to that demand.

41. In order to maximise the 'non-energy' opportunities
arising from a barrage, considerable further investment and
promotion against the background of a regional policy will be
required.   Commercial decisions regarding the timing of
commitments   should   ensure   that   opportunities   are
progressively realised. Good transportation is likely to be
a major factor in determining response rates.

42. Government revenues arising from the construction and
operation   phases   of   either   barrage   are   provisionally
estimated to amount to 40-50% of their capital costs and
annual revenues from the sale of electricity, respectively.
These sources of income have not been taken into account in
the economic calculations for each project presented in Paper
10.

43. A better understanding of the implications of a barrage
on regional infrastructure would permit many of the qualified
statements made in this Paper to be viewed more objectively.
To create that understanding requires that a constructive and
interactive approach to planning is adopted. Where outright
constraints need to be applied, they must be identified and
acknowledged as such. Otherwise, a flexible attitude must be
adopted within which potential opportunities are viewed
against a background of existing Structure Plan objectives,
while bearing in mind that the main impact of either scheme
would occur post-2000, when social requirements and attitudes
will differ from those prevailing now. The thinking required
to shape either project deserves to be on the scale of the
widespread   implications   which   it   would   have   on   the
industrial, commercial and social activities of the affected

community. Present indications are that a rational approach to this task will produce generally beneficial solutions, especially if regular and constructive public consultation is an acknowledged part of the planning process.

REFERENCES

1. SEVERN TIDAL POWER GROUP "Tidal Power from the Severn", Rpt to UK Secretary of State for Energy, 1986.
2. OFFICE OF POPULATION CENSUSES AND SURVEYS "Population Projections 1983-2001", (for England and Wales), HMSO, 1986.

ACKNOWLEDGEMENTS

On behalf of the Severn Tidal Power Group, the author wishes to record his appreciation of the advice readily given by the many organisations consulted during the course of the studies to which this Paper makes either direct or indirect reference. However, the interpretation of this advice and the conclusions drawn from it are those of the author alone.

Appreciation for permission from the UKAEA to publish this Paper is also recorded. The Authority part-funded the study reported in Ref 1, on which much of the Paper is based.

15. Steel caissons for tidal barrages

A. W. GILFILLAN, MSc, FRINA, *Manager, Ship Design Group,* and
G. C. MACKIE, MSc MRINA, *Senior Section Head, YARD Ltd,* and
R. P. ROWAN, BSc, MICE, *Associate, Roxburgh and Partners*

SYNOPSIS. YARD LTD and Roxburgh & Partners were commissioned
by the Department of Energy through ETSU to investigate the
feasibility of constructing the caissons for the Severn
Barrage in steel (Ref. 1). This paper presents the work
carried out on the Turbine Caissons and discusses the
selection of steel as a material and corrosion control. The
proposed method of installation and the overall impact on the
cost of the barrage are described. Finally, the Sluice over
Turbine concept is described, and the implications for the
overall cost of the barrage is assessed.

EVOLUTION OF THE STEEL CAISSON DESIGN CONCEPT
1. The initial objective was to evolve steel caisson
designs equivalent in duty to the concrete reference designs
but utilising to best advantage the properties of steel and
modern shipbuilding construction methods. It was therefore
necessary to differentiate between the functional constraints
on the design of the caisson and the material constraints of
a concrete structure.

Steel Equivalent Design
2. In order to draw comparisons between steel and
concrete structures, a steel equivalent of the concrete
reference design (Ref. 2 - Fig. 1 (a)) was prepared on the
basis of a 3 turbine unit, as shown in Figure 1 (b). A
simple costing calculation was made based on the steelweight
and ballast volume, corrected to a base cost of 100 to be
used as a comparison with other arrangements. The ballast
requirement was based on a minimum factor of safety against
sliding or overturning of 1.2 for the worst loading
condition. The resulting comparison between the two
materials can be summarised as follows :-

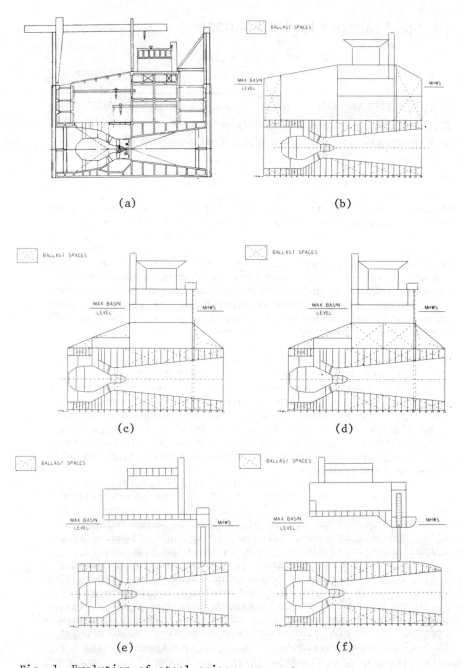

(a)

(b)

(c)

(d)

(e)

(f)

Fig. 1. Evolution of steel caisson concept

	Concrete	Steel
Mass of caisson	76,500 tonnes	26,200 tonnes
Concrete ballast	)	86,400 tonnes
	) 91,200 tonnes	
Water ballast	)	55,100 tonnes
Total mass	167,700	167,700 tonnes
Mass ratio concrete/ steel	3.3	
Strength ratio concrete/steel	0.3	

3. To take advantage of the lighter weight of steel required to achieve the same strength, it is necessary to reduce the buoyancy of the caisson to enable a less dense and therefore less costly ballast material such as water to be used.

Development of an Economic Concept

4. A study of the reference concrete caisson design indicates that much of the internal volume was devoted to ballast spaces and the removal routes for the complete bulb turbine and the transformer, operations which would be carried out at intervals of 20 to 30 years. This excess volume penalises the more buoyant steel structure and requires much more ballast to hold it in place. Schemes for reducing the internal volume were therefore investigated.

5. The Arrangement, shown in Figure 1 (c), has the rectangular wing ballast spaces and maintenance/removal area replaced by sloping wing ballast tanks and watertight access to the turbine. Total removal of the turbine and supporting structure would be by floating crane barge, which could also be used to position the gates used to enable the draught tube to be dewatered.

6. The reduction in volume was still insufficient to enable the ballast requirements to be met by water ballast alone. One solution to this problem is to increase the percentage of internal spaces given over to water ballast by moving the transformers up one level and flooding the original transformer space. This is shown as Figure 1 (d).

7. An alternative approach is to reduce the underwater volume to the absolute minimum consistent with structural integrity of the draught tube and access requirements to the turbine machinery. In the arrangement shown in Figure 1 (e), the machinery spaces are sited above the highest waterline supported on double plate buttresses between each draught tube. A logical development of this arrangement is the sluice over turbine concept, shown in Figure 1 (f), and as described in more detail later.

Table 1. Comparison of turbine caisson options

Arrangement No. See Figure		1 1 (b)	2 1 (c)	3 1 (d)	4 1 (e)
Steel	(tonnes)	20060	17230	17230	17300
Machinery & outfit	(tonnes)	6140	6140	6140	6140
Concrete ballast	(tonnes)	86400	87600	64460	69940
Water ballast	(tonnes)	55100	34400	57540	33060
Total mass	(tonnes)	167700	145370	145370	126440
Relative cost	index	100	90	88	89

8. Table 1 compares the principal particulars of the turbine caisson options. Although the relative cost of arrangement No. 4 is marginally higher than arrangment No. 3 due to increase in steelweight and solid ballast requirements, the layout offers considerable benefits in heavy plant maintenance and removal. Arrangement No. 4 was therefore selected for more detailed development.

MATERIAL SELECTION AND CORROSION CONTROL
9. An instinctive engineering response to the idea of immersing a mild steel structure in sea water for a period of 120 years with no removal for maintenance is that it is impossible to guarantee that structure. However, this reaction is based on experience with more conventional structures such as ships and offshore platforms where a 20 year design life, minimal corrosion margin and a planned corrosion control system are built into the design. By increasing the corrosion margin, improving the corrosion control system and in certain key areas adopting special steels it is possible to arrive at a structure that at the end of its 120 years life will be strong enough to sustain the loadings it was designed for from the outset.

Corrosion Processes
10. The severity of the corrosive attack on steel in seawater varies mainly according to the temperature, salinity, oxygen content and velocity of the water. Pollutant in the water such as ammonia can lead to increased attack on steel. The oxygen content and salinity of the water at the barrage will be little different from that of sea water. Effluent discharged into the Severn estuary may affect the growth of marine fouling on the structure. Marine fouling is beneficial to the barrage as it reduces the attack on steel by reducing the velocity of oxygen carrying seawater and acts as a diffusion barrier to oxygen. However, marine fouling is unlikely to occur in the area of highest flow rate - the draught tube.

11. Water low in oxygen but flowing at high velocity can be as corrosive as water, high in oxygen, but moving more slowly. Where the water flows at high velocity through the draught tubes and sluices of the barrage the rate of corrosion will be much greater than that associated with normal tidal movement. In the region of highest flow rate adjacent to the turbine blades, the corrosion rate of mild steel would be 6 times that of the immersed tidal zone. This increased rate of attack is due to the removal of the rust barrier by the high velocity water, the increased availability of oxygen and possible erosion from the silt content in the moving water.

12. Second in risk to the draught tubes and sluices is the splash zone of the caissons which extends from 2 m above the high water spring tide level to the low water neap tide level. The structure in this area is continuously wet with well aerated seawater and the rate of corrosion can be twice as fast as that of the fully immersed zone. However, the splash zone is generally available for maintenance work to the protective coatings.

13. The fully immersed zone includes the external walls of the caisson in contact with seawater and those areas beneath the caisson and between the adjacent caissons which are grouted with concrete. Although the corrosion rate is less in this area, access for maintaining a paint system could only be achieved at great cost with the use of a cofferdam system and this was not considered realistic for this project.

14. Steel embedded in concrete has a very low corrosion rate (less than 10 m/year) unless cracks in the concrete permit water penetration. This can lead to carbon dioxide penetration with a lowering of the pH locally by carbonation. Unless the steel is first coated before being embedded in concrete, then diffusion of oxygen and chlorides through the concrete onto the exposed steel will lead to localised corrosion with considerable pressure being exerted by the corrosion products. Galvanic corrosion can also occur as steel in seawater is anodic to steel in concrete, thus a galvanic cell will be set-up between the exposed ends of the caissons and the bottom and side plating protected by the concrete grouting.

15. Very heavy corrosion of steel structures in sea water can result from stray electrical current. The generating plant and transmission lines are potential sources of leakage, though this applies mainly to a d.c. system. Damage from strong currents can be minimised by having a large unpainted area in contact with the seawater to reduce the current density, or by operating a system of impressed current cathodic protection with a large number of well distributed anodes and reference electrodes. The later system is proposed for the barrage.

Protective Measures

16. Maintenance of protective coatings on the totally immersed structure is not possible and some form of cathodic protection is essential. An initial protective coating – zinc silicate paint followed by a primer and coal tar epoxy would be applied at the construction side. However for the life of the Severn Barrage total loss of the protective coating would have to be assumed and provision made in the cathodic protection fit for 100% bare steel protection. A combination of sacrificial anodes and impressed current would be used, the sacrificial anodes offering initial protection after launch and up to such time as power is available for commissioning the impressed current system.

17. In the splash and tidal zones corrosion rates are high and maintenance is difficult although not impossible. In this region some form of high performance protective coating would be applied to the whole zone. Monel sheathing has been used on offshore platforms as monel provides excellent corrosion resistance in aerated seawater. Vulcanised rubber up to 13 mm thick has also been used. Other candidate materials are sprayed metallic coatings such as zinc. Where maintenance is difficult near the low waterline, the cathodic protection would give added protection for most of the tidal cycle.

18. The draught tubes and sluice channels are subject to high speed sea water of up to 10 m/s and 4 m/s respectively combined with intermittent periods of stagnant water. Monel suffers virtually no loss from the velocity effects of sea water but is subject to pitting in quiet sea water and is therefore unsuitable in these positions. An acceptable alternative material is Inconel 625 which is highly resistant to all forms of attack in sea water but costs more than twice that of Monel.

19. The use of alternative low alloy steels was considered for the structure. They perform better in the splash zone but offer no better protection under fully submerged conditions and are more subject to pitting attack than carbon steel. However shipbuilding grade high tensile steel could be used with some benefit in reduced costs as their corrosion properties are similar to mild steel and for the same thickness of steel could suffer increased corrosion before structural failure would occur.

DESCRIPTION OF SELECTED OPTION

Arrangement

20. Figure (2) shows plan, elevation and sectional views of the selected arrangement and indicates four main features :-

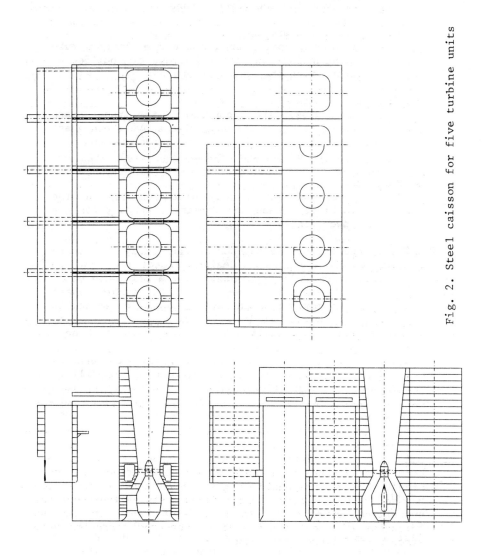

Fig. 2. Steel caisson for five turbine units

(a)   Draught tube structure :-
> a box shaped diaphragm stiffened structure surrounding the draught tube.

(b)   Maintenance dock area :-
> a free flooding area above the draught tube surrounding on three sides by buttresses supporting the machinery space.

(c)   Machinery space :-
> an enclosed space containing the transformers and switchgear and supported above the highest waterline on a box girder structure spanning the top of the buttresses.

(d)   Roadway :-
> a box girder roadway sited on top of the machinery space and forming a conduit for the power lines.

21. The dimensions of the caisson are generally similar to those of the concrete reference design, except the upstream length of the draught tube was reduced and a bellmouth entry introduced. With a minimum submergence of the turbine shaft of 21 metres, and a maximum basin wave height of 4 metres, it is improbable that wave effects will induce air entraining vortices or affect the stability of the flow to the turbine.

22. The distribution of turbines along the barrage is shown in Figure (3). After consideration of stability, towing resistance, and handling requirements, it was decided to adopt a five turbine arrangement per caisson. This reduces the total number of emplacement operations over the 3 turbine caisson concrete proposal.

Structure

23. The objective of the study was to determine the comparative costs of a steel caisson structure. Most of the effort was concentrated on the overall consideration of the scantlings and the evolution of a design which could be produced economically using flat panel lines commonly installed in large shipyards and fabrication yards. For the preliminary design exercise the loadings assumed for the six main structural elements were :-

Draught tube	— hydrostatic head plus wave increment.
Buttresses	— hydrostatic head plus the dead load of box girder structure over.
Sea well	— hydrostatic head plus wave forces.

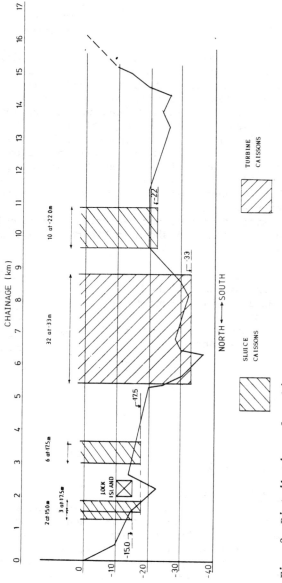

Fig. 3. Distribution of turbine and sluice caissons

Box girder bridge	–	support dead load.
Superstructure	–	Lloyds Rules for ship's superstructures which take account of wind and wave loads.
Roadway	–	supported dead load.

24. Use was made of Lloyd's rules for Mobile Offshore Units to determine permissible stress, levels and plate thicknesses for hydrostatically loaded stiffened panels. The factors of safety were taken as 1.67 for "normal" operating loads and 1.25 for "extreme" load conditions corresponding to 147 N/mm^2 and 196 N/mm^2 respectively. A future detailed design would require the environmental loadings to be agreed by the certification authority and the allowable stresses confirmed. Lloyds grade A, D and E mild steel with a yield stress of 245 N/mm^2 have been assumed, with the grade E being used for the superstructure which has to withstand sub zero temperatures.

25. Detailed weights and centroids were calculated for the structural arrangement shown in Figure (4) and are summarised as follows :-

Structural steel	21176 tonnes
Corrosion allowance	5366 tonnes
Margin (5%) + rounding up	1458 tonnes
Total steel	28000 tonnes
Machinery, equipment and fittings	900 tonnes
Total weight	37000 tonnes

26. An assessment was made of the cost of using high tensile steel for those parts of the structure subject to high static loading, such as the draught tube structure. Based on the use of Lloyds AH34 (yield stress 355 N/mm^2) a saving in weight of 10% was calculated. The cost benefit was partly offset by the need to use additional concrete ballast and the total overall saving in materials and labour was estimated to be 9%.

Machinery Arrangement

27. The study was based on a 9 m diameter bulb turbine with fixed pitch runner blades and output regulated by variable pitch distributor vanes. The turbines are sited with bulb upstream, and horizontal shaft axis. Emergency shut down is normally enacted by the distributor vanes, with the option to install a downstream vertical lift gate, positioned by a travelling gantry crane.

28. Access into the bulb unit is via a vertical shaft of similar dimensions to the concrete design in the upper support strut and a horizontal tunnel from the draught tube maintenance area containing the actuating mechanisms for the distributor vanes. A service lift through the buttress

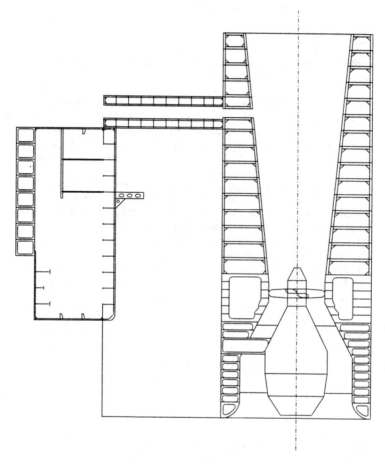

Fig. 4. Turbine caisson structural elevation

provides access to the machinery space and a similar route is taken by the busbar conduits from generator to the first stage transformer.

29. The machinery space is above the highest water level and contains transformers, switchgear and machinery monitoring control equipment.

Turbine Maintenance and Removal

30. Two possible methods are proposed for the installation and removal of bulb turbines at the barrage site within the same caisson arrangement.

31. Removal in the wet envisages the use of a heavy lift crane barge and transport barge as shown in Figure (5). The draught tube is first dewatered by the fitting of closure gates at either end and the necessary disconnections made in the dry. The draught tube is then flooded again and the complete bulb turbine and adjacent draught tube structure removed as a single unit.

32. Removal in the dry is more conventional, and similar to the method used in Reference 2, but precludes the total removal of the bulb unit as a single unit. This method envisages the use of a maintenance dock closure gate, as well as upstream and downstream draught tube gates and the removal of the turbine in major sections using either a smaller floating crane, or a mobile crane on the road over.

33. In either method, the removal of heavier items of equipment such as transformers within the above waterline machinery space would be via a removable soft patch in the box girder floor structure on to a barge positioned in the maintenance dock. Major on site maintenance to items external to the bulb turbine such as the runner blades requires dewatering the draught tube as in the concrete design with dry access from the maintenance area.

Damage Protection

34. Although the flat underside of the machinery support structure has been positioned well above the maximum basin wave crest, there is a possibility of double height waves being generated by reflection and a perforated wall section spans the maintenance dock to remove wave energy and prevent pressure induced vibration damage to the caisson structure.

35. The turbine section of the barrage is well away from the shipping lanes and it is unlikely that the caissons would experience direct impact from a ship at full speed. A more possible scenario is a helpless vessel drifting on the barrage under the influence of tide, wave and wind and the caisson structure should be able to withstand this load without sustaining more than superficial damage. A most important factor is that penetration of the machinery space would not lead to flooding, and water damage to machinery and possible loss of life is less likely than with machinery space sited below the waterline.

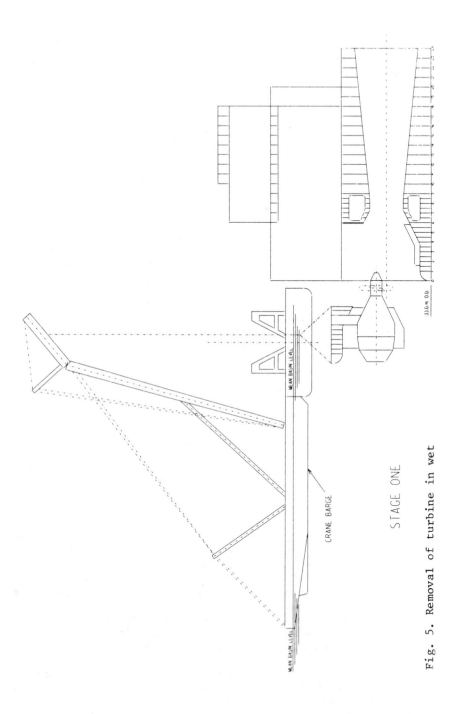

Fig. 5. Removal of turbine in wet

FOUNDATION DESIGN

36. Following a careful study of the work carried out in the concrete reference design, four important factors were identified in respect of the foundations, their preparation and the subsequent installation of the caissons :-

(a) The relative roughness of the pre-dredged rock surface, implying the need for further preparation to provide a sufficiently level caisson underside contact surface.

(b) The very short tidal velocity window.

(c) The relatively light caisson structure, implying a considerably extra amount of secondary weight or downwards force in order to achieve adequate long term stability.

(d) The relatively large number of caisson structures involved, enabling cost effective 'production line' techniques and equipment to be evolved for the offshore operations.

37. Initially, three different schemes were evolved for emplacing and securing the caissons :-

(a) A heavy ballast scheme, in which a high density ballast material is placed in the caisson. A particular problem identified for this scheme was the difficulty of obtaining satisfactory positional accuracy when taken in conjunction with the very short emplacement window.

(b) A template scheme (Figure 6) in which the caisson is locked onto foundation pads pre-installed at seabed level. The pads are secured to the seabed using sufficient rock anchors to obtain satisfactory short term post-emplacement stability. Following emplacement, long term operational stability of the caisson is then achieved by a second set of rock anchors installed at a higher level in the caisson structure. Stabbing guides on the locating pads, together with a re-usable template for positioning each locating pad, enables the caisson to be emplaced with good positional accuracy.

(c) A strong point scheme, in which strong points are pre-installed thereby providing the means of subsequently locating and securing the caissons.

38. The study eventually favoured method (b) for the deeper water turbine caissons on cost and risk grounds, with method (a) being held in reserve in case of rock conditions being unfavourable for installation of rock anchor.

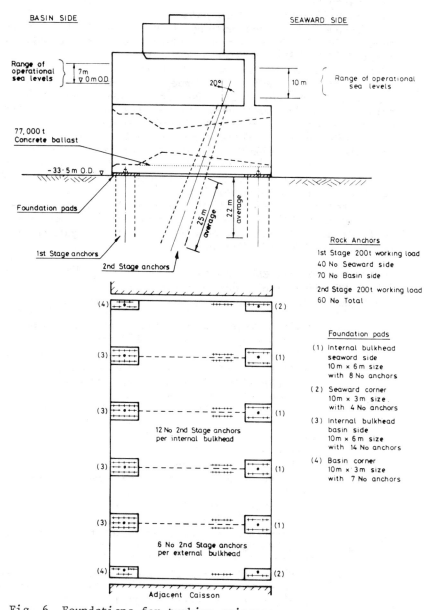

Fig. 6. Foundations for turbine caissons

Emplacement of the sluice caissons could be carried out satisfactorily using method (b) without pre-installation of the stabbing guide locating pads.

39. The use of rock anchors, both to secure the locating pads to the seabed, and to provide long term operational stability to the caisson, was found to be very cost effective. This was primarily because rock anchors use the most efficient restraint force both in direction and magnitude. However, their use in these circumstances is relatively novel, with very little other direct experience being available. Nonethelesss, a lot of the problems likely to be met may be identified and practical answers obtained from other more conventional applications.

40. Problems identified at this stage include the requirement for more careful corrosion protection measures, together with the evolution of remote techniques for the underwater installation of the anchors to the seabed level locating pads. Additionally, the adoption of rock anchors would require further and more detailed investigation of the rock strata under the seabed.

TRANSPORTATION AND EMPLACEMENT

41. The most important problems encountered in evolving the final transportation and installation sequence were :-

(a) The high tidal velocities and short period of slack water.

(b) The confined nature of the last 20 km of approach route to the barrage alignment, particularly when considered in conjunction with the strong tidal currents and the unwieldly nature of the caisson under tow.

(c) The minimisation of horizontal and vertical motions as the caisson is ballasted down onto the locating pads.

(d) The necessity to achieve early post emplacement stability, in order that the worst anticipated environmental conditions over a 3 to 4 year period may be resisted.

42. The transportation and emplacement of the caissons was considered in the following stages :-

(a) Transportation from the various fabrication yards to a deep water site for 1st stage ballasting (possibly in the Firth of Clyde).

(b) Positioning of temporary bulkheads (to improve hydrostatic characteristics during emplacement) and 1st stage ballasting.

(c) Transportation from the 1st stage ballasting site to the Severn Estuary.

(d) Final approach to site location and initial positioning above pre-installed locating pads. This stage is timed to occur within the 3 hour period either side of high water.

(e) Final positioning over the stabbing guides and emplacement during the low water tidal slack period.

(f) Lock onto the stabbing guides to obtain immediate post emplacement short term stability.

(g) Installation of the second stage, long term operational rock anchors, removal of the temporary bulkheads and completion of the structural work including underbase grouting and inter-caisson joints.

43. The fabrication, transportation and installation stages of the caissons were timed to coincide as far as possible, so that storage of the caissons afloat was minimised. The controlling stage in the sequence is emplacement, with neap tides being considered the most appropriate time to emplace. The final programme evolved allowed for 1 turbine caisson to be placed every third neap tide, i.e. every 39 days. However, provision was made in the overall timing for missing one or more tidal windows on up to 30% of the total number of emplacements. This resulted in a overall period of 3.7 years for emplacing 32 number turbine caissons and 2.5 years for 22 number sluice caissons, the combined sequence being carried out in years 9 to 15 of the programme adopted by the Severn Barrage Committee.

COST

44. The costs for the steel caisson study were derived for the same base date as the concrete reference design, i.e. December 1980. The number and diameter of the turbines was the same as for the Bondi Report. However radial gate sluice caissons with an equivalent flow rate to the venturi sluice caissons were costed. The turbine caissons were designed for a constant foundation level and the sluice caissons covered three different foundation levels.

Fabrication

45. The total quantity of steel involved in constructing the caissons is in the order of 1.2 million tonnes which is equivalent to between 15% and 20% of the British Steel annual capacity over a construction period of 6 years. The costs in this study have been based on the list price applicable in December 1980.

46. Caisson fabrication costs vary with the material to be used which in turn is affected by the corrosion protection measures adopted. Two systems of corrosion protection were costed for comparative purposes.

(a) System 1 :
Sacrificial anodes on bare steel immersed structure (short term)
Inconel 625 draught tubes and sluices
Monel splash zone
Zinc sprayed sluice gates
Solvent free epoxy (bottom and ends)
Cathodic impressed current (excluding Inconel 625 areas)

(b) System 2 :
Coal Tar Epoxy on immersed structure
Solvent free epoxy draught tubes and sluices
Zinc sprayed splash zone
Zinc sprayed sluice gates
Solvent free epoxy (bottom and ends)
Cathodic impressed current

System 1 was judged to offer the best protection with low maintenance costs, System 2 offering good protection was less expensive initially but with much higher maintenance costs.

47. Fabrication (including the initial protective coating) and through life maintenance costs are given below for the turbine caissons. The maintenance costs over the 120 year life of the barrage were discounted to present day costs using a 5% discount rate.

	Fabrication Cost	Through Life Maintenance Cost	Total
Mild steel – System 1	£883 m	£ 1.3 m	£884.3m
Mild steel – System 2	£667 m	£33.0 m	£700.0m
High tensile steel – System 2	£612 m	£33.0 m	£645.0m

Offshore Operations

48. The costing of the offshore operations must allow for considerable interaction between costs of materials, labour and plant and the duration of the various operations on a "semi-production line" basis. The relevant plant and labour costs were assesed in two parts, i.e. the period of the operation where full rates were used and the excess time to start of the next sequence where standby rates were used. The approach of continuous charter was retained for all plant except the tugs used to tow the caissons from the ballasting site to the Severn Barrage line.

49. Rock anchors were adopted for securing the caissons. The length of rock anchor involved depended on the type and quantity of the rock assumed to be present under the caissons.

Overall Cost Summary

50. A cost summary indicating the total capital cost for constructing the barrage in mild steel with corrosion protection System 2 is given in Table 2. The steel caisson costs (Reference 1) are compared with the original Bondi Report costs and show a £600 m saving at December 1980 costs with use of steel. Most of this saving is achieved through the reduction in fabrication costs of the caissons.

51. A comparison of the raw material costs for steel or concrete designs indicates that since 1980 the cost of precast concrete has increased at a higher rate than the cost of structural steel. The labour content in the construction of the steel caissons is about 30,000 man years in comparison with 61,000 man years for the concrete caissons. A costing exercise at todays prices would therefore show that the cost difference between steel and concrete has improved in steels favour.

SLUICE OVER TURBINE CAISSON

52. The evolution of the sluice over turbine caisson concept is a natural progression from the selected turbine caisson option and is achieved by substituting a sluice gate for the sea wall. The top surface of the draught tube at 9 m below ordinance datum is the same as the sill height for the proposed radial gate caisson design. To improve the flow to the sluice way a rounded leading edge is built into the top of the draught tube structure. The proposed arrangement of the sluice over turbine arrangement is shown in Figure (7).

53. Reference 2 reviews a selection of gates that could be used to control the level of the barrage and concluded that the vertical lift gate, the radial gate and flap gate were the best alternatives. However, several disadvantages in using these for the sluice over turbine concept were identified and a radial sector gate design was devised to overcome these. The solution proposed is shown in Figure (8), and it is an amalgam of the radial gate design and the rising sector gate as used in the Thames Barrier. Instead of a counterbalance weight 9 combinations of buoyancy and free flood chambers are used to assist the winch in raising the gate to the closed position - a 5 tonne pull is required. When open the gate is stowed flush with the top of the draught tube.

54. The arrangement of machinery, plant maintenance and removal, and installation procedures are generally as described for the turbine caisson.

Table 2. Comparison of costs between steel caisson study and EP46

ITEM NO.	DESCRIPTION	EP46 (VOLUME 1, FIGURE 18)					STEEL CAISSON STUDY					COST DIFFERENCE £M
		COST	C%	E%	PT%	TOTAL £M	COST	C%	E%	PT%	TOTAL £M	
1	Turbines	970	5	7	17.5	1281	970	5	7	17.5	1281	0
2	Transmission	500	0	7	17.5	629	500	5	7	17.5	660	+ 31
3	Caisson Construction Facilities	–	–	–	–	–	50	0	7	17.5	63	+ 63
4	Turbine Caisson Fabrication	1025	10	7	17.5	1418	667	0*	7	17.5	839	)
5	Turbine Caisson Offshore Works						162	10	7	17.5	224	) – 355
6	Sluice Caisson Fabrication	542	19	7	17.5	750	278	0*	7	17.5	350	)
7	Sluice Caisson Offshore Works						46	10	7	17.5	64	) – 336
8	Abutment Caissons	83	10	7	17.5	115	83	10	7	17.5	115	0
9	Ship Locks	300	10	7	17.5	415	300	10	7	17.5	415	0
10	Marine Plant & Misc.	105	10	7	17.5	145	–	–	–	–	–	– 145
11	Embankment	530	10	7	17.5	733	632	10	7	17.5	874	+ 141
12	Prototype Trials	165	–	–	–	165	–	–	–	–	165	0
	TOTAL	4220				5651	3853				5050	– 601

C – Contingency Allowance

E – Engineering Overheads

PT – Post Tender Costs

* Design margin of 5% included in weight estimate

Included in cost of offshore waters

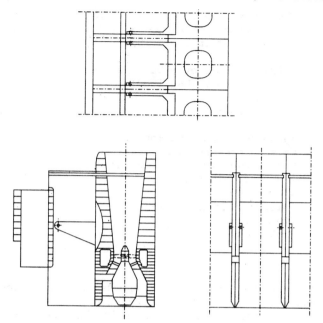

Fig. 7. Sluice over turbine caisson

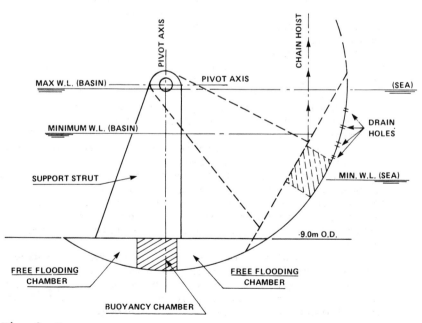

Fig. 8. Buoyant sector gate

327

55. The overall cost of the sluice over turbine caisson concept is summarised in Table (2), and assumes the same contingency as used in Table (3).

Table (3) — Sluice over Turbine Costs
(December 1980)

Item	Cost £M
Turbines & Transmission	1,939
Caisson construction facilities	63
Sluice over turbine caissons	919
Abutment caissons	115
Ship locks	415
Caisson offshore works	224
Embankment	1,086
Prototype	165
Total	4,926

56. The total of £4,926 M represents a small saving of 2½ percent over the steel caisson scheme. However, the sum of £1,086 m for embankment is based, in the absence of any other data, on the rates quoted in Reference 2, and as these appear somewhat pessimistic, a greater saving should be possible.

CONCLUSIONS

57. The study concluded that the construction, installation and maintenance of the steel caissons is technically feasible. The overall cost of the steel option shows a significant potential saving compared with proposals outlined in Reference 2.

58. The proposed steel caissons satisfies the functional requirements of the barrage while remaining a comparatively simple structure to fabricate. The layout of the turbine caisson produces a minimum buoyancy low steelweight design without compromising the provisions for removal and maintenance of the machinery. There are sufficient construction facilities within the U.K. shipbuilding and offshore fabrication industry to construct the caissons within the overall project timescale.

59. Corrosion protection systems are available which would provide a 120 year life comparable with that for concrete. The selected method of corrosion allowance on the steel scantlings, painting and impressed current cathodic protection provides an economic and effective protection system.

60. The combining of the two functions of sluice and turbine housing in one caisson indicates further possible savings, if the problem of embankment costing can be overcome. Alternatively it may be more cost effective to install blank caissons in lieu of embankment.

61. An earlier closure of the barrage would reduce the unit cost of the energy produced. An increase in the number of turbines in each caisson would reduce the number of emplacement operations, and could be accommodated within existing fabrication facilities. Further studies would be required to evaluate the largest size of unit that could be conveniently handled.

62. The rate of inflation for steel construction is less than for concrete construction and it is likely that the cost differential, in favour of steel at todays prices will be greater than indicated in the study. All in all, it is concluded that steel provides an economical means of constructing the barrage and its use should be considered for other locations.

ACKNOWLEDGEMENTS

63. This paper summarises work carried out for the Department of Energy and ETSU, and the authors are grateful to both parties, and to the Directors of YARD LTD and Roxburgh and Partners, for permission to publish this paper.

REFERENCES

1. Steel Caissons for the Severn Barrage. Confidential report to ETSU by YARD LTD and Roxburgh and Partners. June 1984.

2. Tidal Power from the Severn Estuary, Department of Energy Paper No. 46 Vol. I & II.

16. The development of functions relating cost and performance of tidal power schemes and their application to small scale sites

A. C. BAKER, BSc(Eng), FICE, MIWES, CIEE, *Binnie & Partners*

SYNOPSIS. Methods of assessing quickly the energy output and cost of a tidal power scheme, developed during the 1978-81 Severn Tidal Power studies, have been applied to 15 other UK estuaries. The results are combined into various simple cost and energy functions which allow comparison with results from work on foreign schemes, and these generally show good correlation. The cost and energy functions have been applied to some small UK sites and show that several sites could merit more detailed study.

INTRODUCTION

1. The proposed method of construction of the Severn barrage, which was developed during the studies carried out under the direction of the Severn Barrage Committee (ref. 1), could be used for other tidal power schemes having suitable basin areas, tidal ranges and water depths. Indeed, during the 1978-1981 studies, a wide range of different schemes in the Severn estuary was evaluated and compared before the preferred scheme was identified. Energy outputs were assessed quickly and to good accuracy using a desk-top computer and plotter, while the cost of a scheme was estimated by assembling the main components from a set of 'building bricks' comprising turbine caissons, turbines, sluice caissons, embankments, ship locks and transmission. These costs, plus various contingency allowances, were combined with likely construction programmes which were then used together to calculate the unit cost of electricity produced using a standard discount rate of 5%. Optimum schemes were identified largely on the basis of their unit cost of energy.

2. Towards the end of the 1978-81 studies, these quick and reasonably accurate methods of assessing and comparing schemes were used to assess the resource and likely unit cost of energy from six other large UK estuaries (ref. 2), namely Morecambe Bay, Solway Firth and the Dee estuary on the west coast of the UK, and the Humber, Wash and Thames on the east coast. The first five schemes were refined as regards generator capacity and turbine numbers for a subsequent paper (ref. 3).

3. In 1984, after some development of the design of turbine caissons and sluice caissons to suit shallower water, the same methods were applied to seven small tidal power sites and to the Mersey (ref. 4). Subsequently, a comparison was made between the results of these studies and the results of studies carried out abroad of other tidal power sites (ref. 5).

4. This paper presents the results both of this comparison and of the development of various functions relating energy and cost which enable a quick assessment to be made of the main features and economic viability of any tidal power site. Finally, some small UK sites are identified which appear to merit further attention.

PARAMETERS CONSIDERED

5. The parameters considered are divided into two groups, those which define the site and those which define the barrage. Those which define the site are:

- R, the mean tidal range (2 x M_2) in metres at the barrage site. Where the site is remote from a standard tide gauge, careful interpolation or extrapolation is needed

- A, the area of the enclosed basin in km^2. For this study, the area at high water of spring tides has been used, since this is readily measured from Admiralty Charts or from Ordnance Survey maps

- L, the length of the barrage in metres

- H, the maximum depth of water along the line of the barrage, taken as the depth shown on Admiralty Charts plus the mean spring tidal range (2 x (M_2 + S_2)).

6. The remaining parameters which define the barrage are:

- D, the turbine runner diameter in metres and N, the number of turbines

- P, the total installed capacity of the generators, in MW

- E, the annual energy output, in GWh

- T, the total cost of the barrage, including contingencies, engineering and $17\frac{1}{2}$% of post-tender allowance, in January 1983 prices

-    U, the unit cost of electricity from the barrage, in
     p/kWh at January 1983 price levels and using a
     discount rate of 5%. This includes interest during
     construction, annual maintenance at 1% of mechanical
     and electrical plant costs and 0.75% of civil costs,
     and complete replacement of all electrical and
     mechanical equipment after 40 years.

7.   These parameters can be used in isolation or in
combination. For example, AR^2 indicates the energy resource
at the site, AR being the approximate volume of water flowing
in and out each tide, and R indicating the available head
across the turbines. However, since relatively little energy
would be available from turbines operating at heads below 1m,
$A(R-1)^2$ has been found to be more useful for some comparisons.
ND^2 indicates the total swept area of the turbine runners.

UK SITES STUDIED
8.   The parameters discussed above are summarised on Table 1
for the 17 sites in the UK which have been individually
studied on behalf of the Department of Energy. They range
from the outer Severn barrage (the Minehead/Watchet line) with
an installed capacity of 12,000 MW, to 20 MW schemes. The
smaller schemes (Nos. 9 to 15) were not necessarily selected
as good potential sites, but covered a range of basin shapes,
tidal ranges and so forth so that the relative importance of
these factors could be assessed.

9.   Separate studies have been carried out of Strangford
Lough in Northern Ireland (ref. 6) and the Mersey estuary (see
paper at this conference by Haws), Nos. 21 and 17 respectively
on Table 1, although figures for the Mersey are taken from
phase 3 of ref. 4.

OVERSEAS SITES
10.   Tidal power schemes generating electricity have been
built at La Rance in France, Annapolis Royal in Nova Scotia,
Kislaya Guba on the north coast of the USSR and several small
schemes on the south east coast of China. Kislaya Guba is the
only one constructed using caisson technology, but at 400 KW
is only a pilot trial, apparently successful. Similarly, the
Annapolis Royal scheme is a prototype and its operation is
restricted by land drainage requirements, so its performance
and cost are not typical. Adequate details are not yet
available for the Chinese schemes.

11.   La Rance scheme has been well documented but, after 20
years, its equivalent present-day construction cost is not
readily established. However, its then present cost of energy
was stated to the Severn Barrage Committee during a visit in
1980 (ref. 1).

Table 1. Summary of tidal schemes studied

	R	A	L	H	D	N	P	E	C	U
	Mean tidal range (m)	Basin area (m² x 10⁶)	Barrage length (m)	Max water depth (m)	Turbine dia (m)	Turbine No.	Installed capacity (MW)	Annual energy output (GWh)	Capital cost (£M)	Cost of energy (p/kWh)
1 Severn – Inner line	7.76 (7.0)	450	17000	35	9.0	160	7200	12900	6660	3.7
2 Severn – Outer line	7.2 (6.0)	1000	20000	35	9.0	300	12000	19700	10460	4.3
3 Morecambe	6.3	350	16600	30	9.0	80	3040	5400	3610	4.6
4 Solway	5.64 (5.5)	860	30000	28	9.0	180	5580	10050	7480	4.9
5 Dee	5.95	90	9500	29	6.0	50	800	1250	1230	6.4
6 Humber	4.1	270	8300	29	9.0	60	1200	2010	2140	7.0
7 Wash	4.68 (4.45)	590	19600	40	9.0	120	2760	4690	4860	7.2
8 Thames	4.2	190	9000	29	9.0	40	1120	1370	1740	8.3
9 Langstone	3.13	19	550	12	4.0	9	24	53	43	5.3
10 Padstowe	4.75	6	550	9	4.0	6	28	55	35	4.2
11 Hanford	3.0	11	3200	7	4.0	9	20	38	50	8.5
12 L. Etive	1.95	29	350	18	7.5	6	28	55	96	11.7
13 Cromarty	2.75	36	1350	27	7.5	6	47	100	176	11.8
14 Dovey	2.90	13	1300	11	4.0	9	20	45	50	7.2
15 L. Broom	3.15	7	500	27	7.5	3	29	42	90	13.9
16 Milford Haven	4.5	20	1150	27	7.5	6	96	180	270	10.0
17 Mersey	6.45	70	1750	25	7.6	27	620	1320	697	3.6
18 Fundy site A6	9.55	119	5410	42	7.5	53	1643	4530	2250	3.4
19 Fundy site A8	10.05	97	2560	39	7.5	37	1147	3183	1353	2.7
20 Fundy site B9	11.7	282	8000	42	7.5	106	4028	11766	3875	2.2
21 Strangford Lough	3.1	144	1500	27	7.6	30	210	528	544	7.0
22 Garolim Bay (S.Korea)	4.8	100	1850	28	8.0	24	480	893	273	2.1
23 La Rance	8.0	22	750	23.5	5.35	24	240	544	–	3.4
24 Fundy site B9	11.7	282	8000	42	7.5	128	4864	14004	4530	2.2

1. Costs at January 1983 price levels
2. Tidal range in brackets is predicted range after construction of barrage
3. Schemes 1, 2 taken from ref. 1. Schemes 3 to 7 from ref. 3. Scheme 8 from ref. 2. Schemes 9 to 17 from ref. 4. For other schemes see text

12. Useful ·references are found in the reports of detailed feasibility studies of tidal power in the Bay of Fundy (ref. 7) and Garolim Bay in S. Korea (ref. 8). One of the three short-listed Fundy schemes, site B9, has had its performance and cost updated recently (ref. 9).

13. The relevant parameters from La Rance, the Bay of Fundy and Garolim Bay are also listed on Table 1. The capital cost and resulting cost of energy have been converted to UK sterling using the exchange rate current at the time of the base date used for costing purposes in the relevant reports. However, the wide fluctuations in inflation and exchange rates between the dollar and the pound over the last 5 years make precise comparisons of costs impossible. The process used for calculating the unit cost of energy has been rationalised by applying the same method as for the UK sites, as shown on Table 2.

Table 2. Examples of assembly of scheme costs

Site	Padstow	Mersey	Fundy
Scheme No. (Table 1)	10	17	20
Ref. No.	4	4	7
No. of turbines	6	27	106
No. of sluices	6	18	60
Assumed year of first power	1995	1997	1998
CIVIL WORKS			
Years for construction	1992-1995	1992-1997	1992-1998
Cost of lock	1.48	128.0	)
turbine caissons	6.91	150.30	)
sluice caissons	4.30	39.57	) 1616.1
gates	1.22	12.38	)
embankment	2.75	17.28	)
Sub total	16.66	347.53	1616.1
10% contingencies	1.67	34.75	161.6
TURBINES & TRANSMISSION			
Years for construction	1993-1996	1993-1998	1995-2002
Cost of turbines	7.44	130.14	)
transmission	1.90	41.75	) 1304.6
Basic total cost	27.67	554.17	3082.3
7% engineering	1.94	38.79	215.8
17½% post tender costs	5.18	103.77	577.2
TOTAL COST (£M)	34.79	696.73	3875.2*
Cost of energy (p/kWh)	4.15	3.59	2.2

* Based on £1 = $ Canada 1.79 and May 1976 costs, updated to Jan 1983 using UK inflation indices

COMPARISON OF SCHEMES

14. Inspection of the information on Table 1 shows that the
scheme with the lowest unit cost of energy is Garolim Bay in
S. Korea. As will be discussed later, this appears to be an
effect of low labour costs rather than the inherent properties
of the site. Otherwise, as to be expected, the Fundy sites,
with their exceptional tidal ranges, promise the lowest cost
of energy. Thereafter, La Rance, the Severn inner barrage and
Mersey are broadly comparable at 3.4 to 3.7 p/kWh. Of the
small schemes, No. 10, Padstow is the most promising at 4.2
p/kWh.

15. Various parameters are presented in graphical form on
Figures 1 to 6. In Fig. 1, the ratio of the number and size
of turbines (ND^2) to the energy resource $(A(R-1)^2)$ is plotted
against mean tidal range (R).

16. For the two Severn schemes, for which substantial
reductions in tidal range on the seaward side are forecast by
computer models of tidal propagation on the west coast of the
UK, points with and without this reduction are shown. There
is little difference in the relative fit of these points.
Scheme 9 (Langstone Harbour) appears to be under-turbined,
while scheme 15 is over-turbined. La Rance also appears to be
over-turbined by about 20%.

17. In Fig. 2, the ratio of annual energy output (E) to
installed turbine capacity (ND^2) is plotted against mean tidal
range (R). In this curve, the reduced tidal ranges for the
Severn schemes have been shown because these were the basis of
the choice of the turbines. A very consistent relationship is
found for all schemes except La Rance, which has turbines
designed to generate or pump in each direction of flow. This
result is to be expected, because the 1978-81 Severn studies
showed that about 20% more two-way turbines would be needed to
generate the same electricity as one-way turbines. The
updated version of the Fundy scheme B9 has a slightly lower
energy output per turbine than the earlier version, as to be
expected if the number of turbines is increased.

18. In Fig. 3 is plotted the ratio of the energy output of
each barrage (E) to the energy resource AR^2 again against mean
tidal range. This is an indication of the 'efficiency' of
each barrage in extracting the available energy. The points
plotted for the Severn barrages are the present tidal ranges.
The points with tidal range reduction are less consistent.
Schemes 9 and 15 are inconsistent, for the same reasons
discussed in section 16. Otherwise, the ratio $E/R^2 A$ lies in
the range 0.32 to 0.5. The mean is 0.404.

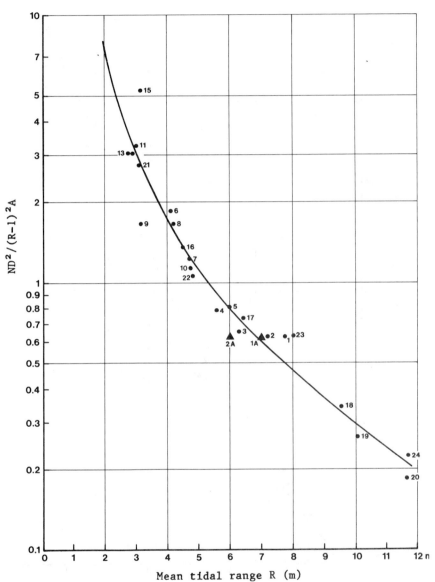

Fig. 1.   Ratio of installed turbine capacity (ND^2) to energy
resource $(R-1)^2$ plotted against mean tidal range (R)

(Note.   1A, 2A = Severn barrages with tidal range reduction)

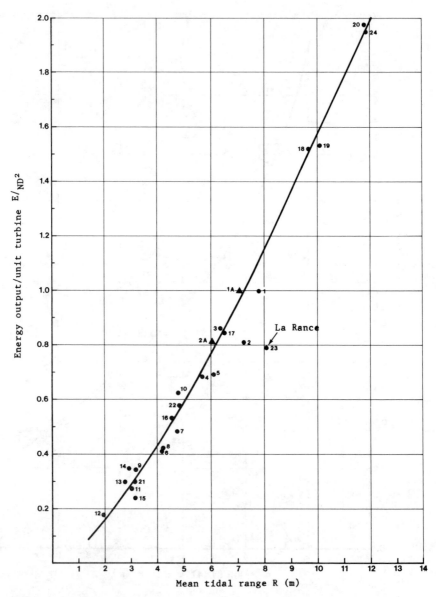

Fig. 2. Ratio of energy output (E) to installed turbine
         capacity (ND^2) plotted against mean tidal range (R)

(Note. 1A, 2A = Severn barrages with tidal range reduction)

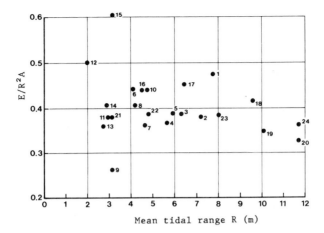

Fig. 3.   Ratio of energy output (E) to energy resource (R^2A) plotted against mean tidal range (R)

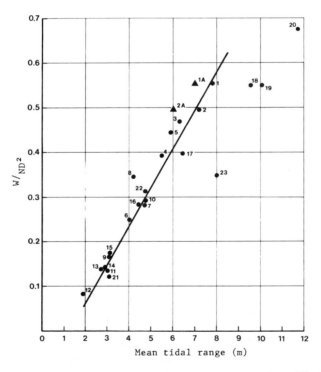

Fig. 4.   Ratio of installed generator capacity (W) to installed turbine capacity (ND^2) plotted against mean tidal range (R)

(Note.   1A, 2A = Severn barrages with tidal range reduction)

339

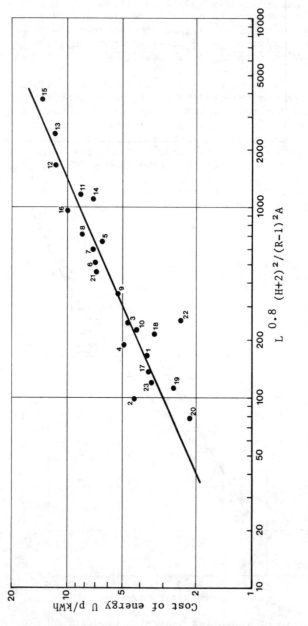

Fig. 5. Cost of energy (U) plotted against energy cost function (C)

19.   In Fig. 4, the ratio of installed capacity (W) to
turbine capacity (ND²) is plotted against mean tidal range.
This is the same ratio as the individual generator capacity
divided by D². A reasonably consistent relationship is found
but the Fundy schemes and La Rance appear to need larger
generators, La Rance by 70% if it were a one-way scheme.
However, the economics of tidal power are not sensitive to the
choice of generator capacity for a particular size of turbine.

20.   In Fig. 5 is plotted the unit cost of energy against
the cost function $C = L^{0.8}(H+2)^2/R-1)^2A$, first published in
ref. 5, now with the refined schemes of ref. 3 and with the
addition of the foreign schemes and Strangford Lough.
Plotting the results on log scales results in a straight line
relationship, with a reasonable amount of scatter. The only
scheme well away from the straight line is Garolim Bay in
Korea, whose cost of energy is about half that to be expected

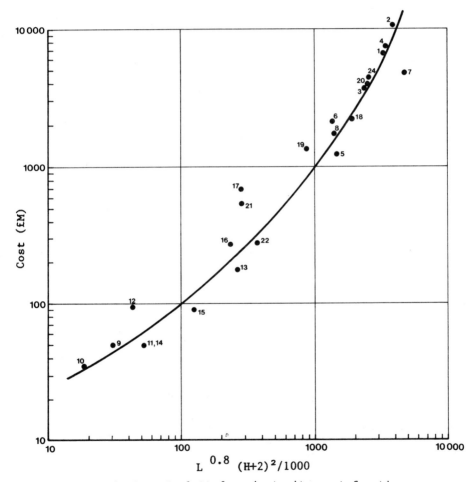

Fig. 6.   Capital cost plotted against site cost function

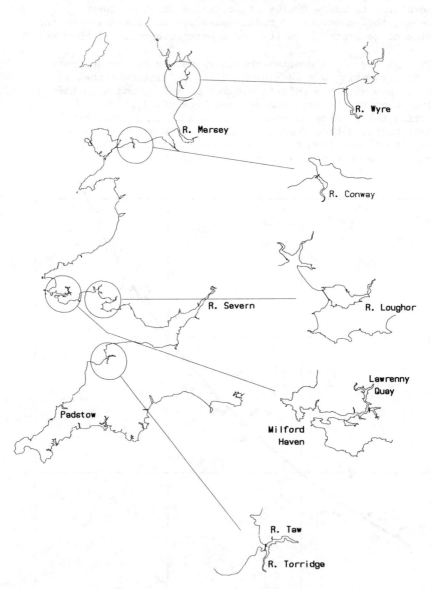

Fig. 7. Locations of sites listed on Table 3

Table 3.   Small scale sites indicating low cost of energy

Location	Mean tidal range (m)	Basin area (km)	Barrage length (m)	Max.water depth (m)	Cost function	Indicated cost of energy (p/kWh)
Taw/Torridge estuary (Appledore)	5.45	13	950	11	158	3.5-4
Wyre estuary (Fleetwood)	6.4	7.6	480	15	182	3.5-4
Conway estuary (N. Wales)	5.2	6.0	225	13.6	175	3.5-4
Loughor estuary (S. Wales)	6.55	46.3	3375	18	277	4.5
Lawrenny Quay (Upper Milford Haven)	4.57	2.82	200	10.5	301	4.5-5

from the other schemes.   Recent work in Korea during a review
of this scheme has shown that construction costs are low in
Korea and this single factor satisfactorily explains the
difference in unit cost.

21.   The results in Fig. 5 show that, if a scheme has a cost
function C below 350 it is probably worth a second look
because the unit cost of energy could be below 5 p/kWh.   If
its cost function is below 200 then the scheme may well be
economic with a cost of energy of 4 p/kWh or less.   The three
Fundy schemes all lie consistently below the line.   This may
be related to the choice of exchange rate but, clearly, they
are outstanding sites.

22.   Finally, Fig. 6 compares the site cost function
$L^{0.8}(H+2)^2$ and the estimated capital cost of each barrage.
Surprisingly, a reasonably consistent relationship is found,
in spite of the cost function taking no account of tidal range
and basin area, factors which define the number and size of
turbines.   Attempts to correlate capital cost against other
simple variables have not produced more consistent
relationships.

APPLICATION TO OTHER SITES
23.   When considering a potential tidal power site, Fig. 5
will provide a quick guide to the likely cost of energy.   If
this is encouraging, then Figure 1 will indicate the turbine
capacity.   As a quick guide, a water depth at low water of
spring tides of at least twice the turbine diameter is needed
to provide adequate submergence and the diameter can be chosen
accordingly.   Many small sites having low cost functions are
too shallow at low water to be practicable.   Having selected a
turbine diameter and calculated the number of turbines from
Fig. 1, Fig. 2 can be used to estimate the energy output of
the scheme.   Fig. 4 indicates the generator capacity and,

finally, Fig. 6 will indicate the approximate capital cost. Capital cost will be sensitive to factors such as the need either for ship locks or for temporary work yards for caisson construction.

24. Using this method, some potentially interesting small UK schemes are listed on Table 3, and their locations are shown in Fig. 7.

ACKNOWLEDGEMENT

25. This paper is largely based on work carried out by Binnie & Partners for the Energy Technology Support Unit (ETSU) at Harwell on behalf of the Department of Energy. Permission to publish this paper is acknowledged with thanks.

REFERENCES

1. SEVERN BARRAGE COMMITTEE. Tidal power from the Severn Estuary, 1981. HMSO Energy Paper No. 46.
2. BINNIE & PARTNERS for Department of Energy. Preliminary survey of tidal energy of UK estuaries. May 1980. Report No. STP 102.
3. WISHART S.J. A preliminary survey of tidal energy from 5 UK estuaries. Proc. 2nd International Symposium on Wave and Tidal Energy. Cambridge,UK. BHRA. Sept 1981.
4. BINNIE & PARTNERS for Department of Energy. Preliminary study of small scale tidal energy. Phases 1 to 3. 1984.
5. BAKER A.C. & WISHART S.J. Tidal power from small estuaries. Proc. 3rd International Symposium on Wave, Tidal, OTEC and Small Scale Hydro Energy. Brighton, UK. BHRA. May 1986.
6. Strangford Lough Tidal Energy. Northern Ireland Economic Council. August 1981.
7. Reassessment of Fundy Tidal Power. Bay of Fundy Tidal Power Review Board and Management Committee. November 1977.
8. SOGREAH for Korea Electric Company. Garolim tidal power plant feasibility studies. November 1981.
9. DELORY R.P. The Annapolis tidal generating station. Proc. 3rd International Symposium on Wave, Tidal, OTEC and Small Scale Hydro Energy. Brighton UK. BHRA. May 1986.

Discussion on Papers 13–16

MR J. C. BENNETT, Ewbank Preece

Paper 13 describes the programme for construction of the
Cardiff-Weston scheme as a seven-year period up to generation
of first power with generation of full power in a further two
years. For the proposed 192 units this implies bringing a unit
into service every three or four days for a period of two
years. This is an impossibly tight programme and is unlikely
to be achieved.

Notwithstanding the proposals, described in other papers, to
carry out the maximum amount of unit construction and
precommissioning work away from the barrage, and the acceptance
that several units will be undergoing final commissioning
trials simultaneously, the implication remains that eight units
will be brought into service each and every month for two
years. There is no precedent anywhere in the hydropower
industry where such a rate of commissioning of mechanical and
electrical plant has been achieved and it would be foolhardy to
imagine that it could be achieved on the Severn tidal scheme.
A more feasible rate of implementation might be three units per
month, but even that would be faster than anything previously
accomplished.

It has been stated elsewhere that the economic viability of
the scheme is sensitive to delays in implementation.
Consequently it is imperative that realistic and practical
programme planning be incorporated in the future studies to be
carried out.

What examples of similar mechanical and electrical
construction and commissioning works have the Authors examined
in arriving at their stated conclusions on the programme?

Mr Buckland's Paper gives estimates of phased costs before
construction and also mentions that there will be an inevitable
need to revise and update estimates for civil and plant costs
during the preconstruction period.

In earlier papers dealing with cost estimates it was noted
that for both Severn options the ratio of civil costs to
mechanical and electrical plant costs is approximately 2:1.
For conventional hydropower developments this ratio is more

usually of the order of 5–6:1. In other words the relative cost of mechanical and electrical plant for the Severn scheme appears to be high. This seems to be inexplicable since, even including the probable high costs of transmission reinforcements, the mechanical and electrical plant proposed is of fairly conventional design, whereas the civil construction works are fairly complex.

Do the Authors have any explanation for this apparent anomaly?

MR R. CLARE

The probable explanation of the different civil/mechanical and electrical cost ratio between tidal and hydropower is that in tidal power there are low head turbines designed to work on a head difference as low as 3 m or even less whereas most hydropower schemes have very high head operating conditions thus resulting in much higher speeds of rotation.

MR J. TAYLOR, Merz & McLellan

I support whole heartedly Mr Bennett's comments on the ambitious commissioning programme. To expect continuous commissioning, and I stress commissioning, of 192 operating sets in the period stated is expecting too much. There will certainly be some overrun.

MR P. C. WARNER, Northern Engineering Industries plc

Both Mr Bennett and Mr Taylor have expressed doubts about the intended commissioning rate and have asked where else in the world it has been achieved. The Severn tidal power project is not regarded as purely reflecting previous experience or having its parameters settled within an envelope of what has been done already. The scale is unprecedented and, even though established technology and world-wide experience will be used as fully as possible, it demands an approach that is appropriately new.

As an illustration, firstly there is the manufacturing challenge of producing three turbine generator sets every month, as the programme requires. The question whether 'existing manufacturing capacity' would be adequate is soon set aside, not just because business and technical evolution produce rapid changes in capacity (which would have been sufficient reason) and it would have been impossible to say what might be 'existing' at the time the production was needed. The more fundamental reason is the necessity for a new production concept to achieve rapid high quality production, and that involves substantial sums spent on equipment for fabrication, machining and handling, as well as design details worked out accordingly. When there are well over £1000 million worth of sets to be produced, significant capital expenditure for the sake of programme integrity is money well spent. All that is explained in the Group's report.

The same principle would apply to commissioning. The final
rate is very high, higher than the manufacturing rate, because
barrage closure must be waited for. The new approach will
involve capital expenditure on commissioning methods
(templates, devices, special instrumentation etc.), none of
which has at this stage been defined. The point to stress is
the need to set that target and to find the methods that will
achieve it. A second consideration in our favour is that,
although complete closure is necessary to create the full load
tidal heads for final commissioning, much work at synchronous
speed can be done with the limited head (1 m or so) that
becomes available relatively early in the installation
programme through partial closure. It is enough for a good
proportion of the commissioning work referred to, but that too
has yet to be worked out in detail, and with the package
concept there are more opportunities than normally for
precommissioning both at the works and immediately after
installation.

PAPER 14

PROFESSOR E. M. WILSON, Salford Civil Engineering Limited
Dr Shaw showed a rates levy of £15 for an operating barrage.
This is equivalent to 0.12p/kWh approximately. This is the
rate applied to public utilities but not to private generation.
If the barrage were to be privately financed, the appropriate
rate might be ten times higher. Has this point been
considered?

MR R. CLARE
Professor Wilson is correct that at present the rates demanded
on privately owned generating plants such as wind turbines or
combined heat and power schemes are about ten times those paid
by the electricity utilities. However, the problem is known to
both the Department of the Environment and the Department of
Energy and steps are being taken to resolve the discrepancy.
Powers exist for the Secretary of State for the Environment to
formulate a rate for private generating plant and an indication
has been given that he would consider exercising this option
for a large plant covering several districts and thus making
rating more comparable with that of the electricity supply
industry.

PAPER 15

DR C. J. BILLINGTON, Steel Construction Institute
I should like to discuss the potential use of steel in barrage
schemes.
    There is no question that the potential for steel is enormous
- should a steel-intensive solution be chosen for the

Severn barrage this implies in excess of 1 million t of steel plate. For comparison at the beginning of October 1986 the achievement of 1 million t of British Steel Corporation (BSC) plate going into the North Sea since the development started more than 15 years ago was celebrated. The Steel Construction Institute has been studying the design work so far carried out by YARD and the Severn Tidal Power Group and expect to be active in the future in developing steel-intensive solutions.

At this symposium concern has been expressed over the durability of steel in the marine environment, made against the background of very few facts. 'Feelings' have been expressed that steel is just not the right material to put into the marine environment and even the YARD Authors start the section of their Paper on material selection and corrosion control with the sentence 'An instinctive engineering response to the idea of immersing a mild steel structure in sea water for a period of 120 years with no removal for maintenance is that it is impossible to guarantee that structure.' However, they then go on to describe how it can be done.

Engineering by instinct is not a very sensible or necessary approach and here are some facts which decribe the corrosion of steel in the marine environment and the experience with properly engineered corrosion protection systems. It is also interesting to note the ready acceptance of steel for components of the barrage systems which can only be made in steel. Fig. 1 shows corrosion rates for the various zones of the structure – atmospheric, splash, tidal immersed and below bed – based on many test samples monitored around the coasts of the UK for many years. The number of data points is shown in the first column of the table, the corrosion rates in the second and the life of the YARD corrosion allowances with no protection, i.e. unprotected bare steel, in the final column. Even in the worst corrosion zone the life of the corrosion allowance (not the structure) is in excess of 100 years and for the majority of the structure it provides the required design life.

Figure 2 provides data on corrosion protection systems recommended for the barrage. In the atmospheric, splash and tidal zones a scaled zinc metal spray is recommended. For this there are many successful case histories with in excess of 30 years' performance. Maintenance involves just further applications of the organic sealing coating.

In the permanent immersion and below bed level zones a coal tar epoxy resin coating together with impressed current corrosion protection is proposed. There have been 35 years of satisfactory experience in the UK and longer in the USA. This is a popular approach for offshore oil platforms which have been dismissed as having a short design life and therefore being of little relevance. However, many platforms are now designed for a much longer life, some up to 60 years, because of the potential for tying in smaller outlying reservoirs in the future and the life of many existing platforms is being extended without difficulty for the same reason. It should

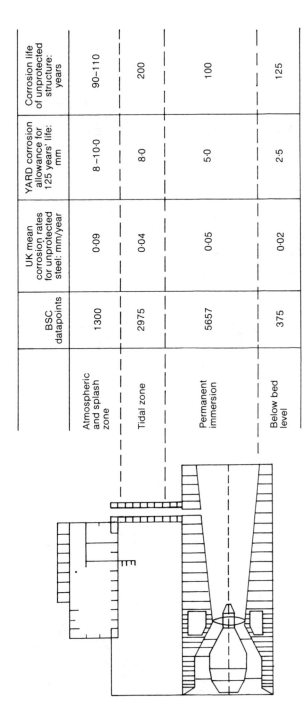

	BSC datapoints	UK mean corrosion rates for unprotected steel: mm/year	YARD corrosion allowance for 125 years' life: mm	Corrosion life of unprotected structure: years
Atmospheric and splash zone	1300	0·09	8–10·0	90–110
Tidal zone	2975	0·04	8·0	200
Permanent immersion	5657	0·05	5·0	100
Below bed level	375	0·02	2·5	125

Fig. 1. Corrosion allowance

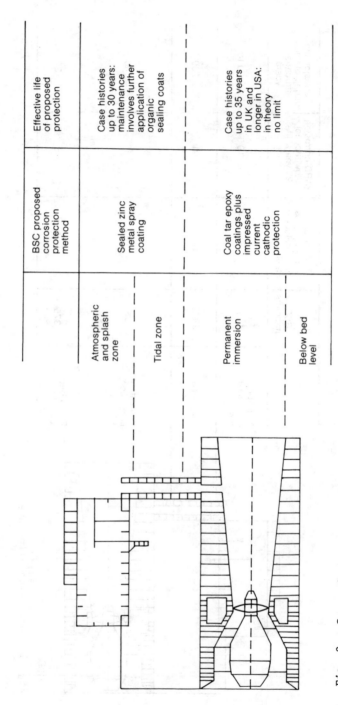

	BSC proposed corrosion protection method	Effective life of proposed protection
Atmospheric and splash zone		
Tidal zone	Sealed zinc metal spray coating	Case histories up to 30 years: maintenance involves further application of organic sealing coats
Permanent immersion	Coal tar epoxy coatings plus impressed current cathodic protection	Case histories up to 35 years in UK and longer in USA: in theory no limit
Below bed level		

Fig. 2. Corrosion protection

also be remembered that these structures are much more
sensitive to corrosion fatigue and therefore many platforms are
designed with the objective of complete corrosion prevention
including localized pitting which would not be necessary for
barrage caissons with a designed-in corrosion allowance.

With regard to costs, the YARD Authors present cost figures
which superficially show a significant saving for a steel
scheme for the Severn. No doubt it will be argued that like is
not being compared with like. The YARD designs in several
respects represent changes in design philosophy (particularly
for maintenance) and improved layout which could perhaps also
be achieved in concrete. In spite of this steel is likely to
remain a strong competitor on economic grounds when detailed
comparisons are made. These comparisons will not necessarily
be like with like but should embody acceptable and agreed
design and operating restraints. The YARD cost comparisons
were made of 1980 prices. Since then there has been a very
significant divergence of cost. Fig. 3 shows a comparison of
typical onshore construction costs in steel and concrete. The
YARD cost comparisons were made on the basis of steel list
prices. Prices for the supply of plate to shipyards over this
period show significant reductions for the materials element
input to the comparison and result in an even greater
divergence of cost. Further savings will be achieved through
automated fabrication where the labour costs might be expected
to be halved.

Faced with such large potential orders for steel the BSC
would probably offer significant discounts over the prices used
for these cost comparisons particularly if plate thicknesses
are matched to economy of production.

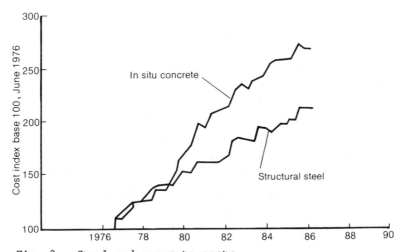

Fig. 3. Steel and concrete costs

Finally the Steel Construction Institute does not view this as a concrete versus steel competition. It may be that the final optimum solution adopts a hybrid or composite approach. However, the Institute will continue to work towards removing some of the myths and instinctive reactions concerning the use of steel in this environment.

MR K. JOHNSON, British Steel Corporation
The YARD Paper mentions two aspects of steel quality that could be expanded on. Firstly, in the use of high strength steels, consider two typical shipbuilding steels (Table 1):

(a) Lloyds A - a mild steel with a yield stress of 235 N/mm^2
(b) Lloyds AH36 - a high strength steel with a yield stress of 355 N/mm^2.

The corrosion properties of these two steels are almost identical: there is no advantage there. However, if a structure is designed at mild steel stresses but then fabricated in high strength steel, a considerable corrosion advantage is gained. A much greater loss in section thickness can be tolerated, simply because of the higher strength of the AH36 material.

From the costs and the yield stress values it can be simply calculated that in this case a 51% increase in permissible corrosion loss for only a 12% increase in material cost is gained.

An alternative way of looking at this is to consider the use of high strength steels as a means of reducing the original plate thicknesses and thereby reducing the initial cost. YARD has estimated a further total overall saving of 9% on caisson costs using this simple device.

Table 1

Steel	Yield stress (up to 50 mm thick): N/mm^2	Basis price: £/t
Lloyds A	235	250
Lloyds AH36	355	280

Secondly, the YARD Paper also mentions the use of special steels in certain key areas of the caissons. However, over the past few years there have been certain steel developments which may be of more general interest for tidal barrage schemes (Table 2). These are low alloy steels and they essentially fall into two categories.

The first category includes steels with improved corrosion performance above the water. These contain additions of elements such as chromium, copper, nickel and phosphorus. Some materials are commercially available, e.g. the Mariner steels, and claims have been made of improved splash zone performance by a factor of about 2.

The second category includes steels with improved corrosion performance below the water. These contain low alloying additions of chromium and aluminium and possibly molybdenum. There are as yet no commercially available materials – but development is active. Claims have been made for improved underwater corrosion resistance by a factor of about 2–4.

Possibly both of these types would be of future interest for use in tidal barrage schemes.

Table 2. Low alloy steels for use in marine environments

Chromium	)	
Copper	)	Improved splash zone corrosion
Nickel	)	performance (x2)
Phosphorus	)	
Chromium	)	Improved sea water immersion corrosion
Aluminium	)	performance (x2–4)
Molybdenum	)	

MR A. W. GILFILLAN, MR G. C. MACKIE and MR R. P. ROWAN
The additional data presented by Dr Billington in Fig. 1 indicate that the corrosion margin alone, without any protective coating system, would offer a reasonable design life for the steel caissons. However, the rates quoted are 'mean corrosion rates'. If these rates were exceeded locally then underwater repair might be necessary to prevent penetration of watertight boundaries or structural failure. To reduce the likelihood of expensive underwater repair operations, the designers used corrosion rates based on a 95% probability of not being exceeded when taken into account with an appropriate protective scheme. This more conservative approach was considered more appropriate at the conceptual design stage, although a detailed analysis of the consequences of full penetration corrosion in such redundant areas as concrete-filled ballast tanks could lead to a reduction in the corrosion margin allowance. Other areas of high corrosion rate, such as the maintenance dock in the splash/intertidal zones, have been made more accessible for maintenance by designing in the ability to de-water them for access in dry conditions.

Mr Johnson's comments that a 51% increase in permissible corrosion loss can be attained by using AH36 steel is erroneous. Classification Society rules normally permit only the square root of the ratio of the yield stresses to be used under bending in determining a reduction in plate thickness as only the extreme fibres are effective. Secondly, only the structural part of the steel plate thickness can be reduced when a high yield strength steel is used. Thus if 60% of the steel plate is structural and 40% is the corrosion margin then the reduction in structural steel thickness that could be used as added corrosion margin is only 60% - 60% x $(235/355)^{1/2}$ or 11%. However, the cost of corrosion margin steel is less than 6% of the total barrage cost and is therefore not worth compromising the design on. In addition, as a philosophical point, a design in mild steel and then fabrication in AH36 steel implies a lack of confidence in the corrosion allowance being applied.

The British Steel Corporation's (BSC's) data on low alloy corrosion-resistant steels are of interest and we would welcome any evidence in support of their claimed low corrosion rates, and of course BSC's cost per tonne should they go into production. Clearly this is an area where closer interaction between the specialist customer and the steel producer is required.

Dr Billington's final graph (Fig. 3) showing the increased differential between steel and concrete structure costs confirms the revised costing work that we are at present carrying out for the Energy Technology Support Unit on the Cardiff-Weston line.

Finally, although the YARD/Roxburgh report has been labelled conservative in its approach to the corrosion and production costs, the total barrage cost as derived shows a significant improvement over the concrete version.

PAPER 16

MR B. MADGE, Polytechnic of Wales
The effect of tidal surges in the Bristol Channel will have very important implications on the barrage operation and the barrage will in turn have a modifying influence on surge effects upstream. What studies have been undertaken to quantify these effects? Fig. 3 of Paper 12 suggests that a value of 1.5 m has been considered as a maximum surge water level superelevation.

At the Polytechnic of Wales, the actual and predicted high and low water tide levels at Newport have been analysed and compared for many years. In 1982, for example, the histogram shown in Fig. 4 indicates that, of the 1380 high/low water values compared, only 11% were as predicted, whereas 67% of the actual levels were 0.2 m different from the predicted levels, and 18% were at variance by at least 0.5 m. Although only the maximum and minimum levels were studied the effect would have been present throughout the whole tidal cycle.

As mentioned, such divergences have two aspects of interest in relation to the barrage.

(a) The enhanced or depleted water surface level will affect the electricity generation time on any day and will make accurate prediction of operating times difficult in the overall context of the Central Electricity Generating Board's network utilization.

(b) The 'damping' effect which the barrage will have on the water level variation behind the structure will be of great benefit to those inhabiting the low-lying areas above the barrage. More emphasis should be placed on this major beneficial outcome of the barrage construction.

Another aspect of the effect of the barrage relates to both the beneficial and adverse influence it will have on the coastal protection regime of the channel/estuary. The wave climate behind the barrage will be dramatically improved, the fetch having been reduced to, at maximum, about 40 km. Only the long period swell waves will pass through the barrage filter. The beneficial outcome of such a situation is obvious. Dr Keiller (Paper 2) and Dr Kirby (Paper 11) have mentioned the modifications which will take place in the sediment regime. Has an investigation been undertaken to ascertain the effect on beaches downstream of the barrage?

Experience gained by engineers in The Netherlands working on the Dutch Delta project should be invaluable in this scheme. For example, the Oosterschelde storm surge barrier used precast pier units, each weighing 18 000 t and costing around £2000 million. Of this, approximately 8% was devoted to research and development and included

(a) the effect of the barrier aperture on the tidal regime in the estuary and the sea side of the structure

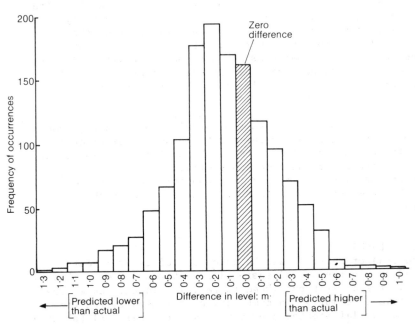

Fig. 4. Frequency of specific differences in actual and predicted levels for 1982 at Newport

(b) the optimal distribution of the opening over the three existing gullies to maintain the present morphological pattern
(c) the changes in salinity and the transport of pollutants
(d) the boundary conditions for the design of the barrier
(e) the detailed flow patterns at the barrier site during different construction phases
(f) the comparison of different gate operation sequences
(g) the effect of translation waves induced by the operation of the gates
(h) storm surge propagation in the case of failure of one or more gates
(i) the effect of the positions of the dams on the tidal amplification and flow velocity in the estuary.

Environmental and ecological impacts of the Delta project are also of value in the current study. Man's obvious interference, with its far reaching consequences, was appreciated in relation to the fishing industry, shipping and recreational facilities. The effect on the bird population was also of great importance, as the area is one of the most valuable winter habitats in Europe for nesting birds from northern regions. Research into the future natural development of the area has produced a wealth of information on the behaviour and ecological requirements of both plant and animal life.

In Dr Shaw's second Paper, the benefits of the barrage to the Severnside area are well illustrated, but it does 'require that opportunities are seen and promoted'. Unfortunately, generally, those living in the immediate vicinity of the barrage and with most to benefit in the long term do not appreciate that such benefits could become a reality. Rather, they feel that the barrage is a marvellous idea which will never come to fruition. It is important that an effective public relations exercise is implemented to convince the general public of the area that the barrage is to be built. It should be possible to form a powerful local lobbying group from such organizations as the county councils, district councils, the Welsh Office, the Welsh Development Agency, the Welsh Tourist Board and local industrialists and entrepreneurs, who could work in conjunction with the Severn Tidal Power Group to promote this excellent, far-sighted project.

MR A. C. BAKER
Figure 3(a) of Paper 12 was based on one of several tests carried out in the one-dimensional computer model described by Dr Keiller and first published in ref. 1. In this case, the surge 'event' comprises a combination of a high spring tide (5.4 m OD at Ilfracombe) plus a 0.48 m surge residual (about a 50 year combined event) plus a fluvial flow of 200 m^3/s. The curves of high water levels along the existing upper estuary

are artificially high because some banks would have been overtopped and this was not taken into account in the model. Nevertheless, this test showed the barrage to have great benefit in reducing water levels in the upper estuary. The value of this benefit in monetary terms will be small in relation to the cost of the barrage (ref. 2).

The effects of the frequent but relatively small differences between the levels of actual and predicted high water on barrage operation and output have not been addressed yet. Because the barrage would not start generating until two hours or more after high tide, the starting time could be maintained, if this was important, with very little penalty in terms of energy output. Otherwise, the starting time could be adjusted to take into account the actual time and height of high water, with the prediction of time of start-up being progressively refined, to take into account the effects of atmospheric conditions on water levels, during the previous day or so.

Turning to the question of coastal defences and wave attack, Mr Madge suggests that conditions behind the barrage will be dramatically improved. This may be the case near the barrage, but further up the estuary there will be no benefit (ref. 2), partly because mean water levels will be higher and there will be a long stand at high water, and partly because at present long fetch waves do not penetrate much beyond the Holm Islands (ref. 3). Downstream of the barrage, the wave regime will be altered by reflection off the barrage, where there are vertical concrete walls, and by the discharge from the turbines. The effects on beaches should be slight except perhaps close to the barrage. This aspect has not yet been studied in detail.

The wide practical experience gained at the Oosterschelde scheme is certainly of value.

DR T. L. SHAW
Figure 3 of Paper 12 is taken from the Severn Barrage Committee's report. It shows how normal high and low water levels on spring and neap tides would be changed by the barrage. It also shows the existing tide-plus-surge profile, extreme levels which need not be reached if the barrage is used to exclude surge effects.

In their recent report, the Severn Tidal Power Group (STPG) suggested that pumping around high tide could increase high water levels in spring tides by up to 0.5 m at the barrage and by about 1.5 m in neap tides. This information is not shown on fig. 3. Mr Madge may be confusing reference to these additions to high water levels, which are not expected to affect highest spring tide conditions.

The data given by Mr Madge for the differences between predicted tide heights and timings are unlikely to affect normal barrage operations to any significant degree, although the occasional extreme variations, of over 0.5 m in either direction, would be significant.

The STPG has considered possible changes in seaward wave and current climates and how beaches may be affected. Much more work on this is planned as part of future studies of the project.

MR J. TAYLOR, Merz & McLellan

The purpose of Paper 16 is to set down a method of ranking potential tidal basins relying only on outline data. Schemes which have been studied in greater detail are used to calibrate the results.

The idea is sound and it is hoped that it will encourage people to concentrate on the best sites. The method is presented to public view because the Author is confident that the method is sound and 'fail safe' and that the conclusions are not so sensitive to coarseness in the original assumptions as to invalidate them. To create this confidence the following points should be amplified.

(a) The difficulty in obtaining more data may be the reason for the simple definition of the energy resource of the site as AR^2 where A is the high spring tide basin area and R the mean tidal range, i.e. A is the maximum area. The layman might think that the slopes of the beds of these large estuaries show a degree of similarity such that the mean area of the basin could be estimated. In this case the volume of water stored will be the mean area multiplied by the mean tidal range (AR) and the potential energy would be proportionally less.

(b) Building blocks form the basis of the capital cost estimate. The turbine caisson for a scheme with a maximum depth of 15 m is surely less costly than a turbine caisson for 30 m depth. Could the Author explain the technique that he uses for 'scaling' the building blocks to suit individual schemes?

(c) It might have been expected that the capacity factors of the generators would be related to the time available for generation. This in turn would be related to the tidal range. However, if the annual capacity factors derived from table 1 of the Paper are compared, for scheme 24, 33% is obtained, for scheme 17, 24%, for scheme 1, 20%, for scheme 2, 18% and for scheme 14, 26%.

Could the Author explain the relatively low values for schemes 1 and 2; does it suggest that too much capacity is proposed?

MR A. C. BAKER

Because the tidal range behind an ebb generation barrage is about half that on the seaward side, low water in the basin will be slightly above mean sea level. Consequently, both the mean area and the maximum area are at extremes of the working range, but the maximum area is readily measured from maps or charts. The schemes which have been studied in detail have all had their energy outputs based on properly measured area/depth relationships. Thus there should be no inherent inaccuracy in using basin area at high water spring tides for assessing the energy output of a 'new' estuary from a graph prepared on the same basis from the results of detailed studies.

For scaling the main components, such as turbine caissons, outline designs have been prepared and principal quantities calculated for turbines of 9 m, 7.5 m, 6 m and 4 m runner diameter, most attention having been paid to the first because it was the size selected for the Severn barrage during the Bondi Committee's work. The water depths at the prospective barrage sites in other estuaries have been used to determine the runner diameters, on the assumption that the minimum water depth should be at least twice the runner diameter to avoid undue problems of cavitation. Thus each size of caisson has been used for a relatively narrow range of water depths. In some small estuaries, such as Padstow and Hamford Water (table 1 in the Paper), even 4 m dia. turbines could not be accommodated without some prior dredging for their caissons. In broadly the same way, sluice caissons and embankments have been designed for a range of water depths.

Mr Taylor's last point concerns the capacity factors of the various schemes. In tidal power schemes, as opposed to run-of-river schemes, the total energy output is not sensitive to the generator capacity (see fig. 2 of Paper 5). Thus quite large variations in capacity factor can be expected when other aspects such as the costs of transmission links are taken into account. Mr Taylor suggests that the time available for generation is related to the tidal range. This is not the case because the 12.4 hour tidal cycle is a dominating constraint and, within broad limits, a barrage designed for a site with a large mean tidal range and one designed for a site with a small mean tidal range will generate for the same length of time during spring tides and the same but shorter time during neap tides.

If a barrage is designed with undersized generators then, during high spring tides when the lack of capacity would be expected to result in restrictions in output, the turbines can be started earlier in the cycle. This has the effect of drawing the basin down early and so reduces the maximum head across the turbines which occurs at low tide, which is when generator capacity is important. Because the generating period is longer, the total energy output is not greatly reduced. During middle range tides there need be no reduction in output while during neap tides the greater efficiency at low heads of turbines designed to suit small generators can result in a slight increase in energy output. This effect is mentioned in paragraph 10 of Paper 2. Of the schemes listed in table 1, schemes 1 and 2, which are referred to by Mr Taylor as having low capacity factors, are the two Severn barrages studied in detail during the Bondi study (EP46). The generator capacities were selected on the basis of least cost of electricity, taking into account transmission costs as well as the costs of the barrage components. Scheme 24 is the latest version of Fundy scheme B9, where power transmission over very long distances can be expected to push the optimum generator size downwards and thus to increase the capacity factor. This effect can be seen for the three Fundy sites on fig. 4 in the Paper.

REFERENCES
1. BINNIE & PARTNERS. Severn tidal power: one-dimensional model studies of the Severn Estuary. Binnie & Partners, London, October 1980.
2. BINNIE & PARTNERS. Severn tidal power: land drainage and sea defence. Binnie & Partners, London, August 1981.
3. HYDRAULICS RESEARCH STATION. Severn tidal power: hindcasting extreme waves. Hydraulics Research Station, Wallingford, March 1981, Report EX 978.

17. The Annapolis Royal tidal power station

E. M. WILSON, PhD, FICE, FASCE, *Professor of Civil Engineering, University of Salford, and Director, Salford Civil Engineering Limited*

SYNOPSIS   The Annapolis tidal power plant is the first tidal station to be built in North America.   Its principal purpose is to demonstrate the operation of a Straflo rim-generator turbine of large diameter.   The paper briefly describes the power plant, its construction and its first two years of operation.

1.   The tidal power station at Annapolis Royal, Nova Scotia, is the first North American hydro-electric plant to use tidal flows.   It is also the first such plant to use a rim-generator straight-flow turbine.

2.   The many studies of tidal-power development in the Bay of Fundy had suggested that there might be cost savings in the use of rim-generator rather than bulb-turbines.

3.   The rim-generator turbine was invented by Leroy Harza, the founder of the Harza Engineering Company in Chicago, but no American examples of the design were built.   Escher Wyss of Zurich and Ravensburg built more than 70 small machines of the type between 1938 and 1953 and, after further development work (some of it by English Electric in the mid-1960s) they and their licencees, ACEC, have constructed a further 10 such units in the 1980s with runner diameters up to 3.7m.

4.   All the tidal power studies of recent years have been postulating the use of runner diameters of 7.5m and upwards, and so the Tidal Power Corporation of Nova Scotia decided to build such a machine (registered now as Straflo) in 1980, to demonstrate its practicability at this stage.

5.   A turbine throat diameter was set at 7.6m which was the approximate size of the largest bulb units extant or planned and in the range of sizes used by tidal power engineers in their feasibility studies.

6.   The site chosen was a small island in the Annapolis River estuary where a tidal surge control barrage had been

constructed some 15 years earlier. This barrier was a rockfill bund across the main channel, with a set of sluice gates
provided in an excavated channel. The island site lies
between the sluiceway and the barraged main channel.

7. The head pond has a surface area of about 12km² at a
level of 7.5m above chart datum, and is large enough to serve
two 7.6m turbines but since the purpose of the installation
was principally demonstration rather than power production,
the lower cost option of one machine was selected.

8. The power house housing the turbo-generator was constructed
by diverting the road crossing the island, excavation,
construction in the dry, backfilling, replacing the road and
excavating the intake and discharge canals.

9. The turbine centre line is 8.39m below GSCD, which is
about 16m below ground level. The power house is 46.5m
long with a 15.5m square intake and a draft tube exit of
14.5m x 11.1m. Between these the intake passage and draft
tube transform to circles at the turbine. Both intake and
draft tube have central piers to reduce the size of the
stop logs.

10. The turbine has both upstream and downstream bearings:
the upstream is housed in the intake dividing pier and the
downstream, which is a guide bearing, is supported by struts
from the draft tube walls.

11. There are no gates. Flow control is by the distributor.
Malfunction of it could entail turbine runaway until the water
levels upstream and downstream equalize, when stoplogs would
be introduced.

12. Stainless steel was used for turbine blading, rotor rim,
the facing of sealing surfaces and guide vane closing edges,

13. Sealing is by special hydrostatic seals which are supplied
with pressurized sea water, two-thirds of which is
recirculated. Consumption is 13.5 l/s of sea water filtered
to 40 microns. There is also an independent pneumatic
standstill seal.

14. Generator poles are bolted to the outer face of the
peripheral ring beam which surrounds the runner. The poles
are supplied with excitation current through the runner blades
and turbine shaft. The generator is air-cooled and generates
at 4160v.

15. Initial problems arose from inadequate encapsulation of
the poles, which allowed sealing water to penetrate to them.
A change in design of labyrinth plates for sealing water

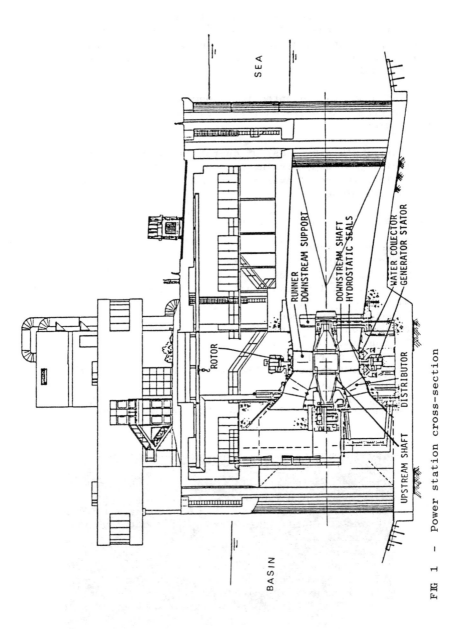

FIG 1 — Power station cross-section

collection and two coats of coal-tar epoxy enamel on the poles were necessary to cure this.

16.   The plant was commissioned in August 1984 and a major inspection was made in June 1985.   The hydrostatic seals were in good condition, there was no sign of corrosion or cavitation damage on the blades.   The cathodic protection was obviously working very well.

17.   The plant was accepted into commercial service in 1985 and the owner has declared itself extremely pleased with the machine.

18.   During its first year of operation the plant had an availability of 99%, missing only 8 tides from a possible 728.   This record of availability has continued during its second year of service.

19.   The main parameters of the machine are

Diameter of runner		7.6m
Number of blades		4
Number of guide vanes		18
Normal operating head range	(approx)	1.4m to 6.8m
Maximum operating head		7.1m
Rated operating head		5.5m
Turbine output at rated head		17.8MW
Turbine maximum output		19.9MW
Discharge at rated head		$378m^3/s$
Turbine efficiency at rated head & load		89.1%
Rotational speed		50rpm
Runaway speed		98rpm
Generator output, (70°C rise, 90% pf)		19,100kVA

PLATE 1 -   Aerial view of the Annapolis tidal plant
(Courtesy Nova Scotia Power Corporation)

Generator output, (85°C rise, 90% pf)   21,250kVA

Generator efficiency   96.4%

Rotor fly-wheel effect   857,000kg/m³

20. A cross-section of the power station is shown in Fig 1 and a photograph of the station in Plate 1.

21. The information in this paper has been derived entirely from References 1 and 2, to which interested readers are referred.

REFERENCES

1. DELORY R. P.   The Annapolis Tidal Generating Station. Proc. Water for Energy Conf. pp 125-132. B.H.R.A. May 1986.

2. HOELLER H. K., HEARN W.G . and GEORGESCU M. A. Operation of the Annapolis Tidal Straflo Unit. Proc. Waterpower '85, pp 316-325, A.S.C.E., Las Vegas, Sept 1985.

18. The way forward

R. CLARE, BSc, DIC, FICE, MIHT, *Sir Robert McAlpine & Sons Ltd*

INTRODUCTION

1. On 9th July the Secretary of State announced a further programme of £5.5M of tidal power studies. A grant of £400,000 would go to the Mersey Barrage Company who, it is understood, will add a similar figure from their shareholders to fund an £800,000 feasibility study. A further £300,000 is allocated for investigation of the more promising smaller schemes, described in an earlier paper to this Symposium by Clive Baker. At the time of writing no decision has been made on how this money will be deployed. The largest amount goes for further work on the Cardiff Weston line in the Severn estuary. This package will total £4.2M and will be jointly and equally funded by the Department of Energy, the CEGB and the Severn Tidal Power Group (STPG). It is timely to consider how this funding will advance the development of tidal power in the U.K. and what more needs to be done to realise the viable schemes.

RECENT PROGRESS

2. This Symposium happens to coincide with the 20th Anniversary of the commissioning of Rance Barrage and is 5 years after the previous ICE Symposium on tidal power, specifically on the Severn project. Since 1981 a feasibility study has been prepared on barrages in the Mersey, and the possibilities in smaller estuaries have been reviewed. Both these parcels of work have been reported to this Symposium. The largest activity within the United Kingdom has, however, been the Study of the barrage schemes in the Severn Estuary by the Severn Tidal Power Group. This was reported to the Secretary of State for Energy earlier this year and published in July. The majority of papers to the Symposium have been devoted to reporting this latter work.

3. Progress with the Severn Tidal Scheme since 1981 may at first sign appear not to be spectacular but, in fact, good advance has been made. Economies in time and cost have been established, contingencies and risk factors have been

assessed with greater care than previously, available environmental data has been assimulated and widely discussed with interested and relevant bodies and a start has been made on regional and planning matters. STPG believe that an improved level of confidence has been established particularly on the environmental side and the remaining uncertainties and risk areas have been highlighted. Thus it is now possible to determine with confidence the way forward. The further phase of study recently announced is seen as an essential and substantial step in that direction.

PRINCIPAL ISSUES TO BE ADDRESSED

4. Two studies in recent years have agreed that an energy generating barrage across the Severn is technically feasible and a similar conclusion has been reached by the study on the Mersey. Hence technical matters are not the most burning problem though extensive work is required to optimise designs, reduce uncertainties and increase the accuracy of cost estimates.

5. More fundamental are such questions as to how should the proposed schemes be moved forward to actual projects, will they be public or privately funded, what planning approval procedures will be adopted and is it advisable to start with a small scheme prior to embarking on the £5,500M Severn project? We must remember the vital matters of the extent and nature of the environmental impact and the public and hence political acceptance which has to be secured.

6. Taking the last question first, Dr. Tom Shaw has dealt in some detail in his papers to this Symposium with Regional and Ecological matters and Dr. Kirby has advised on sedimentation problems. This paper will deal no further with these matters other than to emphasise again the great importance of sufficient investigation and analysis of these matters before completion of the planning stages of a development whether it be the Severn Barrage or the smallest in Clive Baker's list.

7. Michael Buckland's paper has considered the alternative planning procedures that could be adopted for the Severn project. Perhaps it might be true to say that the scale and location of the project could be a major factor in determining the method to be adopted. There must be a great difference between a project like the Severn affecting about 100 local authorities from Parish to County Councils and a small scheme where the construction works could be in the area of one Planning Authority. Whatever the scale of the project, the way forward must include detailed discussion at the earliest stage with the relevant Authorities and that includes the Water Authorities and any affected Ports. The planning procedure could also be affected by the source of funds. Where substantial money has to be raised in the private sector for the pre-contract work, early planning clearance is essential.

8. In the past major civil engineering works privately funded, such as the railways, got their planning permission by Parliamentary Bill. It is unlikely that we will see a modern macro public enquiry financed by the private sector.

9. The recent STPG study has found that the Severn Barrage is unlikely to be fundable in the private sector alone as the paper to this symposium by Dr. John Carr sets out and for the underlying reasons that he gives. It could well be that the same conclusion will be reached for other electricity generating tidal barrages. So the likely outcome is mixed funding unless the project is taken over by the public sector completely. It is therefore necessary in the way ahead to consider how a major project can be jointly funded in the public and private sectors. There are several possibilities and no doubt many variants. The Government and CEGB could participate by taking the risk off the private investor thus making the investment similar to gilts and attractive with a relatively low rate of return. Alternatively, the Government could invest a substantial sum in the project, either directly or in some form of tax relief, against the returns it will get in additional taxation and reduced social security payments thus increasing the return to the equity investor. It is not the objective of this paper to give a treatise on the subject but to point out that there are alternatives that will need examination by financial experts.

DO WE WANT A PROTOTYPE BEFORE THE SEVERN BARRAGE ?

10. All the matters briefly discussed in the previous section have been the subject of other papers to this Symposium but the question "do we want a small scheme before the large one?" has not been addressed.

11. Why should we want a small scheme first? What are the arguments for it? At £5,500M capital cost the Cardiff Weston Barrage on the Severn is a very large project so, the argument goes, we must have a small prototype to see that it works to check methods of construction, to refine design ideas and to check the impact on the environment. If the small scheme can produce some electricity at a useful location, that is all to the good, almost a bonus. But let us examine these points more carefully.

12. Tidal power works, the French have adequately demonstrated it at Rance for more than 20 years. Admittedly their capital cost was high but modern construction methods can reduce that and these methods are tested and proven in the North Sea and the Dutch Delta Works to name but two. Where particular details are to be worked out, such as caisson handling in strong currents and tidal range, then excellent work can now be achieved with scale models in a well equipped testing tank. Other matters such as environmental impact and dredging are site specific and may not be helped by prototype works in different hydrodynamic

conditions and with a seabed of different geology.

13.   No two estuaries are sufficiently similar for one to act as a model for the other.   Indeed, it has been shown at this Symposium that two schemes in the same estuary can have very different effects on the physical and hence ecological environments.   Not only each estuary but also each possible scheme must therefore be assessed in its own right.   This is not an unusual situation for civil engineers to deal with and is not one which cannot be assisted by reference to relevant findings from projects elsewhere.

14.   Uncertainty is not limited to .the environment. Several traditional technical areas remain to be evaluated in detail, as STPG's present studies confirm.   Work on reversible   pump-turbine performance,   optimum   economic machine operation and caisson placing are amongst the topics for further development generally, whilst site specific work will   include   geological surveys followed by   caisson foundation and embankment design.

15.   The main argument against the prototype concept is that it could cause many years of unnecessary delay to development of Severn tidal power whilst contributing little to general knowledge and only insignificantly .to   the country's electricity needs.   The STPG report has shown that an energy generating barrage on the Cardiff Weston line would be fully commissioned by the end of the century if the development is systematically pursued.   This is when the CEGB will be decommissioning many time expired stations and will be requiring replacement capacity.   Why should the country have to suffer such a delay before harnessing one of its greatest natural, non-polluting and truly renewable resources?

16.   This is not to say that the opportunities presented by the smaller estuaries should not be developed in addition to the larger schemes..   Many could provide long term sources of energy and perhaps could be coupled with other benefits such as flood prevention, land drainage schemes and leisure industry developments.

THE FOLLOWING STAGES

17.   The STPG report sets out a pre-construction programme of 6 years based on the development of the project in the private sector with planning by Parliamentary procedures. This programme may require some revision for a mixed financed project or one funded in the public sector but the main elements will still be required.

  (a) Field investigations and design studies to reduce existing uncertainties as far as possible thus reducing contractual risks and   increasing credibility of cost and construction. duration estimates.

  (b) Field   investigations   and data analysis   to establish as far as possible the effects of construction on the environment and how variations in construction methods and sequence might minimise these effects or produce beneficial changes.

(c) Studies to determine regional effects and to value the benefits arising.

(d) Decision to proceed, including securing the necessary finance.

(e) Planning permission required, not only in principle but in detail (e.g. land fall locations, construction sites etc.).

(f) Establishment of the owning/operating body.

(g) Completion of design to the stage where all significant elements can be frozen so that no variation of design shall occur during construction.

(h) Obtaining tenders for the works and selecting principle contractors.

18. The study on which STPG is about to embark will complete item (a) and make substantial progress with (b) and (c), enabling decisions to be taken to proceed with the planning and detailed design stages. For the future the rate of this pre-contract expenditure will continue to increase and hence confidence that the construction will go ahead must grow in parallel.

CONCLUSIONS

19. The future for tidal barrages in the U.K. depends on a broad, comprehensive and integrated approach to the wide range of inter-related issues and interests. These differ little between possible barrage sites. In all cases, a range of Government Departments will claim a direct involvement. In addition to the Departments of Energy and Environment, participation will be demanded by the Ministry of Agriculture, Fishery and Food, the Departments of Transport and Trade and Industry, and, in the case of the Severn Barrage, the Welsh Office. A Solway Barrage would bring in the Scottish Office, and some sites will attract interest from the Ministry of Defence and perhaps the Home Office in regard to certain administrative matters. The case for a co-ordinating body capable of pulling together relevant and balanced information about barrage schemes and advising these and other bodies gets stronger as serious interest in U.K. tidal power grows.

20. A further argument for such centralised guidance is evident in the environmental sector. STPG's studies have clearly shown that much of general value to tidal power can be learned by looking at the environment of various estuaries. The programme of studies now being put in hand in respect of a Cardiff-Weston barrage will therefore assist

the assessment of the environmental impact of other barrage schemes.

21. Expertise demonstrated at the scale of the Severn Barrage will establish the U.K. at the forefront of this technology and has the potential to show those contemplating tidal power developments in other countries that we have understood the way forward.

ACKNOWLEDGEMENT

22. The Author wishes to express his appreciation for help in preparing this paper to Dr. Tom Shaw and Mr. Peter Gibson.

DR K. HOLLER, Sulzer Escher Wyss
Escher Wyss is the only company with bulb and rim generator
experience. Bulb turbine generators up to 9 m are today
proven, reliable technology. Straflo rim generators would be
better technology for tidal applications, because

(a) they have better stability
(b) a single-lift straflo installation requires only 700 t
    compared with 1600 t for a bulb, which means a cost
    reduction
(c) the caisson is shorter and stiffer and there is less
    uplift, which means that more or bigger machines can be
    accommodated in the same channel width
(d) units of 9 m are possible (Annapolis with D = 7.6 m was
    not an extrapolation from D = 3.7 m, but a carefully
    designed new machine).

Two years of experience with no problems at the seal with an
availability of 99% are insufficient to recommend installation
of 192 similar units at the Severn barrage, but with five or
more years of operating experience after 1990 the situation
may be different.
In June 1986 a design review meeting took place between the
client, the consultant, the Canadian government, the civil
contractor, the builder and the equipment supplier. Items for
improvement were identified. If the Annapolis station were to
be built today it would show improvements in

(a) generator insulation
(b) generator pole fixation
(c) the transfer of forces to the construction
(d) water collection around the hydrostatic seal
(e) erection
(f) control
(g) auxiliary equipment.

This revised Straflo design has been chosen for the Bay of

Fundy in a feasibility study conducted by a study group similar to the Severn Tidal Power Group. The Bay of Fundy would be built with variable speed (frequency conversion) and high voltage direct current transmission. The operating regime would be single-effect ebb generation without pumping.

MR H. MILLER, Electrowatt Engineering Services Ltd
The information given by Professor Wilson demonstrates that the trouble-free commissioning and operation of the Annapolis installation is not an extrapolation of the existing next smaller-sized 3.55 m unit at that time but the improved design of large bulb turbine technology. In this context the paper presented by M. Braikevitch of GEC Power Engineering Ltd at the Halifax 1970 conference on tidal power should be noted, where he states that the simplest arrangement of a low head electric unit is that proposed by the American engineer L. F. Harza. It is remarkable that the design developed and advocated by Braikevitch and his engineers - namely of a concrete bulb and with bearings inside the hub of the turbine - is exactly the design on which modern straight flow turbines are being built.

MR J. D. GWYNN, Wilson Energy Associates
Mr Miller has referred to the work done by Braikevitch and his team in the 1960s. I was involved in giving evidence to the then UK Ministry of Technology in support of the English Electric Company's application for grant aid to develop the straight flow turbine with rim generator for large-scale use.
Braikevitch had noted the record of the 73 small machines already built by Escher Wyss which then had run for over 20 years. These machines are still in service over 20 years later - surely a demonstration of the reliability and longevity of the basic concept in this case with generators protected by simple lip seals.
With the development work done since then by Escher Wyss, there is good reason to trust the improved designs for the larger sizes that are now available of which the Annapolis unit is the latest example.

PAPER 18

MR R. CLARE
This symposium has been good and instructive in not concentrating on engineering matters only but addressing legal, planning, regional and ecological factors.
The Bondi Committe report, the Severn Tidal Power Group (STPG) report and the Mersey report all agree that these tidal barrages are technically feasible and while proponents of particular methods of construction or materials to be used will continue to argue the pros and cons for some time, and

quite rightly so, the final conclusions are unlikely to be
more than one percentage point change in the IRR. Although an
economically viable scheme at the public sector rate of return
is essential, no technical factor is likely to arise that will
prevent the construction of one or both of these schemes.
Regional and environmental problems could stop them, however,
and therefore must be given the highest priority in the way
forward. Nor must the time-scale that is necessary for such
work be underestimated. For the Severn scheme it has been
estimated that a period of six years more will be required to
complete the necessary environmental impact analysis, the
planning and the detailed design. Even for a small scheme, it
is doubted whether this period could be sensibly reduced below
four years.

The se points can be illustrated by reference to the future
work of the STPG. Although the STPG consists of six major
engineering companies, it is committed to thorough examination
of regional and environmental factors and the fullest possible
discussion with all local interests and relevant
representative bodies. It is also very conscious of the
Department of Energy's similar commitment and welcomes the
advisory groups that they plan to establish.

The STPG has been very active in establishing in detail its
programme for the next stage of work; it is essential to have
this at the outset. At present it is busy in agreeing the
detail with its partners, the Department of Energy and the
Central Electricity Generating Board. Although not yet
finalized, the possible split of work could be

Geological survey	£800 000
Power system	£500 000
Civil engineering	£800 000
Running costs	£50 000
Hydraulic modelling, water quality, energy capture	£600 000
Regional	£300 000
Ecological	£500 000
Economics and financing	£100 000
Legal - consents and planning	£200 000
Management	£400 000
Total	£4 250 000

MR R. T. P. McLAUGHLIN, George Wimpey PLC
In considering the way ahead, it might be speculated that, if
France had a site as favourable for tidal power as the Severn
Estuary, it would probably have been completed 10 or 15 years
ahead of us. Promotors, contractors, financial institutions,
generating boards and the Government should reflect on the
truth of this and why it should be so.

MR C. J. A. BINNIE, W S Atkins & Partners
This symposium has come at an apposite time because

(a) power/energy demand is increasing

(b) the Central Electricity Generating Board (CEGB) is now having to bring forward several new power generation schemes

(c) there is public concern for nuclear safety, especially after the Chernobyl accident

(d) for oil, there are uncertainties over long-term price and availability

(e) for coal, there are acid rain problems as well as uncertain industrial relations

(f) for hydropower there is little further scope

(g) other renewable energy sources are either small scale or are not cost competitive

(h) the Severn barrage would provide the owner, the CEGB, with direct control over a new and reliable power source

(i) the studies so far show that the barrage would be generally environmentally beneficial and would provide a welcome economic boost to the region.

Therefore now is the time for the tidal power lobby to stand up and to make sure that its benefits are understood.

MR R. J. LIVESEY, Babtie, Shaw and Morton
The subject matter of the symposium may be summarized in terms of financial risk. The lead-in time for the Severn barrage is long so that if it is privately funded the investor's money is at risk a long time before any return is possible.

Reference has been made to the question of world peace as it might affect the project. In the list of government departments to be consulted in relation to the development of the Severn barrage, the Ministry of Defence was omitted.

I have been involved in two public sector projects where the clients set particular goals to be achieved by the completed works. In one case, after the project had been brought to fruition, external local government politics changed the objective and in the other it was the influence of international politics. In both cases the benefits of investment were frustrated and the money need not have been spent.

In contrast there was another project where at the time of conception as part of a larger development into which it was integrated the scheme could not be shown to be viable on its own. Now after several years in operation and with the help of inflation, the revenue from the investment has equalled half the original capital outlay.

During the lead-in time private funding has to face the incalculable risk of third-party effects that could frustrate the completion of the barrage, leading to a loss of the privately invested money.

MR R. CLARE
Mr Livesey is right to draw attention to the question of project risk assessment. Referring to chapter 5 of volume 1

of the Severn Tidal Power Group's report, an initial review
has been made of risk. The difficulty of funding the
pre-contract work is well appreciated and it will be seen by
reference to the report that the rate of pre-contract work
expenditure increases as various stages are passed (e.g. on
Enabling Act) thus increasing the confidence level that the
project will proceed.

MR P. C. WARNER, Northern Engineering Industries plc
The question that the Severn Tidal Power Group (STPG) was set
was whether a power barrage at the Severn could be financed in
the private sector, and STPG members have been explaining that
the treatment of commercial matters was in keeping with that.
On reflection it was a very strange question. To the extent
that there came from it a most fruitful engineering, economic
and environmental exercise that is now a springboard for
further work, putting that question to the STPG has turned out
to have been quite creative, but when thinking about the
future, it should be remembered that it is artificial.
    The circumstances of private sector investment do not apply
readily to a major energy project of this kind, and where it
may have worked abroad it is generally because of special
conditions. These do not need to be debated now, in view of
the simple observation that Sizewell B, which is the energy
project that is attracting priority interest in the UK at
present, is to be funded in the public sector and there is no
sign that anyone is seriously asking whether it should be
attractive to the private investor.
    There are many differences between private and public sector
funding, and if it is chosen to assess some energy projects of
one and some in the other there will be paradoxes of various
kinds, such as the following.

(a) There is the effect of time-scale. If the private
    investor is told, as he is in this case, that an
    attractive return requires many years (30 or more added
    to the construction time), he will want assurances that
    the conditions for which the return has been calculated
    will be stable over that period. Who is there that
    could give such assurances? In public sector energy
    projects, very long lives are commonly assumed: 35
    years' exploitation time for Sizewell B, for instance,
    and similar periods for other power-stations. These are
    quite reasonable on objective grounds, but the strength
    of the public sector system is that it can act without
    asking anyone for formal guarantees that conditions will
    be stable; that is why the STPG said that the Government
    must underwrite that key condition of long-term
    stability for the question to be answered affirmatively.

(b) The second paradox has to do with residual value. The
    barrage, once in place, will stay there and have value
    for a long time, provided that the machinery is brought

up to date at intervals of perhaps every 40 years. Yet the private investor will discount according to the conventions, and no credit can be taken for capital value 40 years hence. There has been an example of this in relation to the English Stones barrage, when first estimates were looked at for silting and it was stated that it was not of great economic importance since it only built up slowly. Fortunately, common sense intervened, the economic analysis was put to one side and it was agreed that a significant risk of silting was an important factor even if it occurred 40 years from now. That was a negative decision, but positive decisions do not seem to operate on a similar common sense basis.

(c) The final example concerns future rises in energy price. The private sector cannot take account of that possibility and prefers to assume that energy prices will stay indefinitely at their present real level. How is that reconciled with the fluctuations that are witnessed (admittedly they are for the price of oil, but that tends to set off prices of the other sources)? One major strength of a tidal project is its imperviousness to fluctuations in future energy prices. So the private investor, if he has been correctly described, is declining to take that significant advantage into account. Once again, on the public side it is different. Nuclear power has some independence of energy price rises (not the total imperviousness of tidal power), and that was reflected in the economic assessment for Sizewell B by assuming a steady increase in energy prices in real terms. Putting a similar factor into the STPG analysis gave, not surprisingly, a significant improvement.

Other examples could be given, but these are sufficient to show why the original question was a strange one. It has catalysed some valuable progress, but the time has now come to put it aside so that the underlying merits of tidal power from the Severn may speak for themselves, and that the project may go ahead in a proper fashion.

Closing address

W. J. CARLYLE, BSc, FICE, FIWES, FGS, *Binnie & Partners*

The Severn Tidal Power Group is on its way with the energy of a Severn spring tide. They have financial support from the Government and the Central Electricity Generating Board, but the initiative is the Group's alone and without it there would be no present activity.

The objective of the recent work was to review and update the scheme as defined by the Severn Barrage Committee and hopefully to find a breakthrough in construction techniques to reduce the capital cost.

A scheme is emerging which consolidates and improves the original without achieving any big breakthrough in cost. Perhaps the most important statement was that of Dr Carr who demonstrated that the value of power would be more dominant in the financial planning than cost overrun or even construction delay. He demonstrated that this must be a public sector project in some form. The rates of return would not be of interest to the private sector in the light of the risk involved.

In contrast those studying the Mersey barrage, admittedly at a lower level of preparation, have claimed a massive reduction in civil engineering costs by an innovative method of diaphragm wall construction. This difference must be resolved if confidence in such civil engineering proposals is not to be eroded.

For the conventional caisson construction it is clear that steel caissons have a part to play in the final selection.